移动互联网开发技术丛书

微课视频版

吴 胜 编著

清華大学出版社
北 京

内容简介

Angular有着广泛的应用。本书由浅入深、循序渐进地介绍Angular(不是Angular.js或称为AngularJS)的应用开发。全书共包括16章,第1章Angular应用开发基础,第2章模板,第3章指令,第4章组件,第5章组件的组合、分解及其应用,第6章路由及其应用,第7章表单及其应用,第8章HTTP客户端服务及其应用,第9章国际化及其应用,第10章动画及其应用,第11章PWA、Service Worker、Web Worker,第12章测试及其应用,第13章高阶技术,第14章最佳实践,第15章综合案例,第16章整合开发。

本书适合作为Angular、Web前端、TypeScript的开发者和学习者(特别是在校学生)阅读和学习的参考书,也可以作为高等学校教材。

图书在版编目(CIP)数据

Angular开发入门与实战:微课视频版/吴胜编著.—北京:清华大学出版社,2023.2
(移动互联网开发技术丛书)
ISBN 978-7-302-62570-4

Ⅰ.①A… Ⅱ.①吴… Ⅲ.①超文本标记语言—程序设计—高等学校—教材 Ⅳ.①TP312

中国国家版本馆CIP数据核字(2023)第008677号

责任编辑:陈景辉
封面设计:刘 键
责任校对:徐俊伟
责任印制:宋 林

出版发行:清华大学出版社
网　　址:http://www.tup.com.cn, http://www.wqbook.com
地　　址:北京清华大学学研大厦A座　　**邮　　编**:100084
社 总 机:010-83470000　　**邮　　购**:010-62786544
投稿与读者服务:010-62776969, c-service@tup.tsinghua.edu.cn
质量反馈:010-62772015, zhiliang@tup.tsinghua.edu.cn
课件下载:http://www.tup.com.cn,010-83470236
印 装 者:三河市铭诚印务有限公司
经　　销:全国新华书店
开　　本:185mm×260mm　　**印　　张**:21.75　　**字　　数**:533千字
版　　次:2023年2月第1版　　**印　　次**:2023年2月第1次印刷
印　　数:1～1500
定　　价:79.90元

产品编号:097735-01

前言

FOREWORD

Angular有着广泛的应用。本书主要介绍Angular的应用开发，不涉及Angular.js(或称为AngularJS)的应用开发。

Angular应用开发的入门学习偏难的原因有四个方面：其一，Angular应用开发时涉及的概念众多(包括组件和模板、样式、指令、服务和依赖注入、Provider、路由、表单、模块、装饰器、响应式编程等)且语法有差异，这些概念、应用开发思想与后端的应用开发有许多相似之处；其二，Angular应用开发时重视模块化开发、设计模式等思想的落实，一个功能点(如组件)包括多个文件(如组件、模板、样式、路由、模块、测试等)；其三，Angular是用TypeScript实现的，Angular的应用开发也使用TypeScript，TypeScript在JavaScript基础上增加了类型系统，语法更为严格，基于TypeScript的Angular应用开发的初学者更容易出错；其四，Angular的类库和工具丰富，即Angular应用开发具有一站式的特点，但是官方文档略显简洁。

Angular应用开发的学习门槛较高的四个原因如下。其一，丰富的概念对于从后端转入前端的开发人员很便利，Angular应用开发人员转到后端开发也很容易，众多概念与后端开发相似的特点便于前后端的分离和整合开发；其二，模块化开发和设计模式等思想的落实，便于大型项目的开发，便于团队分工和合作，便于测试，便于提高开发的质量和效率；其三，TypeScript增加类型系统后能提高代码质量，可以帮助学习者在学会Angular的同时学会TypeScript的应用开发，提升开发技能；其四，Angular的丰富类库和工具，使得Angular具备完备性(一站式的特点)，仅仅用Angular就能较好地进行前端开发，不需要面对同一生态中不同实现(或工具)的问题。

针对Angular应用开发的四个特点，为了降低Angular应用开发的学习难度，本书先针对各个概念由浅入深、按照开发的先后次序组织内容。为了降低学习的难度和精简篇幅，为了减少一个功能点的相关文件数量，本书将模板、样式内容并入组件文件，将路由并入模块文件。本书的TypeScript代码均进行了测试，学习者按照本书的示例来学习Angular和TypeScript的应用开发可以减少出错，并熟悉Angular和TypeScript的应用场景，在熟悉Angular和TypeScript之后就可以独立处理开发中的错误。本书按照开发的需要和学习的难度，有选择、有针对性地介绍Angular的类库和工具。另外，虽然Angular应用开发学习入门偏难，但学会之后，再学习基于JavaScript或TypeScript等应用框架或其他前端框架就可以事半功倍。

全书共有16章。第1章Angular应用开发基础,包括Angular简介、Angular应用项目说明、Angular应用开发步骤、TypeScript基础;第2章模板,包括模板概述、模板绑定、模板变量和模板输入变量、模板的基础应用、模板的综合应用开发;第3章指令,包括指令概述、内置属性型指令、内置结构型指令、自定义属性型指令、自定义结构型指令、指令的基础应用、指令的综合应用开发;第4章组件,包括组件概述、组件样式及其应用、组件生命周期、组件生命周期的综合应用、组件之间的交互及其应用、Angular元素及其应用;第5章组件的组合、分解及其应用,包括内容投影及其应用、视图封装及其应用、依赖注入及其应用;第6章路由及其应用,包括路由概述、路由的应用开发;第7章表单及其应用,包括表单概述、响应式表单、表单验证及实现、动态表单及其构建、表单的综合应用开发;第8章HTTP客户端服务及其应用,包括HTTP客户端服务、拦截机制、HTTP客户端服务的应用、拦截器的应用开发;第9章国际化及其应用,包括国际化概述、翻译、将翻译结果合并到应用中、可选的国际化实践、国际化应用;第10章动画及其应用,包括动画概述、转场动画、路由转换动画、动画的应用开发;第11章PWA、Service Worker、Web Worker,包括PWA概述、Service Worker概述、生产环境下的Service Worker、Service Worker配置、PWA的应用开发;第12章测试及其应用,包括测试概述、TestBed的应用开发、服务测试应用、组件测试应用、Jasmine应用、路由测试应用、异步测试应用、Mock测试应用、测试综合应用;第13章高阶技术,包括Angular统一平台、Angular CLI、Angular语言服务、AOT编译器、Angular应用的运行、Angular库的开发、原理图、Angular发布信息;第14章最佳实践,包括安全的最佳实践、无障碍性、保持最新和属性绑定、惰性加载、令牌、安全的应用开发、无障碍性的应用开发、属性绑定的应用、惰性加载特性模块的应用;第15章综合案例,包括英雄信息、简易通讯录;第16章整合开发,包括与Ant Design of Angular的整合开发、与Spring Boot的整合开发。

本书特色

(1) 易理解。本书避免对官方文档的简单引用,按照学习先后顺序和开发步骤由浅入深地编排知识点,适合于自学和高等院校课程教学的需要。

(2) 内容新。本书使用的Angular版本是14.0.0版,涵盖了新内容。

(3) 全栈式。本书大多引用官方文档的内容;还介绍了与Ant Design of Angular的整合开发,与Spring Boot(使用MySQL 8.x)的整合开发;得益于Angular自身的完备性,参考本书可以较全面地利用Angular进行前端开发。

(4) 示例多。实战案例丰富,涵盖26个知识点示例、两个整合开发案例、两个完整项目案例。

配套资源

为便于教与学,本书配有微课视频(128分钟)、源代码、教学课件、教学大纲、习题答案、教学进度安排。

(1) 获取微课视频方式:读者可以先扫描本书封底的文泉云盘防盗码,再扫描书中相应的视频二维码,观看视频。

(2) 获取源代码和彩色图片方式:先扫描本书封底的文泉云盘防盗码,再扫描下方二维码,即可获取。

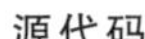

源代码

源代码使用说明

源代码和数据库使用说明

(3) 其他配套资源可以扫描本书封底的"书圈"二维码下载。

读者对象

本书主要面向希望学习 Angular 应用开发的初学者(特别适合高等院校的在读学生)、从事高等教育的专任教师以及广大从事 Web 前端开发的专业人员。

本书的主要内容参考了 Angular 官方文档,在参考文献中已经列出。在此向 Angular 开发者和官方文档的作者表示衷心的感谢和深深的敬意。本书的编写还参考了其他相关资料,在此表示衷心的感谢。

限于时间仓促和个人水平有限,书中难免存在疏漏之处,欢迎读者批评指正。

编　者

2023 年 1 月

CONTENTS

第 1 章

Angular应用开发基础

1.1 Angular 简介

1.1.1 定义

Angular 是用 TypeScript 构建的应用设计框架与开发平台，用于创建高效、复杂、精致的单页面应用(Single Page Application,SPA)。它包括：一个基于组件的框架，可以用于构建可伸缩的 Web 应用程序(简称应用)；一组集成良好的库，涵盖路由、表单管理、客户端与服务器通信等功能；一套开发工具，可帮助开发、构建、测试和更新代码。Angular 让更新更容易，可以用最小的成本升级到最新的 Angular 版本。组件是 Angular 应用的主要构造块。视图用于控制屏幕上的一块区域的显示。用户可以与组件交互，组件之间可以交互，组件和服务之间可以调用。

1.1.2 特点

1. 跨平台

Angular 是渐进式应用(Progressive Web App,PWA)，能充分利用现代 Web 平台的各种能力，提供 App 式体验，具有高性能、离线使用、免安装等特点。可以借助来自 Ionic、NativeScript 和 React Native 中的技术与思想，构建原生移动应用。结合访问原生操作系统 API(Application Programming Interface,应用编程接口)的能力，创造能在桌面环境下安装的应用，实现横跨 Mac、Windows 和 Linux 平台的 Web 应用开发。

2. 高性能

Angular 会把模板(即一块 HTML 代码)转换成针对 JavaScript 虚拟机进行高度优化的代码，轻松获得框架提供的高生产率，同时又能保留手写代码的优点。在服务端渲染应用的首屏(即一个页面)像只有 HTML 和 CSS 的静态页面那样快速展现，支持 Node. js、.NET、PHP 以及其他服务器，为通过搜索引擎优化(Search Engine Optimization,SEO)来优化站点铺平了道路。借助组件路由器，Angular 可以实现快速加载。自动代码拆分机制可以仅仅加载那些用于渲染所请求页面的代码，从而实现高性能。

3. 高生产率

通过简单而强大的模板语法，可以快速创建用户界面(User Interface,UI)视图。可以借助命令行工具 Angular CLI 快速进入构建环节、添加组件和测试，然后立即部署。在常用集成开发环境(Integrated Development Environment,IDE)和编辑器中获得智能代码补全、实时错误反馈及其他反馈等特性。

4. 全流程的开发

通过 Angular 中直观简便的 API 创建高性能、复杂编排和动画时间线(只要非常少的代码)。能使用 Karma 进行单元测试。通过支持 ARIA(Accessible Rich Internet Applications,可访问富互联网应用)的组件、开发者指南和内置的一体化测试基础设施,创建完备、无障碍的应用。

1.1.3 发展简史

AngularJS(或 Angular.js)由 Misko Hevery 等人于 2009 年创建,后被 Google 公司收购。AngularJS 是用 JavaScript 编写的用于构建用户界面的前端框架,是一种前端 MVC 的架构框架。AngularJS 于 2012 年发布 1.0 版,2013 年发布 1.2 版,2014 年发布 1.3 版,2015 年发布 1.4 版,2016 年发布 1.5 版和 1.6 版,2018 年发布 1.7 版。2018 年 7 月,AngularJS 进入了 3 年的长期支持期。2020 年发布 1.8 版。

2016 年 9 月,Google 公司发布了 Angular 2。Angular 2 是采用 TypeScript 语言对 AngularJS 的重写。为了区分,AngularJS(或 Angular.js)是指 1.x 版本,2.0 及以上版本都叫作 Angular。Angular 的版本号包括三个部分:major.minor.patch。如版本 13.3.2 表示主版本号是 13,小版本号是 3,补丁版本号是 2。本书主要介绍 Angular 14.0.0 及以上版本的应用开发,因此,除非明确指出 Angular 版本之外,本书中的 Angular 均是指 14.0.0 及以上版本的 Angular。2017 年 3 月发布 Angular 4(跳过 Angular 3.x)。2017 年 11 月发布 Angular 5。2018 年 5 月发布 Angular 6。2018 年 10 月发布 Angular 7。从 Angular 7 开始,Angular Core 和 Angular CLI 的主要版本已对齐(即两个版本号一致)。这意味着,在开发 Angular 应用时使用的@angular/core 和 Angular CLI 的版本相同。2019 年 5 月发布 Angular 8。2020 年 2 月发布 Angular 9。此后,Angular 所有主版本的一般支持周期是 18 个月。如 2020 年 6 月发布 Angular 长期支持(Long Term Support,LTS)版本 10.0.0 版。2020 年 11 月发布长期支持版本 11.0.0 版,2021 年 5 月发布长期支持版本 12.0.0 版,2021 年 11 月发布长期支持版本 13.0.0 版。2022 年 5 月发布长期支持版本 14.0.0 版,预计 2023 年 11 月不再为 14.0.0 版提供支持。2022 年 10 月发布长期支持版本 14.2.7 版。2022 年 11 月发布长期支持版本 15.0.0 版。

1.1.4 核心概念

1. Angular 应用

Angular 应用程序的基本构造块是模块(NgModule)。模块是组织应用和使用外部库扩展应用的最佳途径,它能把组件、指令(为 Angular 应用中的元素添加额外行为的类)等打包成内聚的功能块,每个模块聚焦于一个特性区域、业务领域、工作流或通用工具。Angular 应用是由一组模块构成的,其中至少会有一个用于引导应用的根模块,通常还会有很多特性模块。NgModule 专注于某个应用领域、某个工作流或一组紧密相关的功能。NgModule 可以将其组件和一组相关代码(如服务)关联起来,形成功能单元。Angular 定义的 NgModule 和 JavaScript 的 ECMAScript 6(简称为 ES6,又称为 ECMAScript 2015)的模块不同且有一定的互补性。

2. 组件

组件是构成应用的砖块,用于定义视图。视图是一组可见的屏幕元素,Angular 可以根据程序逻辑和数据来选择和修改它们。每个应用都至少有一个根组件。组件通常包括用 TypeScript 编写的带有@Component()装饰器的组件类(简称为组件)。组件类中的内容包括一个 CSS 选择器,用于定义如何在模板中使用组件;一个 HTML 模板(简称为组件模板),用于指示 Angular 如何渲染此组件;一组可选的 CSS 样式,用于定义模板中 HTML 元素的外观。

3. 组件模板

组件模板把普通 HTML 和 Angular 指令与绑定标签(Markup)组合起来。这样,Angular 就可以在渲染 HTML 之前先修改 HTML。

4. 指令

指令是为 Angular 应用中的元素添加额外行为的类。使用 Angular 内置指令,可以管理表单、列表、样式以及其他想让用户看到的内容。当组件的状态更改时,Angular 会自动更新已渲染的 DOM(Document Object Model,文档对象模型)。Angular 还支持属性(Property)绑定,以帮助设置 HTML 元素的属性和特性(Attribute)的值,并将这些值传给应用的视图逻辑。可以声明事件监听器来监听并响应用户的操作,例如按键、鼠标移动、单击和触摸等。

5. 组件使用服务

该服务会提供那些与视图不直接相关的功能。服务提供者可以作为依赖被注入组件中,这能让代码更加模块化、高效和可复用。模块、组件和服务都会使用装饰器来标出它们的类型并提供元数据,以告知 Angular 该如何使用(处理)它们。

6. 依赖项

利用依赖注入(Dependency Injection)可以声明 TypeScript 类的依赖项,Angular 的依赖注入框架会在实例化某个类时为其提供依赖项,而开发者无须操心如何实例化它们。依赖项是指某个类执行其功能所需的服务或对象。依赖项注入是一种设计模式;在这种设计模式中,类会从外部源请求依赖项而不是创建它们。这种设计模式能帮助开发人员写出更灵活、更加可测试的代码。为了代码的清晰性和可维护性,可以在单独的文件中定义组件和服务。如果将组件和服务合并在同一个文件中,则必须先定义服务,再定义组件。服务类的元数据提供了一些信息,Angular 要用这些信息来让组件通过依赖注入使用服务。

7. 视图

Angular 应用的组件通常会定义很多视图,并进行分级组织。Angular 提供了路由(router)服务来定义视图(即组件)之间的导航路径。路由器提供了先进的浏览器内导航功能。

1.2 Angular 应用项目说明

微课视频

1.2.1 创建项目 angularcliex1

可以参考电子资源附录 A 或其他资源来完成 WebStorm 的安装,双击图标可以启动它,如图 1-1 所示(首次使用时没有项目信息)。

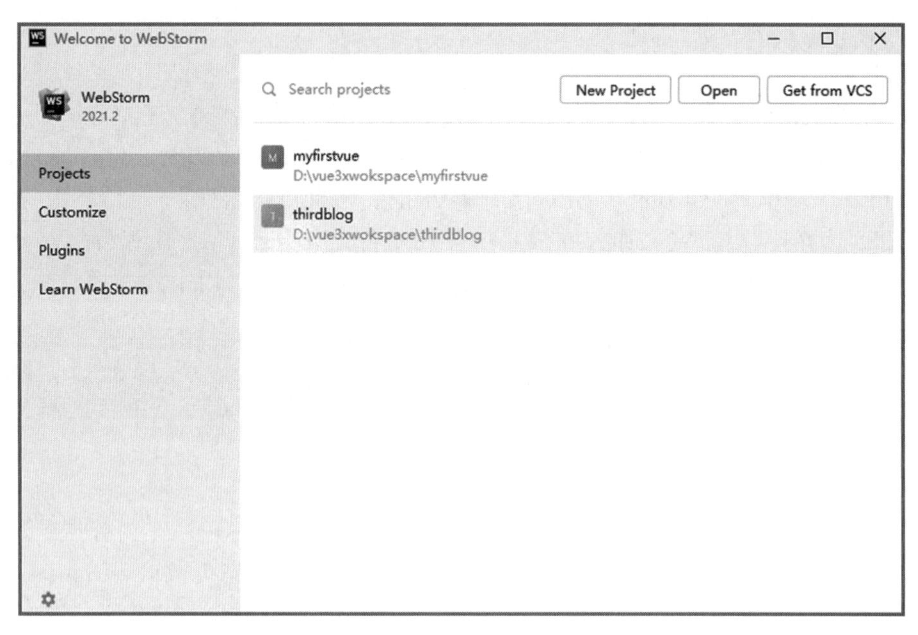

图 1-1 WebStorm 启动后的欢迎界面

在如图 1-1 所示的欢迎界面中单击 New Project 按钮进入项目创建界面。选择 Angular CLI 类型的项目,如图 1-2 所示。创建项目时可以修改项目的位置(Location),如图 1-2 所示中将项目所在的位置(简称地址)修改为 D:\angularjsworkspace\angularcliex1;也可以采用默认地址;地址的最后一个"\"后面的内容为项目名称(如图 1-2 所示的项目名称为 angularcliex1)。

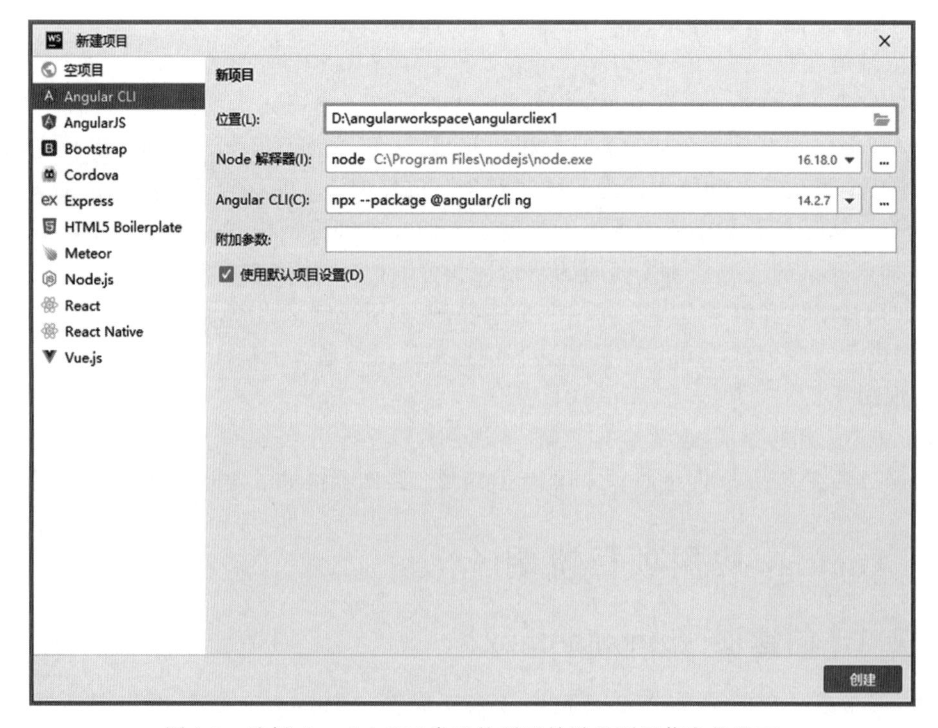

图 1-2 选择 Angular CLI 类型的项目并设置项目信息的界面

1.2.2　项目目录和文件说明

创建完项目 angularcliex1 后，项目目录和项目结构如图 1-3 所示。对图 1-3 所示中的 src\app 目录、src\assets 目录和 src\environments 目录进行展开的结果如图 1-4 所示。

图 1-3　创建新项目后的目录和文件

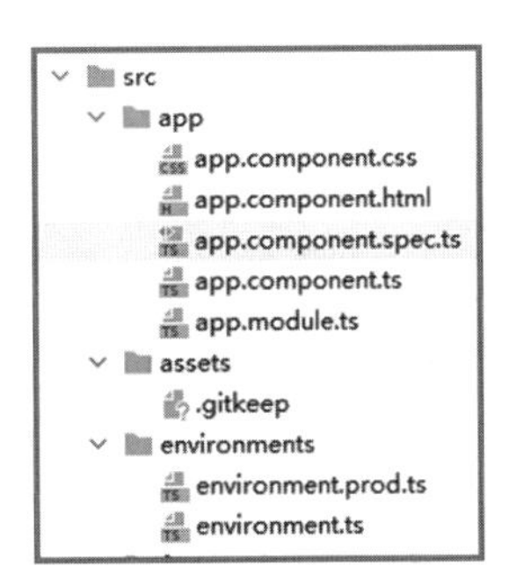

图 1-4　对图 1-3 所示中的 src\app 目录、src\assets 目录和 src\environments 目录进行展开的结果

项目中部分目录、文件(如图 1-3 和图 1-4 所示)的说明如表 1-1 所示。

表 1-1　项目中部分目录、文件的说明

目录或文件	说　明
node_modules	第三方依赖包存放目录
src	源代码目录
. browserslistrc	配置浏览器兼容性的文件
. editorconfig	统一编译器中的代码风格的文件
. gitignore	Git 中忽略文件列表的文件
angular. json	Angular 的配置文件
karma. conf. js	自动化测试框架 Karma 的配置文件
package. json	标准的 NPM 工具的配置文件，包括应用的依赖包、执行命令等
package-lock. json	依赖包版本锁定文件
README. md	项目说明的 MakeDown 文件
tsconfig. app. json	应用的 TypeScript 的配置文件
tsconfig. json	整个工作区的 TypeScript 配置文件
tsconfig. spec. json	用于测试的 TypeScript 配置文件
app	应用源代码目录，包含应用的组件和模块等
assets	应用资源目录，用来存储静态资源(如图片)

续表

目录或文件	说　　明
environments	环境配置目录，Angular 支持多环境开发，可以在不同的环境(开发、测试、生产环境)共用一套代码
favicon. ico	浏览器的图标文件
index. html	应用的根 HTML 文件，程序启动时访问该文件对应的页面
main. ts	应用的入口点文件，Angular 通过该文件来启动项目(扩展文件名 ts 表明是用 TypeScript 编写的文件)
polyfills. ts	兼容不同浏览器的脚本加载文件，用来导入一些必要库，为了让 Angular 能在老版本浏览器中正常运行
styles. css	全局的样式 CSS 文件
test. ts	测试入口文件
app. component. css	app 组件的样式 CSS 文件
app. component. html	app 组件的 HTML 文件
app. component. spec. ts	app 组件的测试类文件
app. component. ts	app 组件的组件类文件
app. module. ts	app 组件的模块类文件
environments. prod. ts	生产环境配置文件
environments. ts	开发环境配置文件

1.2.3 运行项目说明

单击 WebStorm 项目中 Angular CLI Server 后面的三角形按钮(图 1-5 所示中右上角矩形框内)来运行程序，如图 1-5 所示；该操作简称运行程序。运行程序时，会自动在控制台中执行 ng serve 命令，如图 1-6 所示。成功运行程序后，显示 Compiled successfully，并提示可

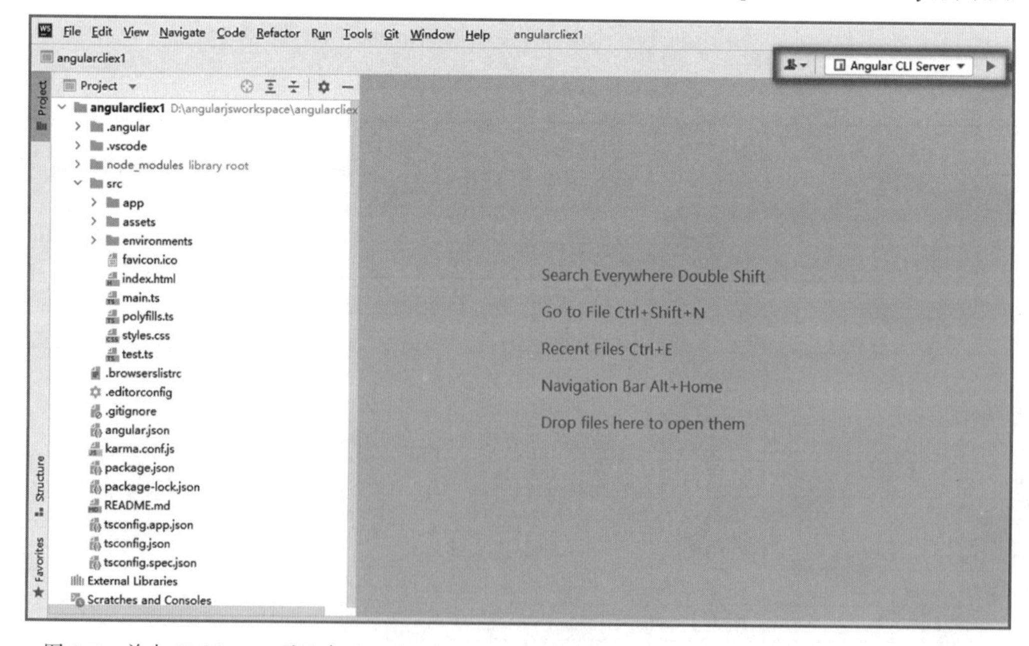

图 1-5　单击 WebStorm 项目中 Angular CLI Server 后面的三角形按钮来运行程序的操作过程界面

以通过浏览器访问 http://localhost:4200/,如图 1-6 所示。在浏览器地址栏中输入 localhost:4200,结果如图 1-7 所示。图 1-7 所示中的火箭飞船(Rocket Ship)小图标后显示 angularcliex1 app is running!,其中 angularcliex1 是项目名称。

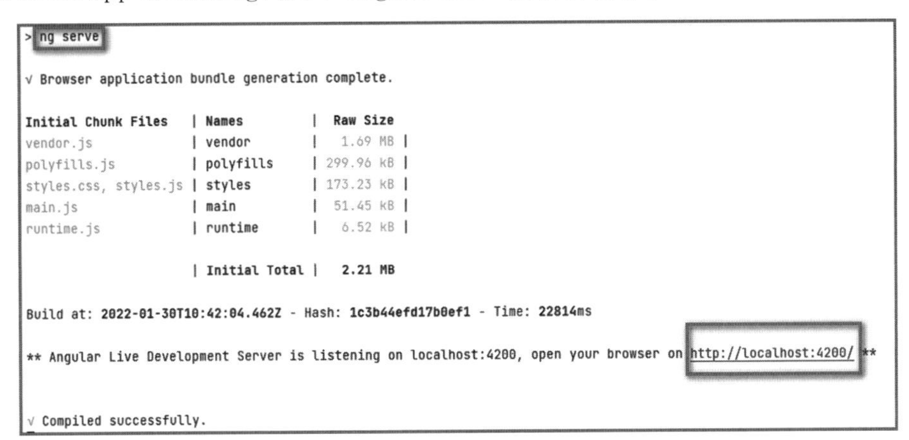

图 1-6 在控制台中执行 ng serve、显示 Compiled successfully 和访问 http://localhost:4200/等信息

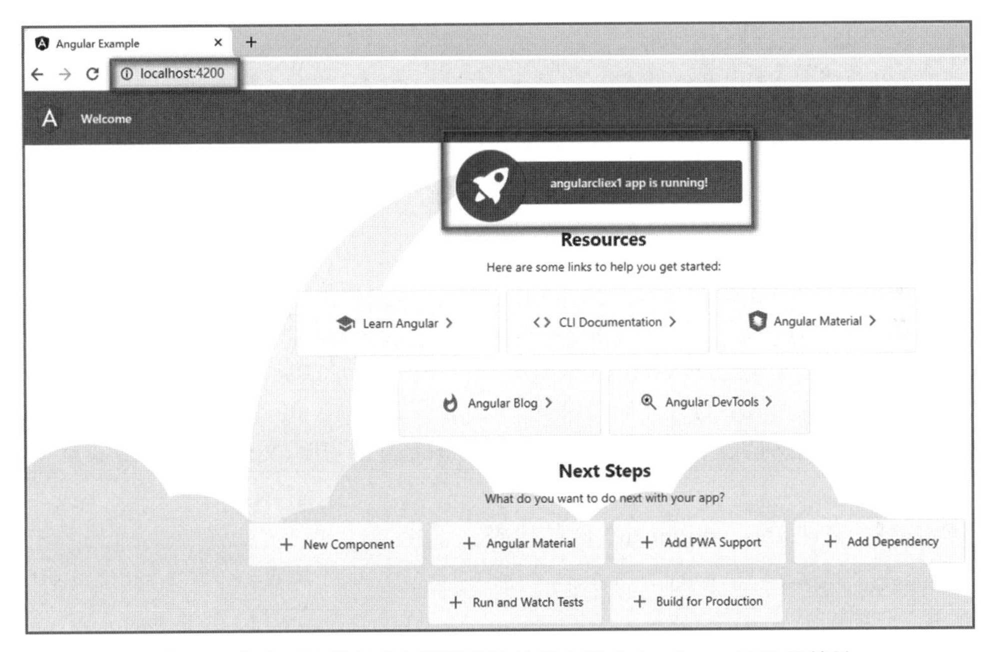

图 1-7 成功运行程序后在浏览器地址栏中输入 localhost:4200 的结果

1.2.4 app 模块中的文件代码和关系说明

项目 angularcliex1 中独立的样式文件 app.component.css 内容为空,因为该示例中将样式内容写入文件 app.component.html 中了。样式内容可以包含在一个独立的样式文件中(称为外部样式),如本例中样式文件 app.component.css。

对项目 angularcliex1 中独立的模板文件 app.component.html 进行简化,代码如例 1-1 所示。模板内容可以包含在一个独立的文件中(称为外部模板),如例 1-1 中模板文件 app.component.html。

【例 1-1】 模板文件 app. component. html 简化后的代码。

```
<!-- CSS 信息 -->
<style>
  :host {
font-family: -apple-system, BlinkMacSystemFont, "Segoe UI", Roboto, Helvetica, Arial, sans-serif, "Apple Color Emoji", "Segoe UI Emoji", "Segoe UI Symbol";
    font-size: 14px;
    color: #333;
    box-sizing: border-box;
    -webkit-font-smoothing: antialiased;
    -moz-osx-font-smoothing: grayscale;
  }
<!-- 省略其他 CSS 信息 -->
</style>
<!-- 省略其他信息 -->
<div class="content" role="main">
  <!-- 省略其他信息 -->
<div class="card highlight-card card-small">
 <!-- 省略其他信息 -->
<svg id="rocket" xmlns="http://www.w3.org/2000/svg" width="101.678" height="101.678" viewBox="0 0 101.678 101.678">
      <!-- 火箭飞船小图标 -->
      <title>Rocket Ship</title>
  <!-- 省略其他信息 -->
    </svg>
    <span>{{ title }} app is running!</span>
<!-- 省略其他信息 -->
</div>
<!-- 省略其他信息 -->
</div>
```

例 1-1 中有两处加粗代码,包括火箭飞船小图标内容和{{title}} app is running!。对比例 1-1 中两处加粗代码和运行程序的结果,可以知道例 1-1 中{{title}}的取值尚待进一步明确。{{title}}的取值 angularcliex1 在组件文件 app. component. ts 中指明,文件 app. component. ts 代码如例 1-2 所示。

【例 1-2】 文件 app. component. ts 的代码。

```
import {Component} from '@angular/core';
//用@Component()装饰器声明组件
@Component({
  //命名的选择器,可以在 HTML 文件中引入该选择器标签,如在 index.html 文件中引入
  selector: 'app-root',
  //引用与本文件相同目录下的文件 app.component.html 作为外部模板
  templateUrl: './app.component.html',
  //引用与本文件相同目录下的文件 app.component.css 作为外部样式
  styleUrls: ['./app.component.css']
})
//导出组件
export class AppComponent {
  //title 的数值
  title = 'angularcliex1';
}
```

这里用到的入口文件是 main. ts,它无法单独(或直接)地处理组件信息,而是需要将组件封装成模块。NgModule(模块)可用于组织代码,也可以用于把关系比较紧密的组件组织到一起。NgModule 可以控制组件、指令、管道等的可见性,同一个 NgModule 中的组件默认互相可见。对于外部组件来说,只能看到 NgModule 导出(Export)的内容。如果定义的 NgModule 不导出任何内容,那么外部文件中即使导入(Import)了这个模块,也无法使用模块中定义的任何内容。NgModule 是脚手架工具@angular/cli 打包的最小单位。打包时,@angular/cli 会检查所有模块和路由配置。如果配置了异步模块,@angular/cli 会自动把模块切分成独立的块(Chunk)。模块是路由进行异步加载的最小单位,而不是组件。不过,模块里面可以只放一个组件。

app 组件在模块文件 app. module. ts 中定义模块,模块文件 app. module. ts 代码如例 1-3 所示。

【例 1-3】 模块文件 app. module. ts 的代码。

```
import {NgModule} from '@angular/core';
import {BrowserModule} from '@angular/platform-browser';
import {AppComponent} from './app.component';
//模块
@NgModule({
  //declarations: 用来放组件、指令、管道的声明
  declarations: [
    AppComponent
  ],
  //imports: 用来导入外部模块
  imports: [
    BrowserModule
  ],
  //providers: 需要使用的服务都放在这里
  providers: [],
  //bootstrap: 设置启动组件
//这个配置项是一个数组,可以指定多个组件作为启动点,往往指定一个组件为启动点
  bootstrap: [AppComponent]
})
export class AppModule { }
```

描述(或称为测试)文件 app. component. spec. ts 是 app 组件的测试类文件,可以暂时不用考虑。

Angular 应用开发时,一个组件往往可以包含样式文件 *. component. css、模板文件 *. component. html、测试文件 *. component. spec. ts、组件类文件 *. component. ts、模块文件 *. module. ts。为了知识点的介绍、代码的简化和篇幅的节省,除非为了说明外部样式和外部模板等内容的需要,后面章节中将外部样式(文件 *. component. css 中)内容、外部模板(文件 *. component. html 中)内容并入组件文件 *. component. ts 中(对应称为内部样式、内部模板)。除了探讨测试等内容的需要外,后面章节中将忽略测试文件 *. component. spec. ts。于是,后面章节的应用开发中将主要讨论组件文件 *. component. ts、模块文件 *. module. ts 等内容。读者熟练掌握了 Angular 应用开发知识后,将内部样式、内部模板内容独立成外部样式文件、外部模板文件很容易;也能较快速地掌握测试文件的编写方法。

1.2.5 文件 main. ts 和 index. html 的说明

项目 angularcliex1 应用的入口文件是 main. ts,其代码如例 1-4 所示。

【例 1-4】 文件 main. ts 的代码。

```
//引入依赖
import {enableProdMode} from '@angular/core';
import {platformBrowserDynamic} from '@angular/platform-browser-dynamic';
//自动生成的引入 app 组件中的依赖
import {AppModule} from './app/app.module';
//引入环境依赖
import {environment} from './environments/environment';
//启用生产环境
if (environment.production) {
  enableProdMode();
}
platformBrowserDynamic().bootstrapModule(AppModule)     //设置启动模块
  .catch(err => console.error(err));
```

platformBrowserDynamic()方法是浏览器平台的工厂方法,返回浏览器平台的实例,然后对根模块进行初始化,链式地将所有的依赖的模块都给加载进来。每个应用都是通过模块的 bootstrapModule()方法创建的。catch()方法捕获错误并输出到控制台。

项目 angularcliex1 应用的根 HTML 文件 index. html 的代码如例 1-5 所示。

【例 1-5】 文件 index. html 的代码。

```
<!doctype html>
<html lang="en">
<head>
  <meta charset="utf-8">
  <title>Angular Example</title>
  <base href="/">
  <meta name="viewport" content="width=device-width, initial-scale=1">
  <link rel="icon" type="image/x-icon" href="favicon.ico">
</head>
<body>
  <!-- app-root 标签和组件文件 app.component.ts 中的选择器(selector)保持一致 -->
  <app-root></app-root>
</body>
</html>
```

1.2.6 配置文件说明

项目 angularcliex1 自动生成的项目文件 package. json 的代码如例 1-6 所示。

【例 1-6】 项目文件 package. json 的代码。

```
{
  "name": "angularcliex1",
  "version": "0.0.0",
  "scripts": {
    "ng": "ng",
    "start": "ng serve",
    "build": "ng build",
```

```
    "watch": "ng build --watch --configuration development",
    "test": "ng test"
  },
  "private": true,
  "dependencies": {
    "@angular/animations": "～14.2.7",
    "@angular/common": "～14.2.7",
    "@angular/compiler": "～14.2.7",
    "@angular/core": "～14.2.7",
    "@angular/elements": "～14.2.7",
    "@angular/forms": "～14.2.7",
    "@angular/material": "^14.2.7",
    "@angular/platform-browser": "～14.2.7",
    "@angular/platform-browser-dynamic": "～14.2.7",
    "@angular/router": "～14.2.7",
    "@ngx-translate/core": "^14.2.7",
    "@ngx-translate/http-loader": "^7.0.0",
    "angular-in-memory-web-api": "^0.13.0",
    "element-angular": "^0.7.6",
    "jasmine-marbles": "^0.8.4",
    "lodash-es": "^4.17.21",
    "ng-translation": "^0.0.3",
    "ng-zorro-antd": "^14.2.7",
    "ng2-bootstrap": "^1.6.3",
    "ngx-bootstrap": "^8.0.0",
    "rxjs": "～7.5.0",
    "tslib": "^2.3.0",
    "zone.js": "～0.11.4"
  },
  "devDependencies": {
    "@angular-devkit/build-angular": "～14.2.7",
    "@angular/cli": "～14.2.7",
    "@angular/compiler-cli": "～14.2.7",
    "@types/jasmine": "～4.0.0",
    "@types/node": "^12.11.1",
    "jasmine-core": "～4.3.0",
    "karma": "～6.4.0",
    "karma-chrome-launcher": "～3.1.0",
    "karma-coverage": "～2.2.0",
    "karma-jasmine": "～5.1.0",
    "karma-jasmine-html-reporter": "～2.0.0",
    "typescript": "～4.7.2"
  }
}
```

在例 1-6 所示的文件 package.json 中，依次配置了项目的名称（name 字段）、版本（version 字段）、可以使用 npm 调用的脚本或封装的命令（scripts 字段）、可见性（private 字段）依赖的基础包（dependencies 字段）、开发时所依赖的工具包（devDependencies 字段）。其中，依赖的基础包包括动画库@angular/animations、提供常用的服务、管道和指令库@angular/common、模板编译器@angular/compiler、内核@angular/core、表单库@angular/forms、与 DOM 和浏览器有关的库@angular/platform-browser、路由器@angular/router 等。

项目 angularcliex1 自动生成的项目文件 angular.json 的代码如例 1-7 所示。

【例 1-7】 项目文件 angular.json 的代码。

```
{
  "$schema": "./node_modules/@angular/cli/lib/config/schema.json",
  "cli": {
    "analytics": false
  },
  "version": 1,
  "newProjectRoot": "projects",
  "projects": {
    "angularcliex1": {
      "projectType": "application",
      "schematics": {
        "@schematics/angular:application": {
          "strict": true
        }
      },
      "root": "",
      "sourceRoot": "src",
      "prefix": "app",
      "architect": {
        "build": {
          "builder": "@angular-devkit/build-angular:browser",
          "options": {
            "outputPath": "dist/angularcliex1",
            "index": "src/index.html",
            "main": "src/main.ts",
            "polyfills": "src/polyfills.ts",
            "tsConfig": "tsconfig.app.json",
            "assets": [
              "src/favicon.ico",
              "src/assets"
            ],
            "styles": [
              "src/styles.css"
            ],
            "scripts": []
          },
          "configurations": {
            "production": {
              "budgets": [
                {
                  "type": "initial",
                  "maximumWarning": "500Kb",
                  "maximumError": "1Mb"
                },
                {
                  "type": "anyComponentStyle",
                  "maximumWarning": "2Kb",
                  "maximumError": "4Kb"
                }
              ],
              "fileReplacements": [
                {
                  "replace": "src/environments/environment.ts",
                  "with": "src/environments/environment.prod.ts"
                }
```

```
              ],
              "outputHashing": "all"
            },
            "development": {
              "buildOptimizer": false,
              "optimization": false,
              "vendorChunk": true,
              "extractLicenses": false,
              "sourceMap": true,
              "namedChunks": true
            }
          },
          "defaultConfiguration": "production"
        },
        "serve": {
          "builder": "@angular-devkit/build-angular:dev-server",
          "configurations": {
            "production": {
              "browserTarget": "angularcliex1:build:production"
            },
            "development": {
              "browserTarget": "angularcliex1:build:development"
            }
          },
          "defaultConfiguration": "development"
        },
        "extract-i18n": {
          "builder": "@angular-devkit/build-angular:extract-i18n",
          "options": {
            "browserTarget": "angularcliex1:build"
          }
        },
        "test": {
          "builder": "@angular-devkit/build-angular:karma",
          "options": {
            "main": "src/test.ts",
            "polyfills": "src/polyfills.ts",
            "tsConfig": "tsconfig.spec.json",
            "karmaConfig": "karma.conf.js",
            "assets": [
              "src/favicon.ico",
              "src/assets"
            ],
            "styles": [
              "src/styles.css"
            ],
            "scripts": []
          }
        }
      }
    }
  },
  "defaultProject": "angularcliex1"
}
```

在例 1-7 所示的文件中，$ schema 字段(或称为属性)指向一个 JSON Schema 文件(该文件描述了 angular.json 所有的字段以及约束)、cli 字段是对 Angular CLI 的配置、newProjectRoot 字段指定由 Angular CLI 创建的新的内部应用和库放置的位置(默认值为 projects)、projects 字段包含了工作空间中所有项目的配置信息、projectType 字段指定项目类型(application 或者 library，前者可以在浏览器中独立运行，而后者则不行)、schematics 字段定义一组原理图(schematic，它可以为该项目自定义 ng generate 命令的默认选项)、architect 字段指定了项目的各个构建器目标配置默认值。其他字段代表的含义可以参考本书后面的介绍、官方文档或其他资源。

1.2.7 项目启动过程

Angular 项目(应用)的启动过程如下所述。

(1) 找到入口文件。Angular CLI 会根据配置文件 angular.json 的 main 字段找到项目的入口文件 main.ts。src 目录是 Angular 应用的根目录。src 目录下的 index.html 是 Angular 启动时加载的页面。src 目录下的 main.ts 是 Angular 启动时加载的脚本(负责引导 Angular 的启动)。

(2) 找到主模块。在文件 main.ts 中找到声明指定的主模块，默认的主模块是 AppModule (如例 1-3 模块文件 app.module.ts)。Angular 导入 platformBrowserDynamic 模块来启动应用，导入生成的整个应用的主模块 AppModule，导入环境配置文件 environment。

(3) 找到主组件。在模块文件 app.module.ts 中找到指定的主组件，默认的主组件 AppComponent(如例 1-2 组件文件 app.component.ts)。在模块文件 app.module.ts 使用 @NgModule()装饰器声明了一个模块类，并向外暴露了 AppModule。

(4) 找到组件的组或内容。在主组件文件 app.component.ts 中找到指定的选择器、模板和样式等，假如有外部模板、样式的话，需要加入外联的模板、样式。从 Angular 核心模块中引入 Component 组件模块，指明了选择器、模板、样式。

(5) 渲染页面中组件对应的区域。将组件渲染到文件 index.html 的选择器中，即打开文件 index.html 对应的页面。在将组件渲染到选择器之前，会先显示选择器本身的内容，直到将选择器中的内容替换为主组件中模板的内容。

微课视频

1.3 Angular 应用开发步骤

1.3.1 创建项目并修改文件 index.html

参考 1.2 节的方法在目录 d:\angularworkspace 下创建项目 firstangular。

修改文件 index.html，修改后的代码如例 1-8 所示。对比例 1-5 和例 1-8 代码(不考虑注释功能代码)，主要差别在一行代码，该行代码标签不同，代表了相关组件(渲染到该标签处的组件)采用了不同的选择器。

【例 1-8】 修改文件 index.html 后的代码。

```
<!doctype html>
<html lang="en">
<head>
```

```
    <meta charset = "utf - 8">
    <title> Angular Example </title>
    <base href = "/">
    <meta name = "viewport" content = "width = device - width, initial - scale = 1">
    <link rel = "icon" type = "image/x - icon" href = "favicon.ico">
  </head>
  <body>
    <!-- 下面一行代码,修改选择器 -->
    <root></root>
  </body>
</html>
```

1.3.2 创建组件文件

在 src 目录下创建子目录 pageex1,在 src\pageex1 目录下创建文件 pageex1. component. ts。一般来说,创建 HTML(或 TypeScript 等)文件之后都需要修改文件的代码;为了叙述的简便,将创建文件并修改代码的过程简称为创建文件。文件代码如例 1-9 所示。对比例 1-2 和例 1-9,可以发现内部模板、内部样式(即直接将模板、样式写入组件中)与外部模板、外部样式(即将独立的模板文件、样式文件导入组件中)的差异。

【例 1-9】 创建文件 pageex1. component. ts 的代码,定义一个组件并将其导出。

```
import {Component} from '@angular/core';
@Component({
  selector: 'root',
//请注意内部模板、内部样式与外部模板、外部样式的语法差异
//内部模板,模板是一块 HTML
  template: `<h1>{{ title }} app is running!</h1>
             <p> hi </p>`,
  //内部样式
  styles: ['p {font - family: "Times New Roman";font - size: larger;color: blue}']
})
export class FirstPage {
  //文本插值允许将文本、数据(如动态字符串值)合并到 HTML 模板中,用来替换模板中的变量
  title = 'firstpage';
}
```

1.3.3 创建模块文件

在 src\pageex1 目录下创建文件 pageex1. module. ts,代码如例 1-10 所示。

对比例 1-3 和例 1-10,可以发现两个文件代码相似,都是声明了组件、指明了根组件(启动组件)。

【例 1-10】 创建文件 pageex1. module. ts 的代码,设置启动组件。

```
import {NgModule} from '@angular/core';
import {FirstPage} from './pageex1.component';
import {BrowserModule} from "@angular/platform - browser";
@NgModule({
  //导入、声明模块组件
  declarations: [
    FirstPage
  ],
```

```
  imports: [
   BrowserModule
  ],
  bootstrap:  [FirstPage]
})
export class Pageex1Module {}
```

1.3.4 修改文件 main.ts

修改目录 src 下的文件 main.ts,代码如例 1-11 所示。对比例 1-4 和例 1-11 可以发现,两个文件代码相似,都指明了根模块(启动模块)。

【例 1-11】 修改文件 main.ts 的代码,设置启动模块。

```
//引入依赖
import {enableProdMode} from '@angular/core';
import {platformBrowserDynamic} from '@angular/platform-browser-dynamic';
//引入 pageex1 依赖
 import { Pageex1Module } from './pageex1/pageex1.module';
//引入环境依赖
import { environment } from './environments/environment';
//启用生产环境
if (environment.production) {
  enableProdMode();
}
platformBrowserDynamic().bootstrapModule(Pageex1Module)
   .catch(err => console.error(err));
```

1.3.5 运行项目

保持其他文件不变后运行程序,成功运行程序后,在浏览器地址栏中输入 localhost:4200,结果如图 1-8 所示。

图 1-8 成功运行程序后在浏览器地址栏中输入 localhost:4200 的结果

1.3.6 Angular 应用开发的一般步骤

在学习语法阶段,开发的一般步骤如下所述。

(1) 创建服务。如果不需要服务,则跳过此步骤。

(2) 创建组件。如果需要用到样式,则给出样式相关内容。

(3) 创建模块。

(4) 修改路由信息。如果不需要修改路由信息,则跳过此步骤。

(5) 修改文件 main.ts 和 index.html。如果不需要修改,则跳过此步骤。

(6) 运行项目。

1.4 TypeScript 基础

1.4.1 说明

TypeScript 是由微软开发的开源编程语言,是 JavaScript 的一个超集,支持 ES6。TypeScript 的设计目标是用于开发大型应用,它可以编译成纯 JavaScript,编译出来的 JavaScript 可以在任何浏览器上运行。

TypeScript 扩展了 JavaScript 的语法(即符合 JavaScript 语法的代码就符合 TypeScript 语法),因此现有的 JavaScript 代码可与 TypeScript 一起工作而无须任何修改。TypeScript 通过类型注解提供编译时的静态类型检查。TypeScript 可处理已有的 JavaScript 代码,并只对其中的 TypeScript 代码进行编译。

1.4.2 应用示例

在 src 目录下创建子目录 testtypescript,在 src\testtypescript 目录下创建文件 tsexample.component.ts,代码如例 1-12 所示。

【例 1-12】 创建文件 tsexample.component.ts 的代码,演示 TypeScript 基础语法的应用。

```
import {Component} from '@angular/core';
import {log} from "ng-zorro-antd/core/logger";
/* 组件、组件装饰器 */
@Component({
  selector: 'root',
  template: `
    <div>TypeScript 示例</div>
    <hr>
    <div>常见类型示例</div>
    <div>布尔类型: {{a}}</div>
    <div>数值类型: {{count}}</div>
    <div>数组类型: {{list[0]}}、{{list[1]}}、{{list[2]}}</div>
    <div>枚举类型: {{color}}</div>
    <div>元组类型: {{x}}、{{x[0]}}、{{x[0]}}</div>
    <div>任意值类型: {{y}}</div>
    <div>null、undefined 等多种类型(用|分隔): {{b}}</div>
    <!-- 下面一行注释实现了 void 用法,不断循环 -->
    <!-- div>void 类型: {{hello()}}</div -->
    <div>对象类型: {{student}}、{{student.getSno()}}、{{student.getName()}}</div>
    <div>接口类型: {{teacher}}、{{teacher.getName()}}、{{firstName}}、{{secondName}}</div>
    <hr>
    <div>装饰器示例 1: {{msg}}</div>
    <div>装饰器示例 2: {{p}}、{{fName}}、{{sName}}</div>
    <hr>
    <div>泛型示例 1: {{newarr[0]}}、{{newarr[1]}}、{{newarr[2]}}</div>
    <div>泛型示例 2: {{arradd[0]}}、{{arradd[1]}}、{{arradd[2]}}、{{arradd[3]}}</div>
    <hr>
    <div>属性 get/set 示例: {{monkey.age}}</div>
```

```
  `,
})
export class TsexampleComponent {
  //变量名：变量类型 = 取值
  a: boolean = false;
  count: number = 5;
  list: number[] = [1, 2, 3];
  color: Color = Color.Black;
  x: [String, number] = ["hello", 1];
  y: any = 1;
  b: number | undefined | null = 1;
  c: any;
  student: Student = new Student("001", "zsf");
  fullName: FullName = {
    firstName: "张",
    secondName: "三丰"
  }
  teacher: Teacher = new Teacher(this.fullName);
  firstName: string = this.getFirstName(this.teacher.getName());
  secondName: String = this.getSecondName(this.teacher.getName());
  mda: MethodDecoratorApp = new MethodDecoratorApp();
  msg: String = this.mda.testMessage("方法装饰器");
  p: PersonComponent = new PersonComponent("类装饰器", "Angular + TypeScript");
  fName: String = this.p.firstName;
  sName: String = this.p.secondName;
  arrays: number[] = [31, 24, 6];
  m: MinHelp<number> = new MinHelp(this.arrays);
  newarr: number[] = this.m.sort();
  arradd: number[] = this.m.add(18);
  monkey : Animal = new Animal();
  hello(): void {
    alert("hello");
  }
  getFirstName(fullName: FullName): string {
    return fullName.firstName;
  }
  getSecondName(fullName: FullName): String {
    return <String> fullName.secondName;
  }
}
//枚举
enum Color {
  Red,
  Black,
  Blue,
}
//类
class Student {
  sno: string | undefined;
  name: String | undefined;
  //构造方法(或称为构造函数)
  constructor(sno: string, name: String) {
    this.sno = sno;
    this.name = name;
```

```
  }
  getSno(): String {
    if (this.sno == undefined) {
      return "000";
    } else {
      return this.sno;
    }
  }
  getName(): String {
    if (this.name == undefined) {
      return "zsf";
    } else {
      return this.name;
    }
  }
}
//接口
interface FullName {
  firstName: string;
  secondName?: String; //?表示可选
}
class Teacher {
  name: FullName | undefined;
  defaultFullName = {
    firstName: "z",
    secondName: 'sf'
  }
  constructor(name: FullName) {
    this.name = name;
  }
  getName(): FullName {
    if (this.name == undefined) {
      return this.defaultFullName;
    } else {
      return this.name;
    }
  }
}
//方法装饰器
class MethodDecoratorApp {
  @log
  testMessage(arg: string) {
    return arg;
  }
}
//组件装饰器
@Component({
  selector: 'person',
  template: ``
})
export class PersonComponent {
  constructor(
    public firstName: String,
    public secondName: String
  ) {
  }
```

```
}
//泛型
class MinHelp<T> {
  arrays: T[] = [];                          //原始数组
  newarr: T[ ] = [];                         //排好序的数组
  arraysadd: T [] = [];                      //增加一个元素之后的数组
  constructor(arrays: T[]) {
    this.arrays = arrays;
  }
  //排序
  sort(): T[] {
    for (var j = 0; j < this.arrays.length - 1; j++) {
      var min = this.arrays[j];
      for (var i = j + 1; i < this.arrays.length; i++) {
        if (min > this.arrays[i]) {
          let tempt = this.arrays[i];
          this.arrays[i] = min;
          min = tempt;
        }
      }
      this.newarr[j] = min;
    }
    this.newarr[this.arrays.length - 1] = this.arrays[this.arrays.length - 1];
    this.arrays = this.newarr;
    return this.newarr;
  }
  //增加一个元素后并排好序
  add(element: T): T[] {
    let k;
    let flag = 0;
    for (k = this.arrays.length; k > 0; k--) {
      if (this.arrays[k - 1] >= element) {
        this.arraysadd[k] = this.arrays[k - 1];
      } else {
        this.arraysadd[k] = element;
        flag = k;
        break;
      }
    }
    for (let i = 0; i < flag; i++) {
      this.arraysadd[i] = this.arrays[i];
    }
    return this.arraysadd;
  }
}
class Animal {
  get age(): number {
    return this._age;
  }
  set age(value: number) {
    this._age = value;
  }
  private _age: number = 1;
}
```

在 src\testtypescript 目录下创建文件 tsexample.module.ts,代码如例 1-13 所示。

【例 1-13】 创建文件 tsexample. module. ts 的代码,设置启动组件。

```
import {NgModule} from '@angular/core';
import {PersonComponent, TsexampleComponent} from './tsexample.component';
import {BrowserModule} from "@angular/platform-browser";
@NgModule({
  //导入、声明模块组件
  declarations: [
    TsexampleComponent,
    PersonComponent,
  ],
  imports: [
   BrowserModule
  ],
  bootstrap:  [TsexampleComponent]  //启动处(组件)
})
export class TsexampleModule {}
```

修改目录 src 下的文件 main. ts,代码如例 1-14 所示。文件 main. ts 主要用来设置启动模块,本书后面章节对此不再说明。

【例 1-14】 修改文件 main. ts 的代码,设置启动模块。

```
import {enableProdMode} from '@angular/core';
import {platformBrowserDynamic} from '@angular/platform-browser-dynamic';
import {TsexampleModule} from "./testtypescript/tsexample.module";
import {environment} from './environments/environment';
if (environment.production) {
  enableProdMode();
}
platformBrowserDynamic().bootstrapModule(TsexampleModule)
   .catch(err => console.error(err));
```

保持其他文件不变下成功运行程序后,在浏览器地址栏中输入 localhost:4200,结果如图 1-9 所示。

图 1-9 成功运行程序后在浏览器地址栏中输入 localhost:4200 的结果

习题 1

一、简答题

1. 简述对 Angular 的理解。
2. 简述对 Angular 特点的理解。
3. 简述对 Angular 核心概念的理解。
4. 简述 Angular 应用开发的一般步骤。

二、实验题

1. 实现 Angular 的应用开发。
2. 实现 TypeScript 的应用开发。

第2章

模　板

2.1　模板概述

2.1.1　模板含义

Angular模板是一种HTML代码，与同常规HTML一样，可以在浏览器中渲染成视图或用户界面(UI)，但Angular模板比常规HTML功能更丰富。在Angular模板中，可以使用Angular特有语法。例如，可以通过内置的模板函数、变量、事件监听和数据绑定等功能来动态获取和设置DOM中的值，可以根据应用逻辑、应用状态和DOM数据来修改HTML内容。模板可以使用数据绑定来协调应用视图或用户界面与DOM中的数据，使用管道(在模板中声明显示内容的转换逻辑)在内容被显示出来之前对其进行转换，使用指令可以把程序逻辑应用到要显示的内容上。

由于Angular模板只是整个网页的一部分而不是整个网页，因此不需要包含诸如<html>、<body>等标签。为了消除脚本注入攻击的风险，Angular不支持模板中使用<script>标签。Angular会忽略<script>标签，并向浏览器控制台输出一条警告。

2.1.2　模板分类

可以采用引用外部模板或直接在组件内部编写内部模板这两种方式为组件定义模板。但注意，不能在组件中同时引用外部模板和编写内部模板代码。在引用外部模板时，通过templateUrl属性将外部模板文件名添加到@Component()装饰器中；在定义内部模板时，可以用template属性将内容模板代码添加到@Component()装饰器中；要想让模板代码跨越多行，可以使用反引号(`)把模板代码包裹起来。

可以在Angular应用中将SVG(Scalable Vector Graphics，可缩放矢量图形)文件用作模板。当使用SVG文件作为模板时，可以像HTML模板一样使用指令和实现数据绑定。使用这些功能可以动态生成交互式图形。

2.1.3　模板语句

模板语句可用于响应用户事件，如可以通过显示动态内容或提交表单之类的动作吸引用户。响应用户事件是Angular单向数据流的一个特点。可以在单个事件循环中更改应用中的任何内容。模板语句是一段带有Angular特点的HTML语句，支持使用类似于

JavaScript 的语法。在模板语句中可以使用赋值(=)和带有分号(;)的串联表达式,不允许使用 new、递增和递减运算符、赋值运算符(如+=)、按位运算符(如|和 &)、管道操作符(|)等 JavaScript 和模板表达式语法。

模板语句具有上下文(即语句所属应用中的特定部分)。模板语句只能引用模板语句所属的上下文中的内容,通常是组件实例。模板语句所属的上下文还可以引用模板自身的上下文属性。如代码< button (click) = "onSave($event)"> Save </button >中处理单击(click)事件的 onSave()方法将模板语句自身的 $event 对象(属性)用作参数。$event 的上下文是< button >语句模板。

模板上下文中的参数名称(标识符)优先于组件上下文中参数的名称。可以使用方法调用或基本属性赋值来让模板语句保持最小化。模板语句的上下文可以是组件实例或模板。因此,模板语句无法引用全局名称空间中的任何内容(如 Window 或 Document 等)。模板语句也不能调用 console. log()或 Math. max()等方法。

2.1.4 文本插值与模板表达式

在 Angular 模板中,可以通过用文本插值(简称为插值)的方法,动态更改用户界面中要显示的内容。也可以用插值的方法将动态字符串值合并到模板中,从而实现用户界面的动态更新。默认情况下,插值使用双花括号{{和}}作为定界符。在本书 1.3 节中的 pageex1 组件说明了文本插值的用法,例如,在组件中给出数据(如在本书 1.3 节中文本 firstpage)替换模板中的占位变量{{title}}。

模板语句是处于开始标签、结束标签之间的内容。模板表达式会产生一个值,它出现在双花括号{{和}}中,如{{1+1}}。Angular 解析表达式(如{{1+1}})时,先对表达式进行求值并将求值结果(即 2)转换成字符串(即"2"),再将字符串值赋给绑定目标(如可以是 HTML 元素、组件或指令)的某个属性。插值可以将模板表达式嵌入被标记的文本中。

模板表达式的解析器与模板语句的解析器有所不同。许多 JavaScript 表达式都是合法的模板表达式,但不能使用那些具有或可能引发副作用的 JavaScript 表达式作为模板表达式。不能在模板表达式中使用赋值(=,+=,-=)、运算符、new、typeof 或 instanceof 等。在模板表达式中可以使用";"或","串联起来的表达式、自增和自减运算符(++和--)、一些 ES6(及 ES6 以后)版本的运算符。模板表达式不支持位运算(如|和 &)、新的模板表达式运算符(如|和?. 以及!)。

模板表达式具有上下文(表达式所属应用中的特定部分)。通常,此上下文是组件实例。模板表达式可以引用模板语句上下文中的属性,如输入或引用变量。模板表达式不能引用全局命名空间中的任何内容(如 Windows 或 Document 等),也不能调用 console. log()或 Math. max()等方法。它们只能引用模板表达式上下文中的成员。模板表达式求值的上下文是模板变量、指令的上下文对象(如果有的话)以及组件成员的并集。

使用模板表达式时,尽量遵循以下规则。

(1) 使用短表达式。尽可能地使用属性名称或方法调用。将应用和业务逻辑保留在组件中(以用作属性名称或方法调用),更便于开发和测试。

(2) 快速执行。Angular 会在每个变更检测周期之后执行模板表达式。许多异步活动都会触发变更检测,如解析 promise、HTTP 结果、计时器事件、按键和鼠标移动。

(3) 表达式应尽快完成，以提升用户体验，尤其是在速度较慢的设备上。如计算值需要很多资源时，要考虑使用缓存。

(4) 没有可见的副作用。根据 Angular 的单向数据流模型，除了目标属性的值之外，模板表达式不应更改应用状态(用户界面)。读取组件值也不应更改其他显示值。用户界面应在整个渲染过程中保持稳定。

(5) 使用幂等表达式。幂等表达式能减少副作用，提高 Angular 的变更检测性能。幂等表达式总会返回完全相同的内容，除非其依赖值之一发生了变化；即在单独的一次事件循环中，被依赖的值不应该改变。例如，如果幂等的表达式返回一个字符串或数字，连续调用它两次会返回相同的字符串或数字。如果幂等的表达式返回一个对象(包括 Date 或 Array)，连续调用它两次会返回同一个对象的引用。

2.1.5 管道

Angular 管道在模板中声明用户界面中要显示的转换逻辑。Angular 自带了很多内置管道，如 date 和 currency，开发者也可以自定义新管道。可以创建自定义管道来封装那些内置管道没有提供的转换。自定义管道时，在带有@Pipe()装饰器的类中，定义一个转换函数，用来把输入值转换成供应用界面显示用的输出值。

可以在模板表达式中使用管道操作符(|)来指定值的转换方式。如代码“<p>生日是{{birthday|date}}</p>”中，birthday 值通过管道操作符(|)流向 date 管道。还可以用可选参数微调管道的输出。如代码“<p>{{amount|currency: 'EUR'}}欧元</p>”中将数值 amount 转换成欧元(EUR)表示的 currency。如果管道能接收多个参数，就用冒号(:)分隔这些值。如代码“{{ amount | currency:'EUR': '兑换后的欧元数量是'}} ”中把第二个参数(字符串'兑换后的欧元数量是')添加到输出字符串中。可以使用任何有效的模板表达式作为管道参数，如字符串字面量或组件的属性。有些管道需要至少一个参数，并且允许使用更多的可选参数，如代码{{slice:1:5}}中会创建一个新数组或字符串(以第 1 个元素开头，并以第 5 个元素结尾)。

2.2 模板绑定

2.2.1 属性绑定

DOM 是 HTML 和 XML 的编程接口，DOM 定义了访问和操作 HTML 页面的方法。一个 HTML 页面对应一个 DOM。在 HTML 标签中出现的往往是特性(attribute)，如代码<input id="name" value="Tom"/>中，input 标签有 id 和 value 两个特性。浏览器会将 HTML 代码解析为 DOM 对象，会将 HTML 标签特性(简称 HTML 特性)解析为 DOM 对象属性(简称 DOM 属性)。

DOM 属性(property)以 DOM 元素为对象，其附加内容在 DOM 中定义，如 childNodes 等。一般来说，DOM 属性与 HTML 特性并不是一一对应的，少量的 DOM 属性既是 DOM 属性也是 HTML 特性，如 id 等。非标准特性也不会自动映射为 DOM 属性。通常，HTML 特性代表初始值(初始化后不改变原值)，而 DOM 属性代表着当前值(随着属性值的变化而动态改变)。HTML 特性不区分大小写，而 DOM 属性区分大小写。

Angular应用中的属性绑定是指通过方括号([])将模板视图中的DOM属性与组件中的属性进行绑定。属性绑定可设置HTML元素或指令的属性值。使用属性绑定,可以执行切换按钮、以编程方式设置路径以及在组件之间共享值等功能。属性绑定可以将组件的属性(源属性)值赋值给目标元素(即HTML元素或指令等)的属性,目标元素的属性(简称为目标属性)就是要对其进行赋值的DOM属性。例如,代码< img [src]="itemImageUrl">中的目标属性是src属性,等号后面字符串中的内容是组件的属性itemImageUrl(源属性)。

属性绑定时,目标属性放在方括号内意味着将等号右侧字符串中的内容看作动态表达式(变量)进行求值。如例2-1中的代码< a [href]="neteasy">用中括号来表明属性href,就是根据变量neteasy的值表示跳转路径,即http://localhost:4200/net </a >。如果不使用方括号,Angular就会将等号右侧字符串中的内容看作字符串字面量(常量),并将此属性设置为该静态值。如例2-1中的代码< a href="neteasy">没有用中括号来标识属性href,就用字符串常量值(neteasy)表示跳转路径,即http://localhost:4200/neteasy </a >。

【例2-1】 创建文件bindexamplehome.component.ts的代码,演示本章的基础知识点。

```
import {Component} from '@angular/core';
import {Item} from "./item";
@Component({
  selector: 'root',
  template: `
  <h5>示例</h5>
  <button (click) = "deleteHero()"> Delete </button>
  <button (click) = "onSave( $ event)">保存</button>
  <p>生日是 {{birthday | date}}</p>
  <p>{{amount | currency:'EUR'}}欧元</p>
  <p>{{amount | currency:'EUR':'兑换后的欧元数量是'}}</p>
  <p>新数组{{a |slice:1:5}} </p>
  <img [src] = "itemImageUrl">
  <hr>
  <a [href] = "neteasy">用中括号来标识属性,就根据变量值表示跳转路径,即 http://localhost:
  4200/net </a>
  <hr>
  <a href = "neteasy">没有用中括号来标识属性,就用字符串表示跳转路径,即 http://localhost:
  4200/neteasy </a>
  <button [disabled] = "isUnchanged"> Disabled </button>
  <p [ngClass] = "classes">[ngClass] binding to the classes property making this blue </p>
  <app - item - detail [childString] = "parentString" [childItem] = "parentItem"> </app - item -
detail>
  <!-- 下面 4 行代码演示插值、property 绑定实现相同效果 -->
  <p><img src = "{{itemImageUrl}}"> 是 <i>插值</i> image.</p>
  <p><img [src] = "itemImageUrl"> 是 <i>property 绑定</i> image.</p>
  <p><span>"{{interpolationTitle}}"是 <i>插值</i> title.</span></p>
  <p>"<span [innerHTML] = "propertyTitle"></span>" 是 <i>property 绑定</i> title.</p>
  <button [attr.aria - label] = "actionName">{{actionName}} with Aria </button>
  <button [attr.aria - details] = "actionName2">{{actionName2}} with Aria </button>
  <table border = "1">
    <tr><th colspan = "2">横跨两列示例</th></tr>
    <tr><th colSpan = "2">横跨两列示例</th></tr>
    <tr><td>第 1 列</td><td>第 2 列</td></tr>
    <tr><td [colSpan] = "1 + 1"> colSpan 属性绑定</td></tr>
    <tr><td [attr.colspan] = "1 + 1"> colspan 特性绑定</td></tr>
```

```
  </table>
  <p class="classExpression">红色文本</p>
  <p class="{{cexpression}}">红色文本</p>
  <p [class]="cexpression">红色文本</p>
  <p [class.sale]="onSale" [class]="classExpression">类绑定实现样式覆盖(蓝色覆盖红色)
</p>
  <p [attr.class]="attributeClass" [class]="classExpression">特性绑定实现样式覆盖(蓝色
覆盖红色)</p>
  <p [class]="ce1">红色居中文本</p>
  <p [class]="ce2">文本</p>
  <p [class]="ce3">文本</p>
  <p [style.font-size]="bigfont" [style.width.%]="width">用中线格式编写样式特性</p>
  <p [style.font-size]="smallfont" [style.width.%]="width">用 camelCase 格式编写样式特
性</p>
  <p [style]="stringStyle">字符串样式特性</p>
  <p [style]="objectStyle">对象样式特性</p>
  <div [style.color]="color"> red </div>
  <!-- 绑定到 EventDetailComponent 的 deleteRequest 事件 -->
  <!-- deleteRequest 事件触发时 Angular 就会以 currentItem 为参数调用其父组件的 deleteItem(). -->
  <app-event-detail (deleteRequest)="deleteItem($event)" [item]="currentItem"></app-
event-detail>
  <app-directive-detail></app-directive-detail>
  <hr>
  <h4>Result: {{currentItem.name}}</h4>
  <!-- 用户做出更改时会引发 input 事件,该事件允许代码监听这些更改 -->
  <!-- 这个绑定会在一个上下文中执行该语句,此上下文中包含 DOM 事件对象 $event. -->
  <input [value]="currentItem.name"
         (input)="currentItem.name=getValue($event)">
  <app-doublebind-detail></app-doublebind-detail>
  <app-template-detail></app-template-detail>
  `,
  styles:['.special  { color :blue} ' +
  '.classExpression {color:red;}' +
  '.sale {color:blue;}' +
  '.attributeColor {color:blue;}' +
  '.fontEx {font-size:32px;}' +
  '.centerEx {text-align:center;}']
})
export class BindexamplehomeComponent {
  birthday: string = "20220303";
  amount: number = 100;
  a: any = [0,1,2,3,4,5,6];
  itemImageUrl: any = "../../assets/logo.png";
  neteasy: string = "net";
  isUnchanged: boolean = true;
  classes: string = 'special';
  parentString: string = '从父组件传给子组件的信息';
  parentItem: Item[] = [
    {
      id:21,
      name:'张三丰'
    }
  ];
  interpolationTitle: string = 'inter';
```

```
    propertyTitle: string = 'div';
    actionName: string = 'hi';
    actionName2: any = 'hello';
    onSale: any =  true ;
    classExpression: any = ' { color:red } ';
    attributeClass: any = 'attributeColor';
    cexpression: any = 'classExpression';
    ce1: any = 'classExpression fontEx centerEx ';   //多个类,用空格分开
    ce2: any = {fontEx: true, centerEx: false};      //类名作为键名的对象
    ce3: any = ['fontEx', 'centerEx'];               //类名的数组
    bigfont: any = '35px';
    smallfont: any = '15px';
    width: any = '50';
    stringStyle: any = "width: 150px; height: 40px; background - color: cornflowerblue;";
    objectStyle: any = {width: '150px', height: '40px', backgroundColor: 'blue'};
    color = 'red'
    currentItem = {id:21,name: 'teapot'};
    deleteItem(item: Item) {
      alert(`Delete the ${item.name}.`);
    }
    deleteHero() {
      alert("单击 Delete 按钮时,就会调用组件中的 deleteHero()方法");
    }
    onSave($event: MouseEvent) {
      alert($event.relatedTarget);
    }
    //调用 getValue($event.target) 来获取更改后的文本,并用它更新 name 属性
    //在模板中,$event.target 的类型只是 EventTarget
  //getValue()方法把 EventTarget 转为 HTMLInputElement 类型,以允许对其 value 属性进行类型安全的访问
    getValue(event: Event): string {
      return (event.target as HTMLInputElement).value;
    }
  }
```

例 2-1 中要根据布尔值禁用按钮 button 的功能,可以将其 DOM 的 disabled 属性所对应的源属性 isUnchanged 设置为 true。代码<button [disabled]="isUnchanged">Disabled</button>中属性 isUnchanged 的值是 true,会禁用该按钮(即该按钮显示为灰色且无法对其进行单击等操作)。

要设置指令(指令是为 Angular 应用中的元素添加额外行为的类)的属性,可以将指令放在方括号中,如例 2-1 中代码[ngClass]后跟等号和一个源属性 classes。要使用该属性 classes,必须在组件中声明它,如语句 classes = 'special '。Angular 会将 special 类应用到<p>元素,以便可以通过 special 来应用 CSS 样式。

要设置自定义组件的属性,可以将目标属性放在方括号[]中,且在其后面跟着等号和双引号(双引号中设置源属性)。要使用目标属性(如例 2-1 代码<app-item-detail [childString]="parentString"[childItem]="parentItem"></app-item-detail>中 childString)和源属性(如例 2-1 代码中的 parentString),必须在它们各自的类中声明它们。目标属性要在自定义的被调用组件(如例 2-2)中定义,源属性要在调用自定义组件的组件(如例 2-1)中定义。自定义组件(如例 2-2)代码中包含一个带有@Input()装饰器的

childString 属性,如语句@Input() childString = '',才能让数据流入其中。源属性的数据类型必须和目标属性的数据类型一致。例如,childString 的类型为字符串,parentString 也必须为字符串。

【例 2-2】 创建文件 item-detail.component.ts 的代码,定义组件 app-item-detail。

```
import {Component, Input} from '@angular/core';
import {Item} from "./item";
@Component({
  selector: 'app-item-detail',
  template: `
    <hr>
    <p>{{childString}}</p>
    <p>id:{{childItem[0].id}}, name: {{childItem[0].name}}</p>
  `,
})
export class ItemDetailComponent {
  @Input() childString = '';
  @Input() childItem :Item[] = [];
}
```

通过遵循一些规则,可以使用属性绑定来最大限度地减少错误并让代码保持可读性。如果模板表达式改变了所绑定的其他东西的值,那么这种更改就会产生副作用。Angular 可能显示也可能不显示更改后的值。如果 Angular 确实检测到了这个变化,就会抛出一个错误。

模板表达式返回的类型应该和目标属性所期望的值类型保持一致。例如,如果目标属性需要一个字符串,就返回一个字符串;如果需要一个数字,就返回一个数字;如果需要一个对象,就返回一个对象。

通常,插值和属性绑定可以达到相同的结果。将数据值渲染为字符串时,可以使用任一种形式,插值形式更易读。但是,要将元素属性设置为非字符串数据值时,必须使用属性绑定。

2.2.2 特性绑定

Angular 的特性(attribute)绑定可直接设置 HTML 特性值。使用特性绑定,可以提升无障碍性、动态设置应用样式以及同时管理多个 CSS 类或样式。Angular 应用开发时优先使用属性绑定。当元素没有可绑定的属性时,可以使用特性绑定。例如,ARIA(Accessible Rich Internet Applications,可访问的富互联网应用)和 SVG 只有特性,必须使用特性绑定。

特性绑定的语法类似于属性绑定,但不是直接在方括号之间放置元素的属性,而是在特性名称前面加上前缀 attr 和一个点(.)。然后,使用解析为字符串的表达式设置特性值。如代码<button [attr.aria-label]="actionName">{{actionName}} with Aria</button>中特性值为 aria-label,表达式为 actionName。

特性绑定和属性绑定最容易混淆的地方是属性 colSpan 和特性 colspan(大写字母 S、小写字母 s 不同)。特性 colspan 可以编程方式让表格保持动态。根据应用中用来填充表的数据量,某一行要跨越的列数可能会发生变化,如例 2-1 中代码“<tr><td [colSpan]="1+1">colSpan 属性绑定</td></tr>”。colSpan 属性能起到同样的作用,如例 2-1 中代码“<tr><td [attr.colspan]="1+1">colspan 特性绑定</td></tr>”。

2.2.3 类绑定

可以用类绑定从元素的 class 特性中添加和删除 CSS 类名称。创建单个类绑定时使用前缀 class 和点(.)后加 CSS 类的名称，如[class. sale]="onSale"。onSale 为真值时添加类，在表达式为假值时(undefined 除外)删除类。绑定到多个类时要使用[class]来设置表达式，如[class]="classExpression"。表达式 classExpression 可以取用空格分隔的类名字符串、以类名作为键名并将真或假表达式作为值的对象、类名的数组。具体示例可以参考例 2-1 中的代码。对于对象格式，Angular 会在其关联的值为真时才添加类。对于任何类似对象的表达式(如 Object、Array、Map 或 Set)，必须更改对象的引用 Angular 才能更新类列表。在不更改对象引用的情况下只更新其 attribute 是无效的。

2.2.4 样式绑定

可以用样式绑定来动态设置样式。创建对单个样式的绑定时使用前缀 style 和点(.)后加 CSS 样式特性名称，如[style. width]="width"。Angular 会将该特性(如 width)设置为绑定表达式的值，这个值通常是一个字符串。还可以添加单位扩展(表示单位的扩展内容)。例如，添加的单位扩展为 em 或%时，它的值应属于数字类型。

在绑定多个样式时，可直接绑定到[style]特性，如[style]="styleExpression"。styleExpression 可以是样式的字符串列表、一个对象(其键名是样式名，其值是样式值)。当把[style]绑定到对象表达式时，该对象的引用必须能改变，这样 Angular 才能更新这个类列表。在不改变对象引用的情况下更新其属性值是无效的。注意，不支持把数组绑定给[style]。

一个 HTML 元素可以将其 CSS 类列表和样式值绑定到多个源(如来自多个指令的宿主绑定)。当有多个相同的类名或样式属性的绑定时，Angular 使用一组优先规则来解决冲突并确定最终将哪些类或样式应用于元素。

类绑定或样式绑定越具体，其优先级就越高。绑定到具体类(如[class. foo])将优先于不特定[class]的绑定，绑定到特定样式(如[style. bar])将优先于不特定[style]的绑定。元素可以在其声明的模板中绑定，在其匹配的指令中进行宿主绑定，在其匹配的组件中进行宿主绑定。模板绑定是最具体的(具有最高优先级)，因为它们会直接且排他地应用于元素。指令宿主绑定被认为不太具体(优先级低于模板绑定)，因为指令可以在多个位置使用。指令通常会增强组件的行为，因此组件的宿主绑定具有最低的优先级。此时，class 和[class]具有相似的特异性，但是[class]绑定更优先一些，因为它是动态的。可以用 undefined 值来把高优先级的样式“委托”给较低优先级的样式。将样式属性设置为 null 可以确保样式被删除，而将其设置为 undefined 将导致 Angular 回到该样式的次高优先级绑定。

模板绑定中的属性绑定，如< div [class. foo]="hasFoo">或< div [style. color]="color">；映射表达式绑定，如< div [class]="classExpr">或< div [style]="styleExpr">；静态值绑定，如< div class="foo">或< div style="color: blue">。

指令(组件)宿主绑定中的属性绑定，如 host: {'[class. foo]': 'hasFoo'}或 host: {'[style. color]': 'color'}；映射表达式绑定，如 host: {'[class]': 'classExpr'}或 host: {'[style]': 'styleExpr'}；静态值绑定，如 host: {'class': 'foo'}或 host: {'style': 'color: blue'}。

优先级从高到低为模板绑定(属性绑定、映射表绑定、静态值绑定)、指令宿主绑定(属性绑定、映射表绑定、静态值绑定)、组件宿主绑定(属性绑定、映射表绑定、静态值绑定)。

在某些情况下,需要根据在 host 元素上以 HTML 特性的形式设置的静态值来区分组件或指令的行为。例如,可能有一个指令需要知道< button >或< input >元素的 type 值。@Attribute()参数装饰器可以通过依赖注入将 HTML 特性的值传递给组件(或指令)构造函数。注入的值将捕获指定 HTML 特性的当前值。将来对属性值的修改不会影响最初注入的值。例如,RouterOutlet 指令,该指令利用@Attribute()装饰器检索每个路由插槽上的唯一名称。要持续跟踪特性的值并更新关联的属性时,使用@Input()装饰器。若要将 HTML 特性的值注入组件或指令的构造函数中,则需要使用@Attribute()装饰器。

2.2.5 事件绑定

通过事件绑定可以监听并响应用户操作,如鼠标移动等操作。Angular 事件绑定语法由等号左侧括号内的目标事件名(如 click)和等号右侧引号内的模板语句(如 onSave()方法)组成,如< button(click)="onSave()"> Save </button >。Angular 还支持被动事件侦听器。

指令通常使用 Angular 的 EventEmitter 引发(raise)自定义事件。指令创建一个 EventEmitter 并将其对外暴露为属性。然后,指令调用 EventEmitter. emit(data)方法发出事件,传入消息数据,该消息数据可以是任何东西。父指令通过绑定到该属性来监听事件,并通过传入的 $ event 对象接收数据。

为了确定事件的目标,Angular 会检查目标事件的名称是否与已知指令的事件属性匹配。如果目标事件名称未能匹配元素上的事件或指令(含自定义指令)的输出属性,则 Angular 将报告"未知指令"错误。在事件绑定中,Angular 会为目标事件配置事件处理函数,还可以将事件绑定用于自定义事件。

当组件或指令引发(raise)事件时,处理程序就会执行模板语句。模板语句会执行一个动作来响应这个事件。处理事件的常见方法之一是把事件对象 $ event 传给处理该事件的方法。$ event 对象通常包含该方法所需的信息,例如用户名或图片 URL。

目标事件决定了 $ event 对象的形态。如果目标事件是来自原生 DOM 元素,那么 $ event 是一个 DOM 事件对象,具有 target 和 target. value 等属性。如果目标事件属于某个指令或组件,那么 $ event 就具有指令或组件中生成的状态。在模板中,$ event. target 的类型是 EventTarget。

2.2.6 双向绑定

如果没有框架,开发人员负责把数据值推送到 HTML 控件中,并把来自用户的响应转换成动作和对值的更新。手动实现这些功能会很枯燥、容易出错,且代码难以阅读。Angular 支持双向数据绑定(主要用于模板驱动表单中),这是一种对模板中各个部件与组件中的各个部件进行协调的机制。往模板 HTML 中添加绑定标签可以告诉 Angular 该如何连接它们。

双向绑定为应用中的组件提供了一种共享数据的方式。可以使用双向绑定来侦听事件并在父组件和子组件之间同步更新值。双向绑定将属性绑定与事件绑定结合在一起构成一种单独写法:通过属性绑定设置特定的元素属性,事件绑定侦听元素更改事件。在双向绑定中,数据属性值通过属性绑定从组件流到用户界面中某个元素,用户对用户界面某个元素的修改通过事件绑定流回组件,把属性值设置为最新的值。

Angular 的双向绑定语法是方括号和圆括号的组合[()]。[]进行属性绑定,()进行事件绑定,如代码< app-sizer [(size)]="fontSizePx"> </app-sizer >中[(size)]。为了使双向数据绑定有效,@Output()装饰器修饰的属性名字必须遵循 name+"Change"命名方式进行命名。其中,name 是相应@Input()装饰器修饰的属性名字(name)。例如,如果@Input()装饰器修饰的属性名字为 size,则@Output()装饰器修饰的属性必须为命名为 sizeChange。

双向绑定代码可以拆成单独的属性绑定和事件绑定形式的代码,如代码< app-sizer [(size)]="fontSizePx"> </app-sizer >可以拆成< app-sizer [size]="fontSizePx"(sizeChange)="fontSizePx= $event"> </app-sizer >。其中,$event 变量包含组 sizeChange 事件的数据。当用户单击按钮时,Angular 将$event 赋值给 fontSizePx。

因为没有任何原生 HTML 元素遵循了 x 值和 xChange 事件的命名模式,所以与表单元素进行双向绑定需要使用 NgModel。

2.3 模板变量和模板输入变量

2.3.1 模板变量

模板变量可以帮助用户在模板的一部分使用另一部分的数据。使用模板变量,可以执行某些任务,如响应用户输入或微调应用视图中的表单。模板变量可以引用模板中的 DOM 元素、指令、TemplateRef、Web 组件等。

可以使用井号(#)来声明一个模板变量,如在代码< input #phone placeholder="phone number"/>中就声明了模板变量 phone。在声明了模板变量之后就可以在组件模板中的任何地方引用该变量。

Angular 根据声明模板变量所在的位置给模板变量赋值:在组件上声明变量,该变量就会引用该组件实例。如果在标准的 HTML 标签上声明变量,该变量就会引用该元素。如果在< ng-template >元素上声明变量,该变量就会引用一个 TemplateRef 实例来代表此模板。如果该变量在右侧指定了一个名字,如#var="ngModel",那么该变量就会引用所在元素上具有这个导出(exportAs)名字的指令或组件。在大多数情况下,Angular 会把模板变量的值设置为它所在的元素。如< input #phone placeholder="phone number" />中 phone 引用的是电话号码,代码< button(click)="callPhone(phone. value)"> Call </button >中按钮的 click 事件处理程序会把< input >中的电话号码(即 phone 的值)传给 callPhone()方法。

可以在包含此模板变量的模板中的任何地方引用模板变量;而结构型指令(如 *ngIf 和 *ngFor)同样充当了模板的边界。不能在这些边界之外访问其中的模板变量。同名变量在模板中只能定义一次,这样运行时它的值就是可预测的。在嵌套模板中,内部模板可以访问外模板定义的模板变量。访问父模板中的模板变量是可行的,因为子模板会从父模板继承上下文。但是,从外部的父模板无法访问本模板中的变量。

对于结构型指令,Angular 无法知道模板是否曾被实例化过。Angular 无法访问该值并返回错误。在< ng-template >上声明变量时,该变量会引用一个 TemplateRef 实例来表示该模板。如< ng-template #ref3 > </ng-template > < button(click)="log(ref3)"> Log type of #ref </button >中,单击(click)按钮会调用 log()方法,log()方法把#ref3 的值输出到控制台。因为#ref 变量在< ng-template >上,所以它的值是一个 TemplateRef。

2.3.2 模板输入变量

模板输入变量是可以在模板的单个实例中引用的变量。可以用关键字 let 声明模板输入变量，如 let hero。此变量的范围仅限于可复写模板中的单个实例，可以在其他结构型指令的定义中再次使用相同的变量名；相反，可以通过在变量名称前加上＃来声明模板变量，如＃var。模板变量引用其附加的元素、组件或指令。模板输入变量和模板变量名称具有各自的名称空间，且模板输入变量 hero 和＃hero 中的模板变量 hero 是不同的。

2.4 模板的基础应用

2.4.1 基础代码

在项目 firstrangular 的 src 目录下创建子目录 examples，在项目 src\examples 目录下创建子目录 bindexamples，在 src\examples\bindexamples 目录下创建文件 bindexamplehome.component.ts，代码如例 2-1 所示。在 src\examples\bindexamples 目录下创建文件 item-detail.component.ts，代码如例 2-2 所示。

在 src\examples\bindexamples 目录下创建文件 item.ts，代码如例 2-3 所示。

【例 2-3】 创建文件 item.ts 的代码，定义接口。

```
export interface Item {
  id: number,
  name: string
}
```

2.4.2 事件

在 src\examples\bindexamples 目录下创建文件 event-detail.component.ts，代码如例 2-4所示。并在项目的 src\assets 目录下准备图片文件 logo.png。

【例 2-4】 创建文件 event-detail.component.ts 的代码，演示在组件中直接定义 click 事件方法。

```
import {Component, EventEmitter, Input, Output} from "@angular/core";
import {Item} from "./item";
@Component({
  selector: 'app-event-detail',
  template: `
    <hr>
    <img src="{{itemImageUrl}}" [style.display]="displayNone">
    <!-- 它会显示 item 信息并响应用户操作 -->
    <span [style.text-decoration]="lineThrough">{{ item.id }},{{ item.name }}
    </span>
    <!-- 显示了一个删除按钮,会引发一个报告用户要求删除的事件 -->
    <button (click)="delete()">Delete</button>
  `,
  styles:['.detail {\n' +
  '  border: 1px solid rgb(25, 118, 210);\n' +
  '  padding: 1rem;\n' +
```

```
  '  margin: 1rem 0;\n' +
  '}\n' +
  '\n' +
  'img {\n' +
  '  max - width: 100px;\n' +
  '  display: block;\n' +
  '  padding: 1rem 0;\n' +
  '}']
})
export class EventDetailComponent {
  @Input() item! : Item;
  itemImageUrl: any = "../../assets/logo.png";
  lineThrough = '';
  displayNone = '';
  @Input() prefix = '';
  //组件定义了一个 deleteRequest 返回 EventEmitter 的属性
  @Output() deleteRequest = new EventEmitter < Item >();
  //当用户单击 Delete 时,该组件将调用 delete() 方法
  delete() {
    //delete() 方法让 EventEmitter 发出 Item 对象
    this.deleteRequest.emit(this.item);
    this.displayNone = this.displayNone ? '' : 'none';
    this.lineThrough = this.lineThrough ? '' : 'line - through';
  }
}
```

在 src\examples\bindexamples 目录下创建文件 click. directive. ts,代码如例 2-5 所示。

【例 2-5】 创建文件 click. directive. ts 的代码,演示自定义事件 myClick 的服务。

```
import {Directive, ElementRef, EventEmitter, Output} from '@angular/core';
@Directive({selector: '[myClick]'})
export class ClickDirective {
  @Output('myClick') clicks = new EventEmitter < string >();
  toggle = false;
  constructor(el: ElementRef) {
    el.nativeElement
      .addEventListener('click', (event: Event) => {
        this.toggle = !this.toggle;
        this.clicks.emit(this.toggle ? 'Click!' : '');
      });
  }
}
```

在 src\examples\bindexamples 目录下创建文件 directive-detail. component. ts,代码如例 2-6 所示。

【例 2-6】 创建文件 directive-detail. component. ts 的代码,演示自定义事件 myClick 的应用。

```
import {Component} from "@angular/core";
@Component({
  selector: 'app - directive - detail',
  template: `
    < hr >
    < h4 > myClick is an event on the custom ClickDirective:</h4 >
```

```
    <!-- 若目标事件 myClick 未能匹配元素上事件或 ClickDirective 输出属性,就会报告"未知指
    令"错误 -->
    <button (myClick) = "clickMessage = $ event" clickable> click with myClick </button>
    {{clickMessage}}
  `,
})
export class DirectiveDetailComponent {
  clickMessage = '';
}
```

在 src\examples\bindexamples 目录下创建文件 sizer. component. ts，代码如例 2-7 所示。

【例 2-7】 创建文件 sizer. component. ts 的代码，演示多个事件的用法。

```
import {Component, Input, Output, EventEmitter} from '@angular/core';
@Component({
  selector: 'app - sizer',
  template: '<div>\n' +
    '  <button (click) = "dec()" title = "smaller"> - </button>\n' +
    '  <button (click) = "inc()" title = "bigger"> + </button>\n' +
    '  <label [style.font - size.px] = "size"> FontSize: {{size}}px </label>\n' +
    '</div>',
})
export class SizerComponent {
  @Input() size! : number | string;
  @Output() sizeChange = new EventEmitter<number>();
  dec() {this.resize( - 1);}
  inc() {this.resize( + 1);}
  resize(delta: number) {
    this.size = Math.min(40, Math.max(8, + this.size + delta));
    this.sizeChange.emit(this.size);
  }
}
```

2.4.3 绑定

在 src\examples\bindexamples 目录下创建文件 doublebind-detail. component. ts，代码如例 2-8 所示。

【例 2-8】 创建文件 doublebind-detail. component. ts 的代码，演示双向绑定的用法。

```
import {Component} from "@angular/core";
@Component({
  selector: 'app - doublebind - detail',
  template: `
    <hr>
    <div id = "two - way - 1">
      <app - sizer [(size)] = "fontSizePx"></app - sizer>
      <div [style.font - size.px] = "fontSizePx"> Resizable Text </div>
      <label> FontSize (px): <input [(ngModel)] = "fontSizePx"></label>
    </div>
    <br>
    <div id = "two - way - 2">
      <h2> De - sugared two - way binding </h2>
```

```
    <app-sizer [size]="fontSizePx" (sizeChange)="fontSizePx=$event"></app-sizer>
  </div>
  `,
})
export class DoublebindDetailComponent {
  constructor() { }
  fontSizePx = 16;
}
```

2.4.4 变量

在 src\examples\bindexamples 目录下创建文件 template-detail.component.ts，代码如例 2-9 所示。

【例 2-9】 创建文件 template-detail.component.ts 的代码，演示模板变量和模板输入变量的用法。

```
import {Component, ViewChild} from '@angular/core';
import {NgForm} from '@angular/forms';
@Component({
  selector: 'app-template-detail',
  template: `
    <hr>
    <input #phone placeholder="phone number"/>
    <button (click)="callPhone(phone.value)">Call</button>
    <!-- 将 NgForm 与模板变量一起使用 -->
    <!-- 若没有 NgForm 属性值,itemForm 引用的值将是 HTMLFormElement,也就是<form>元素 -->
    <form #itemForm="ngForm" (ngSubmit)="onSubmit(itemForm)">
      <label for="name">Name</label>
      <input type="text" id="name" class="form-control" name="name" ngModel required/>
      <button type="submit">Submit</button>
    </form>
    <div [hidden]="!itemForm.form.valid">
      <p>{{submitMessage}}</p>
    </div>
    <p>Here's the desugared syntax:</p>
    <pre><code [innerText]="desugared0"></code></pre>
    <pre><code [innerText]="desugared1"></code></pre>
    <h3>Accessing from outside parent template. (Doesn't work.)</h3>
    <div class="example">
      <input *ngIf="true" #ref2 type="text" [(ngModel)]="secondExample"/>
    </div>
    <p>Here's the desugared syntax:</p>
    <pre><code [innerText]="desugared2"></code></pre>
    <h3>*ngFor and template reference variable scope</h3>
    <pre><code [innerText]="ngForExample"></code></pre>
    <h3>Accessing a on an <code>ng-template</code></h3>
    See the console output to see that when you declare the variable on an <code>ng-template
  </code>, the variable refers to a <code>TemplateRef</code> instance, which represents the
  template.
    <ng-template #ref3></ng-template>
    <button (click)="log(ref3)">Log type of #ref</button>
    <ng-template #hero let-hero let-i="index" let-odd="isOdd">
      <div [class]="{'odd-row': odd}">{{i}}:{{hero.name}}</div>
```

```
    </ng-template>
  `,
  styles: ['h3 {\n' +
  '  font-weight: 700;\n' +
  '}\n' +
  '\n' +
  'pre, .wrapper {\n' +
  '  background-color: rgb(240, 250, 250);\n' +
  '  padding: 1rem;\n' +
  '  border: 1px solid #444;\n' +
  '}\n' +
  '\n' +
  'input {\n' +
  '  margin: .5rem;\n' +
  '  padding: .5rem;\n' +
  '}']
})
export class TemplateDetailComponent {
  public firstExample = 'Hello, World!';
  public secondExample = 'Hello, World!';
  public ref2 = '';
//在这种情况下,有一个包含这个<span>的隐式<ng-template>,而该变量的定义在该隐式模板之外
  public desugared0 = `
  <input #ref1 type="text" [(ngModel)]="firstExample" />
    <span *ngIf="'true">Value: {{ ref1.value }}</span>
  `  ;
  public desugared1 = `
  <input #ref1 type="text" [(ngModel)]="firstExample" /><!-- A new template! -->
  <ng-template [ngIf]="true">
    <!-- … and it works   -->
    <span>Value: {{ref1.value}}</span>
  </ng-template>`  ;
  public desugared2 = `<ng-template [ngIf]="true">
    <input #ref2 type="text" [(ngModel)]="secondExample" />
  </ng-template>
  <!-- 尝试在模板外访问 ref2 的操作无效 -->
  <span>Value: {{ ref2?.value }}</span>`;
  public ngForExample = `<ng-container *ngFor="let i of [1,2]">
    <input #ref type="text" [value]="i" />
  </ng-container>
<!-- ref.value 不起作用。*ngFor 将模板实例化了两次, *ngFor 对数组中的两个元素进行迭代,
无法定义 ref.value 指向的是谁。-->
  {{ref.value}}`;
  @ViewChild('itemForm', {static: false}) form!: NgForm;
  private _submitMessage = '';
  get submitMessage() {
    return this._submitMessage;
  }
  onSubmit(form: NgForm) {
    this._submitMessage = 'Submitted. Form value is ' + JSON.stringify(form.value);
  }
  callPhone(value: string) {
    alert(`Calling ${value} ...`);
  }
```

```
  callFax(value: string) {
    console.warn(`Faxing ${value} ...`);
  }
  log(ref3: any) {
    console.warn(ref3.constructor);
  }
}
```

2.4.5 模块

在 src\examples\bindexamples 目录下创建文件 app-bindexample.module.ts,代码如例 2-10 所示。

【例 2-10】 创建文件 app-bindexample.module.ts 的代码,定义路由并声明组件。

```
import {NgModule} from '@angular/core';
import {BrowserModule} from '@angular/platform-browser';
import {BindexamplehomeComponent} from './bindexamplehome.component';
import {RouterModule} from "@angular/router";
import {ItemDetailComponent} from "./item-detail.component";
import {EventDetailComponent} from "./event-detail.component";
import {DirectiveDetailComponent} from "./directive-detail.component";
import {ClickDirective} from "./click.directive";
import {DoublebindDetailComponent} from "./doublebind-detail.component";
import {SizerComponent} from "./sizer.component";
import {FormsModule} from "@angular/forms";
import {TemplateDetailComponent} from "./template-detail.component";
@NgModule({
  imports: [
    BrowserModule,
    RouterModule.forRoot([
      {path: 'bindexample', component: BindexamplehomeComponent},
    ]),
    FormsModule,
  ],
  declarations: [
    BindexamplehomeComponent,
    ItemDetailComponent,
    EventDetailComponent,
    DirectiveDetailComponent,
    ClickDirective,
    DoublebindDetailComponent,
    SizerComponent,
    TemplateDetailComponent
  ],
})
export class AppBindexampleModule { }
```

在 src\examples 目录下创建文件 examplesmodules1.module.ts,代码如例 2-11 所示。

【例 2-11】 创建文件 examplesmodules1.module.ts 的代码,设置启动组件。

```
import {NgModule} from '@angular/core';
import {BindexamplehomeComponent} from './bindexamples/bindexamplehome.component';
import {AppBindexampleModule} from "./bindexamples/app-bindexample.module";
@NgModule({
```

```
  imports: [
    AppBindexampleModule
  ],
  bootstrap: [BindexamplehomeComponent]
})
export class ExamplesmodulesModule1 {}
```

修改目录 src 下的文件 main.ts，代码如例 2-12 所示。

【例 2-12】 修改文件 main.ts 的代码，设置启动模块。

```
import {enableProdMode} from '@angular/core';
import {platformBrowserDynamic} from '@angular/platform-browser-dynamic';
import {environment} from './environments/environment';
import {ExamplesmodulesModule} from "./examples/examplesmodules1.module";
if (environment.production) {
  enableProdMode();
}
platformBrowserDynamic().bootstrapModule(ExamplesmodulesModule1)
   .catch(err => console.error(err));
```

2.4.6 运行结果

保持其他文件不变后运行程序，成功运行程序后，在浏览器地址栏中输入 localhost:4200，部分结果如图 2-1 所示。更多的结果请读者自己参考源代码进行验证。

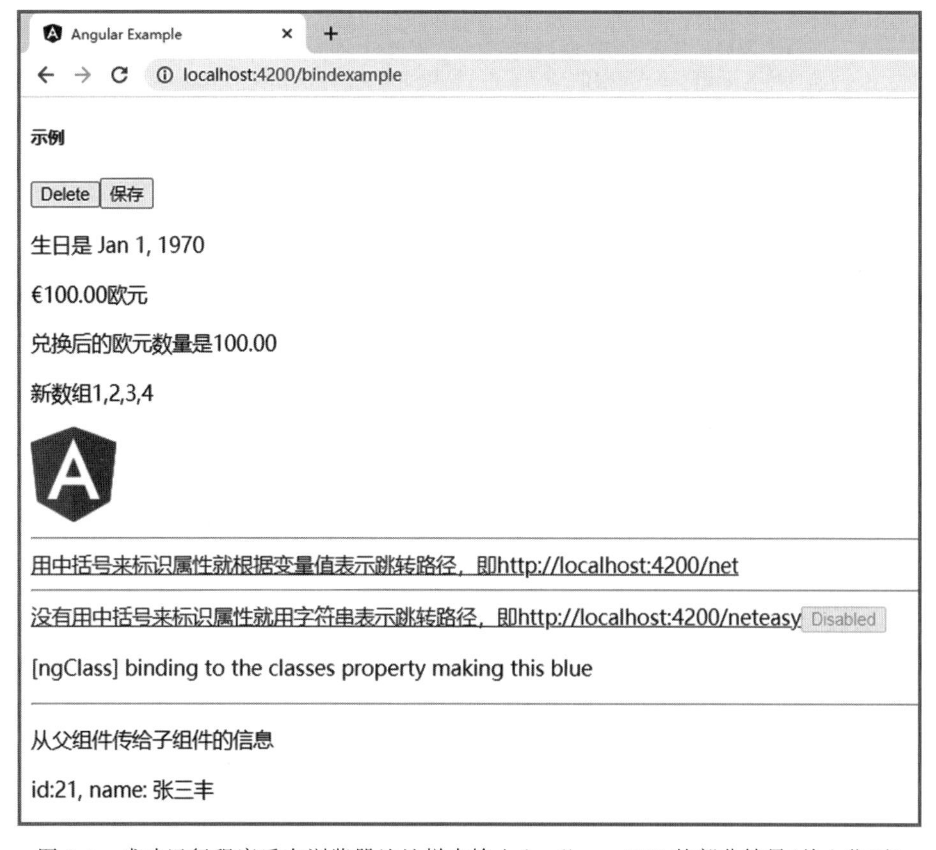

图 2-1　成功运行程序后在浏览器地址栏中输入 localhost:4200 的部分结果(从上往下)

微课视频

2.5 模板的综合应用开发

2.5.1 组件及相关文件

在项目 src\examples 目录下创建子目录 templateexamples，在 src\examples\templateexamples 目录下创建文件 templateexamples.component.ts，代码如例 2-13 所示。

【例 2-13】 创建文件 templateexamples.component.ts 的代码，演示本章内容的应用。

```
import {Component, EventEmitter, Output} from '@angular/core';
@Component({
  selector: 'main-root',
  template: `
    <div>用插值解析表达式 The sum of 1 + 1 is {{1 + 1}}.</div>
    <ul>
      <!-- list 结构数据循环显示 -->
      <li *ngFor="let author of authors">{{ author.value }}</li>
    </ul>
    <button (click)="clickEventMethod()">响应事件模板语句示例</button>
    <div>单一管道示例(日期 {{ birthday }} 格式化) {{ birthday | date :"yy/MM/dd"  }}</div>
    <div>多个管道示例(日期默认格式化,并转换成大写) {{ birthday | date| uppercase  }}</div>
    <div>多个管道示例(日期格式化,并转换成大写) {{ birthday | date: "full" | lowercase }}</div>
    <div>管道示例(货币,欧元) {{ amount | currency:'EUR' }} </div>
    <div>管道示例(把数字 {{ pi }} 转换成带小数点的字符串){{pi | number}}</div>
    <div>管道示例(把数字 {{ a }} 转换成百分比字符串): {{a | percent}}</div>
    <div>插值实现属性设置示例(img 元素的 src 属性)<img src=" {{ itemImageUrl }} "></div>
    <div>属性绑定示例(img 元素的 src 属性)等效实现<img [src]="itemImageUrl"></div>
    <div>属性绑定示例(绑定到表格 colspan)
      <tr>
        <td [attr.colspan]="1 + 1">One-Two</td>
      </tr>
    </div>
    <div>样式绑定示例</div>
    <nav [style]='navStyle'>
      <a [style.text-decoration]="activeLinkStyle">Home Page</a>
      <a [style.text-decoration]="linkStyle">Login</a>
    </nav>
    <div>双向绑定示例
      <button (click)="dec()" title="smaller">-</button>
      <button (click)="inc()" title="bigger">+</button>
      <label [style.font-size.px]="size">FontSize: {{size}}px</label>
    </div>
    <div>模板变量示例
      <input #phone placeholder="phone number"/>
      <button (click)="callPhone(phone.value)">Call</button>
    </div>
    <hr>
  `,
  //内部样式
  //注意内部模板和内部样式的语法差异
  styles: ['div {font-family: "Times New Roman";font-size: larger;color: blue}']
})
export class TemplateexamplesComponent {
  //文本插值允许将文本、数据(如动态字符串值)合并到 HTML 模板中,用来替换模板中变量
```

```
    authors: any = [{value: '司马迁'}, {value: '班固'}, {value: '司马光'}];
    clickEventMethod() {
      alert('单击按钮时 Angular 就会调用组件中 clickEventMethod()方法');
    }
    birthday: any = '2022 - 02 - 03';
    amount: number = 100;
    pi: number = 3.14159265359;
    a: number = 0.259;
    itemImageUrl: String = '../assets/logo.png';
    navStyle: String = 'font - size: 1.2rem; color: cornflowerblue;';
    linkStyle: String = 'underline';
    activeLinkStyle: String = 'overline';
    //sizeChange 事件是一个@Output()装饰器,它允许数据从 sizerComponent 流出到父组件
    @Output() sizeChange = new EventEmitter<number>();
    //inc()方法和 dec()方法分别使用 + 1 或 - 1 调用 resize()方法
    //dec()方法用于减小字体大小,inc()方法用于增大字体大小
    //它使用新的 size 值引发 sizeChange 事件
    dec() {
      this.resize( - 1);
    }
    inc() {
      this.resize( + 1);
    }
    size: number = 12;
    //使用 resize() 在最小/最大值的约束内更改 size 属性的值,并发出带有新 size 值的事件
    resize(delta: number) {
      this.size = Math.min(40, Math.max(8, + this.size + delta));
      this.sizeChange.emit(this.size);
    };
    phone: String = '12345678901';
    callPhone(number: any) {
      alert(number)
    }
  }
```

在 src\examples\templateexamples 目录下创建文件 svg.component.svg,代码如例 2-14 所示。

【例 2-14】 创建文件 svg.component.svg 的代码,定义 svg 文件。

```
<svg>
  <g>
    <rect x = "0" y = "0" width = "100" height = "100" [attr.fill] = "fillColor" (click) =
"changeColor()" />
    <text x = "120" y = "50">click the rectangle to change the fill color</text>
  </g>
</svg>
```

在 src\examples\templateexamples 目录下创建文件 svg.component.css,代码如例 2-15 所示。

【例 2-15】 创建文件 svg.component.css 的代码,定义 svg 文件对应的样式。

```
svg {
  display: block;
  width: 100%;
}
```

在 src\examples\templateexamples 目录下创建文件 svgtemplates. components. ts，代码如例 2-16 所示。

【例 2-16】 创建文件 svgtemplates. components. ts 的代码，演示在组件中使用 svg 模板。

```
import {Component} from '@angular/core';
@Component({
  selector: 'svg-root',
  templateUrl: './svg.component.svg',
  styleUrls: ['./svg.component.css']
})
export class SvgtemplatesComponents {
  fillColor = 'rgb(255, 0, 0)';
  changeColor() {
    const r = Math.floor(Math.random() * 256);
    const g = Math.floor(Math.random() * 256);
    const b = Math.floor(Math.random() * 256);
    this.fillColor = `rgb(${r}, ${g}, ${b})`;
  }
}
```

在 src\examples\templateexamples 目录下创建文件 templatehome. component. ts，代码如例 2-17 所示。

【例 2-17】 创建文件 templatehome. component. ts 的代码，演示两个组件组合成一个组件。

```
import {Component} from '@angular/core';
@Component({
  selector: 'root',
  template: `
    <main-root></main-root>
    <svg-root></svg-root>
  `,
  styles: []
})
export class TemplatehomeComponent {}
```

2.5.2 模块创建

在 src\examples\templateexamples 目录下创建文件 app-template. module. ts，代码如例 2-18 所示。

【例 2-18】 创建文件 app-template. module. ts 的代码，定义路由并声明组件。

```
import {NgModule} from '@angular/core';
import {BrowserModule} from "@angular/platform-browser";
import {RouterModule} from "@angular/router";
import {TemplatehomeComponent} from "./templatehome.component";
import {SvgtemplatesComponents} from "./svgtemplates.components";
import {TemplateexamplesComponent} from "./templateexamples.component";
@NgModule({
  imports: [
    BrowserModule,
    RouterModule.forRoot([
      {path: 'template', component: TemplatehomeComponent},
    ]),
  ],
```

```
  declarations: [
    SvgtemplatesComponents,
    TemplateexamplesComponent,
    TemplatehomeComponent
  ],
})
export class AppTemplateModule { }
```

修改 src\examples 目录下的文件 examplesmodules1.module.ts，代码如例 2-19 所示。

【例 2-19】 修改文件 examplesmodules1.module.ts 的代码，定义启动模块。

```
import {NgModule} from '@angular/core';
import {AppTemplateModule} from './templateexamples/app-template.module';
import {TemplatehomeComponent} from './templateexamples/templatehome.component';
@NgModule({
  imports: [
    AppTemplateModule
  ],
  bootstrap: [TemplatehomeComponent]
})
export class ExamplesmodulesModule1 {}
```

2.5.3 模块的综合应用运行结果

保持其他文件不变后运行程序，成功运行程序后，在浏览器地址栏中输入 localhost:4200，结果如图 2-2 所示。对图 2-2 所示中的按钮操作、文本框等功能的操作请读者自己参考源代码进行操作。

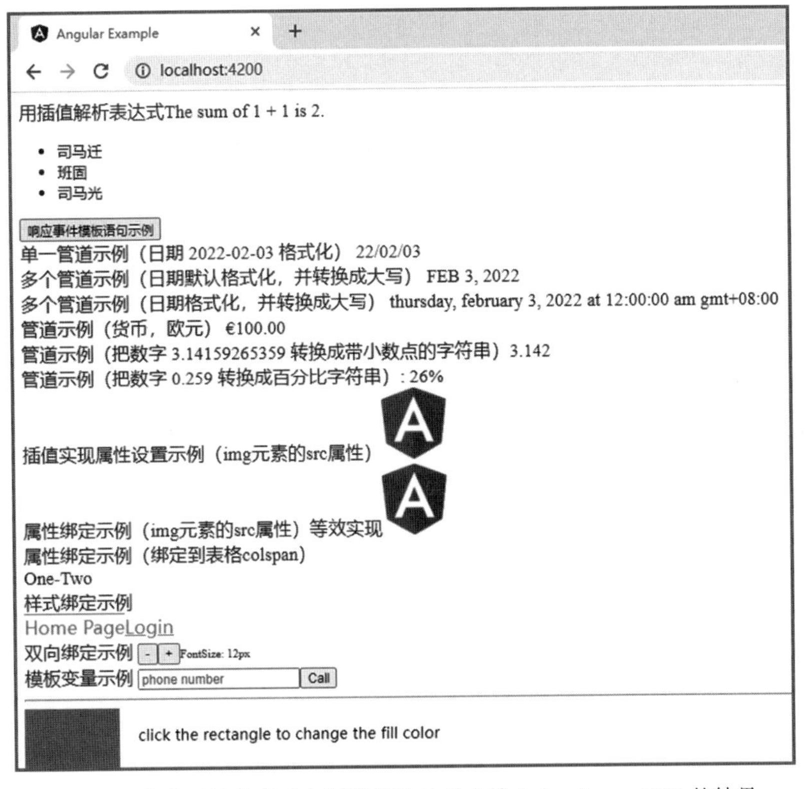

图 2-2 成功运行程序后在浏览器地址栏中输入 localhost:4200 的结果

习题 2

一、简答题

1. 简述对模板的理解。
2. 简述对各类绑定的理解。
3. 简述对模板变量的理解。
4. 简述对 SVG 模板的理解。

二、实验题

实现模板的应用开发。

第 3 章

指　令

3.1 指令概述

3.1.1 指令含义

Angular 的模板是动态的，当 Angular 渲染它们时，会根据模板中的指令给出的指示对 DOM 进行转换。指令是为 Angular 应用中的元素添加额外行为的类，是一个带有@Directive()装饰器的类。从技术角度上说，组件也是指令。由于组件对 Angular 应用来说非常重要和独特，因此 Angular 专门定义了@Component()装饰器，它使用一些面向模板的特性扩展了@Directive()装饰器，即组件是带有模板特性的指令。

Angular 本身定义了一系列结构型指令、属性型指令等内置指令，可以管理表单、列表、样式以及用户界面上要让用户看到的其他内容；也可以使用@Directive()装饰器定义自定义指令。

3.1.2 指令类型

指令可以分为组件、属性型指令、结构型指令等不同类型。为了区分，本书将组件和指令分开介绍。属性型指令可以更改元素、组件或其他指令的外观或行为；结构型指令可以通过添加和删除或替换 DOM 元素来更改 DOM 布局。

3.1.3 指令和模板的关系

指令的元数据把它所装饰的指令类和一个选择器(selector)关联起来，selector 用来把关联的指令插入 HTML 模板中。在模板中，指令通常作为特性出现在模板元素(HTML 标签)上，可能仅仅作为名字出现，也可能作为赋值目标或绑定目标出现。在模板中，它们看起来就像普通的 HTML 特性一样。

3.2 内置属性型指令

3.2.1 内置属性型指令说明

属性型指令会监听并修改 HTML 元素和组件的行为、attribute 和 property。许多

NgModule(如 RouterModule 和 FormsModule)都定义了自己的属性型指令。最常见的属性型指令包括添加和删除一组 CSS 类的指令 NgClass、添加和删除一组 HTML 样式的指令 NgStyle、将数据双向绑定添加到 HTML 表单元素的指令 NgModel 等指令。内置指令只会使用公开 API。它们不会访问任何不能被其他指令访问的私有 API。

3.2.2 NgClass 说明

要添加或删除单个类,优先使用类绑定而不是 NgClass。用 NgClass 可以同时添加或删除多个 CSS 类。可以将 NgClass 与表达式一起使用,即在要设置样式的元素上添加[ngClass]并将其设置为等于某个表达式。如代码< div [ngClass]="isSpecial ? 'special' : ''"> This div is special </div>中将 isSpecial 设置为布尔值 true 后,NgClass 就会把 special 类应用于< div >上。

要想将 NgClass 与方法一起使用,需要将方法添加到组件中。如例 3-1 中的 setCurrentClasses()方法使用一个对象来设置属性 currentClasses,该对象根据 canSave、isUnchanged、isSpecial 等组件属性为 true 或 false 来添加或删除三个类。该对象的每个键(key)(如 saveable)都是一个类名。如果键(类名)的取值为 true,则 NgClass 添加该类。如果键(类名)的取值为 false,则 NgClass 删除该类。于是,在模板中把 NgClass 属性绑定到 currentClasses,根据它来设置此元素的 CSS 类。在例 3-1 中,Angular 会在初始化以及发生更改的情况下应用这些类,即在 ngOnInit()方法中进行初始化以及通过单击按钮更改相关属性时调用 setCurrentClasses()方法。

【例 3-1】 创建文件 inside-directive. component. ts 的代码,演示本章的基础知识点。

```
import {Component} from '@angular/core';
import {Item} from "./item";
@Component({
  selector: 'root',//'app-inside-directive',
  template: `
    <div id="ngClass"> NgClass 绑定</div>
    <p> currentClasses is {{currentClasses | json}}</p>
    <div [ngClass]="currentClasses"> This div is initially saveable, unchanged, and special.
</div>
    <ul>
      <li>
        <label for="saveable"> saveable </label>
        <input type="checkbox" [(ngModel)]="canSave" id="saveable">
      </li>
      <li>
        <label for="modified"> modified:</label>
<input type="checkbox" [value]="! isUnchanged" (change)="isUnchanged = ! isUnchanged" id=
"modified"></li>
      <li>
<label for="special"> special: <input type="checkbox" [(ngModel)]="isSpecial" id=
"special"></label>
      </li>
    </ul>
    <button (click)="setCurrentClasses()"> Refresh currentClasses </button>
    <div [ngClass]="currentClasses">
      This div should be {{ canSave ? "": "not"}} saveable,
      {{isUnchanged ? "unchanged" : "modified"}} and
```

```
  {{isSpecial ? "": "not"}} special after clicking "Refresh".</div>
<br><br>
<!-- 使用属性切换 special 类开/关(on/off ) -->
<div [ngClass] = "isSpecial ? 'special' : ''"> This div is special </div>
<div class = "helpful study course"> Helpful study course </div>
<div [ngClass] = "{'helpful':false, 'study':true, 'course':true}"> Study course </div>
<hr>
<div> NgStyle 绑定</div>
<div [style.font-size] = "isSpecial ? 'x-large' : 'smaller'">
  This div is x-large or smaller.
</div>
<h4>[ngStyle] binding to currentStyles - CSS property names </h4>
<p> currentStyles is {{currentStyles | json}}</p>
<div [ngStyle] = "currentStyles">
  This div is initially italic, normal weight, and extra large (24px).
</div>
<br>
<label> italic: <input type = "checkbox" [(ngModel)] = "canSave"></label> |
<label> normal: <input type = "checkbox" [(ngModel)] = "isUnchanged"></label> |
<label> xlarge: <input type = "checkbox" [(ngModel)] = "isSpecial"></label>
<button (click) = "setCurrentStyles()"> Refresh currentStyles </button>
<br><br>
<div [ngStyle] = "currentStyles">
  This div should be {{ canSave ? "italic": "plain"}},
  {{isUnchanged ? "normal weight" : "bold"}} and,
  {{isSpecial ? "extra large": "normal size"}} after clicking "Refresh".</div>
<hr>
<div id = "ngModel"> NgModel 双向绑定</div>
<fieldset><h4> NgModel examples </h4>
  <p> Current item name: {{currentItem.name}}</p>
  <p>
    <label for = "without"> without NgModel:</label>
    <input [value] = "currentItem.name" (input) = "currentItem.name = getValue( $event)"
id = "without">
  </p>
  <p>
    <label for = "example-ngModel">[(ngModel)]:</label>
    <input [(ngModel)] = "currentItem.name" id = "example-ngModel">
  </p>
  <p>
    <label for = "example-change">(ngModelChange) = "...name = $event":</label>
<input [ngModel] = "currentItem.name" (ngModelChange) = "currentItem.name = $event" id =
"example-change">
  </p>
  <p>
    <label for = "example-uppercase">(ngModelChange) = "setUppercaseName( $event)"
<input [ngModel] = "currentItem.name" (ngModelChange) = "setUppercaseName( $event)" id =
"example-uppercase">
    </label>
  </p>
</fieldset>
<hr>
<div *ngIf = "nullCustomer"> Hello, <span>{{nullCustomer.id}}</span></div>
<div *ngIf = "nullCustomer2"> Hello, <span>{{nullCustomer2}}</span></div>
```

```
<div *ngFor="let item of items; let i=index">{{i + 1}} - {{item.name}}</div>
<div *ngFor="let item of items; trackBy: trackByItems">
  ({{item.id}}) {{item.name}}
</div>
<p>
  I turned the corner
  <ng-container *ngIf="item">
    and saw {{item.name}}. I waved
  </ng-container>
  and continued on my way.
</p>
<hr><h2>NgSwitch Binding</h2>
<p>Pick your favorite item</p>
<div>
  <label *ngFor="let i of items">
    <div><input type="radio" name="items" [(ngModel)]="currentItem" [value]="i">
{{i.name}}
    </div>
  </label>
</div>
<div [ngSwitch]="currentItem.feature">
  <app-stout-item    *ngSwitchCase="'stout'"    [item]="currentItem"></app-
stout-item>
  <app-device-item    *ngSwitchCase="'slim'"    [item]="currentItem"></app-
device-item>
  <app-lost-item    *ngSwitchCase="'vintage'"   [item]="currentItem"></app-lost-
item>
  <app-best-item    *ngSwitchCase="'bright'"    [item]="currentItem"></app-best-
item>
  <div *ngSwitchCase="'bright'"> Are you as bright as {{currentItem.name}}?</div>
  <app-unknown-item    *ngSwitchDefault          [item]="currentItem"></app-
unknown-item>
</div>
<hr>
<!-- 当指针悬停在p元素上时,背景颜色就会出现;而当指针移出时,背景颜色就会消失 -->
<p  [appHighlight]="color">Highlight me!</p>
<h2>Pick a highlight color</h2>
<div>
  <input type="radio" name="colors" (click)="color='lightgreen'">Green
  <input type="radio" name="colors" (click)="color='yellow'">Yellow
  <input type="radio" name="colors" (click)="color='cyan'">Cyan
</div>
<p [appHighlight]="color">Highlight me!</p>
<p ngNonBindable>不会显示 2: {{ 1 + 1 }}</p>
<div ngNonBindable [appHighlight]="'yellow'">
  This should not evaluate: {{ 1 +1 }}, but will highlight yellow.
</div>
<hr>
<p *appUnless="condition" class="unless a">
  (A) This paragraph is displayed because the condition is false.
</p>
<p *appUnless="!condition" class="unless b">
  (B) Although the condition is true,
  this paragraph is displayed because appUnless is set to false.
```

```
    </p>
    <p>
      The condition is currently
      <span [ngClass]="{ 'a': !condition, 'b': condition, 'unless': true }">{{condition}}</span>.
      <button
        (click)="condition = !condition"
        [ngClass] = "{ 'a': condition, 'b': !condition }" >
        Toggle condition to {{condition ? 'false' : 'true'}}
      </button>
    </p>
  `,
})
export class InsideDirectiveComponent {
  currentClasses: Record<string, boolean> = {};
  currentStyles: Record<string, string> = {};
  canSave: boolean = true;
  isUnchanged: boolean = true;
  isSpecial: boolean = true;
  item!: Item;
  items: Item[] = [];
  currentItem!: Item;
  nullCustomer: any = {
    id:21,
    name:'zsf'
  };
  nullCustomer2: any;
  color: any = 'red';
  condition: any = false;
  resetItems() {
    this.items = Item.items.map(item => item.clone());
    this.currentItem = this.items[0];
    this.item = this.currentItem;
  }
  ngOnInit() {
    this.resetItems();
    this.setCurrentClasses();
    this.setCurrentStyles();
  }
  setCurrentClasses() {
    this.currentClasses = {
      saveable: this.canSave,
      modified: !this.isUnchanged,
      special: this.isSpecial
    };
  }
  setCurrentStyles() {
    this.currentStyles = {
      'font-style':  this.canSave      ? 'italic' : 'normal',
      'font-weight': !this.isUnchanged ? 'bold'   : 'normal',
      'font-size':   this.isSpecial    ? '24px'   : '12px'
    };
  }
  getValue(event: Event): string {
    return (event.target as HTMLInputElement).value;
```

```
  }
  setUppercaseName(name: string) {
    this.currentItem.name = name.toUpperCase();
  }
  trackByItems(index: number, item: Item): number { return item.id; }
}
```

3.2.3 NgStyle 说明

可以用 NgStyle 根据组件的状态同时设置多个内部样式。要使用 NgStyle，就要向组件添加一个方法。如例 3-1 中，setCurrentStyles()方法基于该组件 canSave、isUnchanged、isSpecial 等属性的状态，用一个定义了 font-style、font-weight、font-size 三个样式的对象设置了 currentStyles 属性。设置元素的样式，需要将 ngStyle 属性绑定到 currentStyles。在例 3-1 中，Angular 会在初始化以及发生更改的情况下应用这些类。完整的示例会在 ngOnInit()方法中进行初始化以及通过单击按钮更改相关属性时调用 setCurrentClasses()方法。

3.2.4 NgModel 说明

可以用 NgModel 指令显示数据属性，并在用户进行更改时更新该属性。导入 FormsModule，并将其添加到 NgModule 的 imports 列表中，如例 3-2 所示。在 HTML 的 <form>标签上添加[(ngModel)]绑定并将其设置为等于属性，如代码<input [(ngModel)]="currentItem.name" id="example-ngModel">中设置数据绑定属性。要自定义配置，可以编写可展开的表单，该表单将属性绑定和事件绑定分开。使用属性绑定来设置属性，并使用事件绑定来响应更改。NgModel 指令适用于值访问器接口 ControlValueAccessor 支持的元素。Angular 为 HTML 表单所有基本元素提供了值访问器接口。要将[(ngModel)]应用于非表单型内置元素或第三方自定义组件，必须编写一个值访问器接口。编写 Angular 组件时，如果根据 Angular 的双向绑定语法命名 value 和 event 属性，则不需要用值访问器接口或 NgModel。

【例 3-2】 创建文件 app-directiveexample.module.ts 的代码，定义路由并声明组件和指令。

```
import {NgModule} from '@angular/core';
import {BrowserModule} from '@angular/platform-browser';
import {RouterModule} from "@angular/router";
import {DirectiveexamplesComponent} from "./directiveexamples.component";
import {FormsModule} from "@angular/forms";
import {InsideDirectiveComponent} from "./inside-directive.component";
import {
  BestItemComponent,
  DeviceItemComponent,
  LostItemComponent,
  StoutItemComponent,
  UnknownItemComponent
} from "./item-switch.component";
import {HighlightDirective} from "./highlight.directive";
import {UnlessDirective} from "./unless.directive";
```

```
@NgModule({
  imports: [
    BrowserModule,
    RouterModule.forRoot([
      {path: 'insidedirective', component: InsideDirectiveComponent},
      {path: 'directiveexample', component: DirectiveexamplesComponent},
    ]),
    FormsModule,
  ],
  declarations: [
    InsideDirectiveComponent,
    StoutItemComponent,
    BestItemComponent,
    DeviceItemComponent,
    LostItemComponent,
    UnknownItemComponent,
    HighlightDirective,
    UnlessDirective,
    DirectiveexampleComponent
  ],
})
export class AppDirectiveexampleModule { }
```

3.3 内置结构型指令

3.3.1 内置结构型指令说明

结构型指令的职责是进行 HTML 布局，并通过添加、移除和操纵它们所附加的宿主元素(DOM 节点)来塑造或重塑 DOM 结构。由于结构型指令会在 DOM 中添加和删除节点，因此每个元素只能应用一个结构型指令。常见的内置结构型指令包括从模板中创建或销毁子视图的 NgIf 指令、为列表中的每个条目重复渲染一个节点的 NgFor 指令和一组在备用视图之间切换的 NgSwitch 指令。

3.3.2 NgIf 说明

在宿主元素上用 NgIf 指令可以通过条件决定是否添加或删除宿主元素。如果 NgIf 指令为 false，则 Angular 将从 DOM 中移除对应的元素及其后代。然后，Angular 会销毁其组件，从而释放内存和资源。要添加或删除元素，需要将 * ngIf 绑定到条件表达式。如代码 <app-item-detail *ngIf="isActive" [item]="item"></app-item-detail>中将 * ngIf 绑定到 isActive，当 isActive 表达式返回 true 值时，NgIf 指令会把 app-item-detail 所在组件添加到 DOM 中。在默认情况下，NgIf 指令会阻止显示已绑定到空值的元素。如要使用 NgIf 指令保护<div>，就需要将代码“*ngIf="yourProperty"”添加到<div>，修改之后的结果可能为<div *ngIf="currentCustomer"> Hello，{{currentCustomer.name}}</div>。如果属性 currentCustomer 为 null，则 Angular 不会显示<div>(注意，是整个<div>都不会显示)。

3.3.3 NgFor 说明

NgFor 指令用于显示条目列表。定义一个 HTML 块，该块用于决定 Angular 如何渲染单个条目，如代码<div *ngFor="let item of items">{{item.name}}</div>中要列出的条目item，把字符串"let item of items"赋给*ngFor。字符串"let item of items"用于指示Angular执行将items中的每个条目存储在局部循环变量item中、让每个条目在每次迭代时的模板HTML中都可用、将"let item of items"转换为环绕宿主元素的<ng-template>、对列表中的每个item重复<ng-template>等操作。Angular会将指令转换为<ng-template>，然后反复使用此模板为列表中的每个item创建一组新的元素和绑定。

要复写某个组件元素，可以将*ngFor应用于其选择器，如代码<app-item-detail *ngFor="let item of items" [item]="item"></app-item-detail>中的选择器为<app-item-detail>。在宿主元素的后代中，该选择器用以访问条目的属性，将item通过绑定传递给<app-item-detail>组件的item属性。

可以在模板输入变量中获取*ngFor的index索引并在模板中使用它。在*ngFor中，添加一个分号(;)和let i=index的简写形式，如代码<div *ngFor="let item of items; let i=index">{{i + 1}} - {{item.name}}</div>即把index赋予一个名为i的变量中，并将其与条目名称一起显示。NgFor指令上下文的index属性在每次迭代中都会返回该条目的从零开始的索引号。

3.3.4 NgIf、NgFor和容器

若要在特定条件为true时重复某个HTML块，则可以将*ngIf放在*ngFor元素的容器元素上。它们之一或两者都可以是<ng-container>，这样就不必引入额外的HTML层次了。

通过跟踪对条目列表的更改，可以减少应用对服务器的调用次数。使用*ngFor的trackBy属性，Angular只能更改和重新渲染已更改的条目，而不必重新加载整个条目列表。如向某组件添加一个trackByItems()方法，该方法返回NgFor指令应该跟踪的值(item.id)；如果浏览器已经渲染过某个id，Angular就会跟踪它而不会重新向服务器查询相同的id。如果没有trackBy属性，就会由触发完全的DOM元素替换。有了trackBy属性，只有修改了id的按钮才会触发元素替换。

Angular的<ng-container>是一个分组元素，它不会干扰样式或布局，因为Angular不会将其放置在DOM中。当没有单个元素承载指令时，可以使用<ng-container>。同时，要有条件地排除<option>，需要将<option>包裹在<ng-container>中。

3.3.5 NgSwitch说明

NgSwitch指令会根据条件显示几个可能的元素中的一个；而Angular只将选定的元素放入DOM。NgSwitch指令包括属性型指令NgSwitch、结构型指令NgSwitchCase和NgSwitchDefault。NgSwitch指令可以更改其伴生指令的行为。当把[ngSwitch]绑定到一个返回开关值(开关值可以是任何类型)的表达式。当[*ngSwitchCase]绑定值等于开关值时，可将其元素添加到DOM中，而在其不等于开关值时将其绑定值移除。当NgSwitchCase没有被选中时，将[*ngSwitchDefault]宿主元素添加到DOM中。NgSwitch指令也同样适用于内置HTML元素和Web Component。

3.4 自定义属性型指令

3.4.1 创建

创建属性型指令的步骤如下。

(1) 先创建指令文件 *.directive.ts 并在模块文件中声明指令类。

(2) @Directive()装饰器的配置属性会指定指令的 CSS 属性选择器(如 appHighlight)。

(3) 从@angular/core 导入 ElementRef,ElementRef 的 nativeElement 属性会提供对宿主元素的直接访问权限。

(4) 在指令的 constructor()方法中添加 ElementRef 以注入对宿主元素的引用,该元素就是指令 CSS 属性选择器的作用目标。向指令类中添加逻辑。

要注意的是,指令不支持名称空间。

3.4.2 应用

要应用自定义属性型指令,可以将元素(如标签< p >)添加到 HTML 模板中,并以伪指令(或称为属性选择器,如 appHighlight)作为属性,如< p appHighlight > Highlight me! </p >,Angualr 会创建指令类的实例,并将元素的引用注入该指令的构造函数中。

属性型指令可以用于处理用户事件,需要添加事件处理程序,每个事件处理程序都带有@HostListener()装饰器。

@Input()装饰器会将元数据添加到指令类用于绑定的属性,将值传递给属性型指令。

要防止在浏览器中进行表达式求值,可以将 ngNonBindable 添加到宿主元素。如果将 ngNonBindable 应用于父元素,则 Angular 会禁用该元素的子元素的任何插值和绑定。ngNonBindable 会停用模板中的插值、指令和绑定。但是,ngNonBindable 仍然允许指令在应用 ngNonBindable 的元素上工作。如代码< div ngNonBindable [appHighlight]="'yellow'"> This should not evaluate: {{1+1}}, but will highlight yellow. </div >中,appHighlight 指令仍处于活跃状态,但 Angular 不会对表达式{{1+1}}求值。

3.5 自定义结构型指令

3.5.1 创建

结构型指令(如 * ngIf)中的星号(*)语法是 Angular 解释并转换为较长形式的简写形式。Angular 将结构型指令前面的星号转换为围绕宿主元素及其后代的< ng-template >。以代码< div *ngIf="hero" class="name">{{hero.name}}</div >为例,它等价于"< ng-template [ngIf]="hero">< div class="name">{{hero.name}}</div ></ng-template >"。Angular 不会创建真正的< ng-template >元素,只会将< div >和占位符{{hero.name}}的内容渲染到 DOM 中。

解析器会将帕斯卡命名(PascalCase)法应用于所有指令(即指令名的每个单词首字母大写),并为它们加上指令的属性名称(例如 ngFor),如 ngFor 的输入特性 of 和 trackBy 会映射为 ngForOf 和 ngForTrackBy。当 NgFor 指令遍历列表时,它会设置和重置上下文对象的属性。

Angular 的< ng-template >元素定义了一个默认情况下不渲染任何内容的模板。使用< ng-template >可以手动渲染内容以完全控制内容的显示方式。如果没有结构型指令,并且将某些元素包装在< ng-template >中,则这些元素会消失。

3.5.2 应用

模板中的结构型指令会根据输入的表达式来控制是否要在运行时渲染该模板。为了帮助编译器捕获模板类型中的错误,应该尽可能详细地指定模板内指令的输入表达式所期待的类型。类型保护函数会将输入表达式的预期类型缩小为可能在运行时传递给模板内指令的类型的子集;可以提供这样的功能(函数)来帮助类型检查器在编译时推断出表达式的正确类型。

为模板中指令的输入表达式提供更具体的类型,要在指令中添加 ngTemplateGuard_xx 属性,其中静态属性名称 xx 就是@Input()装饰器修饰的属性(字段)名字。该属性的值可以是基于其返回类型的常规类型窄化函数,也可以是字符串,例如 NgIf 中的"binding"。例如,NgIf 的实现使用类型窄化来确保只有当 * ngIf 的输入表达式为真时,模板才会被实例化。为了提供具体的类型要求,NgIf 指令定义了一个静态属性 ngTemplateGuard_ngIf: 'binding'。这里的 binding 值是一种常见的类型窄化的例子,它会对输入表达式进行求值,以满足类型要求。

微课视频

3.6 指令的基础应用

3.6.1 基础代码

在项目 src\examples 目录下创建子目录 directiveexamples,在 src\examples\directiveexamples 目录下创建文件 inside-directive. component. ts,代码如例 3-1 所示。在 src\examples\directiveexamples 目录下创建文件 app-directiveexample. module. ts,代码如例 3-2 所示。在 src\examples\directiveexamples 目录下创建文件 item. ts,代码如例 3-3 所示。

【例 3-3】 创建文件 item. ts 的代码,定义并导出条目类 Item。

```
export class Item {
  static nextId = 0;
  static items: Item[] = [
    new Item(
      0,
      'Teapot',
      'stout'
    ),
    new Item(1, 'Lamp', 'bright'),
    new Item(2, 'Phone', 'slim' ),
    new Item(3, 'Television', 'vintage' ),
    new Item(4, 'Fishbowl')
```

```
  ];
  constructor(
    public id: number,
    public name?: string,
    public feature?: string,
    public url?: string,
    public rate = 100,
    ) {
    this.id = id ? id : Item.nextId++;
  }
  clone(): Item {
    return Object.assign(new Item(this.id), this);
  }
}
```

3.6.2 自定义指令

在 src\examples\directiveexamples 目录下创建文件 highlight.directive.ts，代码如例 3-4 所示。

【例 3-4】 创建文件 highlight.directive.ts 的代码，演示自定义属性型指令的方法。

```
import {Directive, ElementRef, HostListener, Input} from "@angular/core";
@Directive({
  selector: '[appHighlight]'
})
export class HighlightDirective {
  //@Input() 装饰器会将元数据添加到此类，以便让该指令的 appHighlight 属性可用于绑定
  @Input() appHighlight = '';
  private highlightColor = 'red';
  @Input() defaultColor = '';
  constructor(private el: ElementRef) {
    //向 HighlightDirective 类中添加逻辑，将背景设置为黄色
    el.nativeElement.style.backgroundColor = 'yellow';
  }
  //在鼠标进入时作出响应，事件处理程序都带有 @HostListener() 装饰器
//要订阅本属性型指令宿主元素(例 3-1 中标签<p>)的事件，可以使用 @HostListener() 装饰器
  @HostListener('mouseenter') onMouseEnter() {
    this.highlight(this.highlightColor || this.defaultColor || 'red');
  }
  //在鼠标离开时作出响应，事件处理程序都带有 @HostListener() 装饰器
  @HostListener('mouseleave') onMouseLeave() {
    this.highlight('');
  }
  //辅助方法 highlight()，该方法会设置宿主元素 el 的颜色
  private highlight(color: string) {
    this.el.nativeElement.style.backgroundColor = color;
  }
}
```

在 src\examples\directiveexamples 目录下创建文件 unless.directive.ts，代码如例 3-5 所示。

【例 3-5】 创建文件 unless. directive. ts 的代码,演示自定义结构型指令的方法。

```
import {Directive, Input, TemplateRef, ViewContainerRef} from '@angular/core';
@Directive({selector: '[appUnless]'})
export class UnlessDirective {
  private hasView = false;
  constructor(
    //在指令的构造函数中将 TemplateRef 和 ViewContainerRef 注入成私有变量
    //TemplateRef 可帮助获取 <ng-template> 的内容,而 ViewContainerRef 可以访问视图容器
    private templateRef: TemplateRef<any>,
    private viewContainer: ViewContainerRef) { }
  //每当条件 condition 的值被更改时,Angular 都会设置 appUnless 属性
//若条件取假值且 Angular 尚未创建视图,则此 setter 方法会导致视图容器从模板创建出嵌入式视图
  //若条件取真值且当前正显示着视图,则此 setter 方法会清除容器,这会导致销毁该视图
    @Input() set appUnless(condition: boolean) {
    if (!condition && !this.hasView) {
      this.viewContainer.createEmbeddedView(this.templateRef);
      this.hasView = true;
    } else if (condition && this.hasView) {
      this.viewContainer.clear();
      this.hasView = false;
    }
  }
}
```

3.6.3 组件

在 src\examples\directiveexamples 目录下创建文件 item-switch. component. ts,代码如例 3-6 所示。

【例 3-6】 创建文件 item-switch. component. ts 的代码,定义组件。

```
import {Component, Input} from '@angular/core';
import {Item} from './item';
@Component({
  selector: 'app-stout-item',
  template: "I'm a little {{item.name}}, short and stout!"
})
export class StoutItemComponent {
  @Input() item!: Item;
}
@Component({
  selector: 'app-best-item',
  template: 'This is the brightest {{item.name}} in town.'
})
export class BestItemComponent {
  @Input() item!: Item;
}
@Component({
  selector: 'app-device-item',
  template: 'Which is the slimmest {{item.name}}?'
```

```
})
export class DeviceItemComponent {
  @Input() item!: Item;
}
@Component({
  selector: 'app-lost-item',
  template: 'Has anyone seen my {{item.name}}?'
})
export class LostItemComponent {
  @Input() item!: Item;
}
@Component({
  selector: 'app-unknown-item',
  template: '{{message}}'
})
export class UnknownItemComponent {
  @Input() item!: Item;
  get message() {
    return this.item && this.item.name ?
      `${this.item.name} is strange and mysterious.` :
      'A mystery wrapped in a fishbowl.';
  }
}
export const ItemSwitchComponents =
    [ StoutItemComponent, BestItemComponent, DeviceItemComponent, LostItemComponent,
UnknownItemComponent ];
```

3.6.4 模块

修改 src\examples 目录下的文件 examplesmodules1.module.ts，代码如例 3-7 所示。

【例 3-7】 修改文件 examplesmodules1.module.ts 的代码，设置启动组件。

```
import {NgModule} from '@angular/core';
import {AppDirectiveexampleModule} from './directiveexamples/app-directiveexample.module';
import {InsideDirectiveComponent} from './directiveexamples/inside-directive.component';
@NgModule({
  imports: [
    AppDirectiveexampleModule
  ],
  bootstrap: [InsideDirectiveComponent]
})
export class ExamplesmodulesModule1 {}
```

3.6.5 运行结果

保持其他文件不变并成功运行程序后，在浏览器地址栏中输入 localhost:4200，部分结果如图 3-1 所示。更多的结果请读者自己参考源代码进行验证。

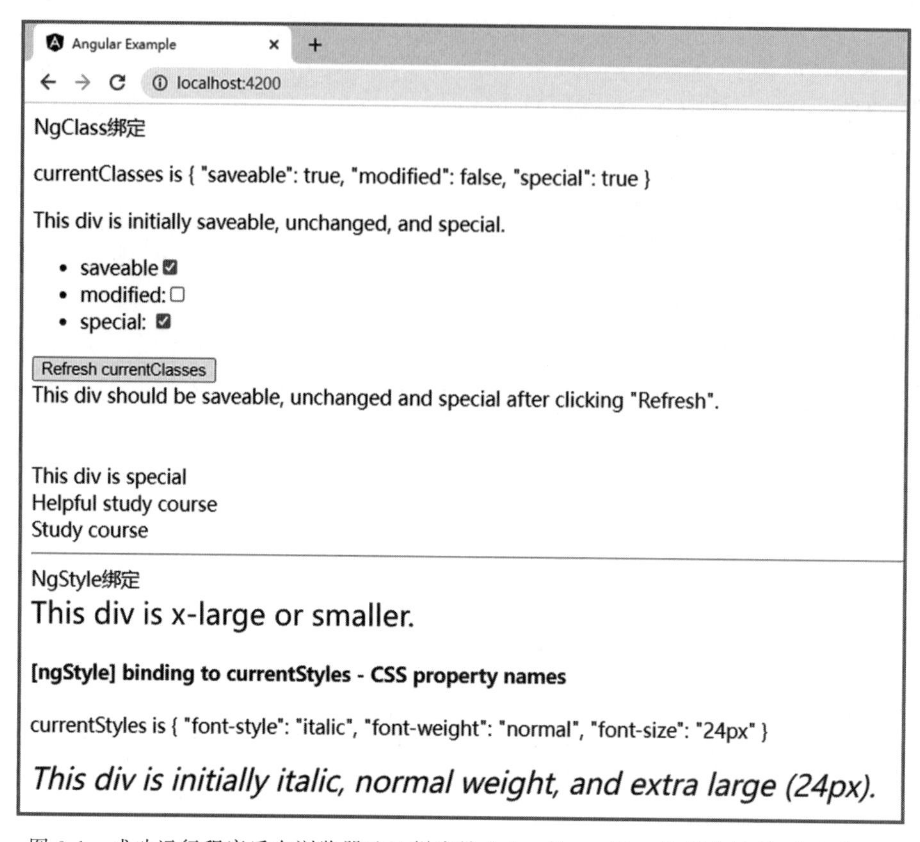

图 3-1　成功运行程序后在浏览器地址栏中输入 localhost:4200 的部分结果(从上往下)

微课视频

3.7　指令的综合应用开发

3.7.1　组件

在 src\examples\directiveexamples 目录下创建文件 directiveexamples. component. ts,代码如例 3-8 所示。

【例 3-8】　创建文件 directiveexamples. component. ts 的代码,定义组件。

```
import {Component} from '@angular/core';
@Component({
  selector: 'root',
  template: `
    NgIf 指令示例(布尔变量 showName 的值为{{showName}} )
    <div *ngIf="showName"> showName 为真时显示</div>
    <hr>
    NgFor 指令示例
    <div *ngFor="let author of authors; let i = index">{{i}} : {{author.name}} {{author.
book}}</div>
    <hr>
    NgSwitch 指令示例
    <div [ngSwitch]="day">
      <b *ngSwitchCase="days.MONDAY">星期一</b>
      <b *ngSwitchCase="days.TUESDAY">星期二</b>
```

```
        <b *ngSwitchCase="days.WEDNESDAY">星期三</b>
        <b *ngSwitchCase="days.THURSDAY">星期四</b>
        <b *ngSwitchCase="days.FRIDAY">星期五</b>
        <b *ngSwitchDefault>休息日</b>
      </div>
      <hr>
NgClass 指令示例(先定义 CSS 样式,再在 ngClass 后面用对象、数组、字符串的方式引入所定义的样式)
      <div [ngClass]="objectStyleFormat">用对象的方式描述要定义的样式(大字)</div>
<div [ngClass]="{'redClassStyle': flag, 'blueClassStyle': !flag}">用对象的方式描述要定义的
样式(不是大字)</div>
      <div [ngClass]="arrayStyleFormat">用数组的方式描述要定义的样式(加粗大字)</div>
      <div [ngClass]="stringStyleFormat">用字符串的方式描述要定义的样式(加粗不是大字)</div>
      <hr>
      NgStyle 指令示例
      <div [ngStyle]="{'font-size':'18px'}">18px 字体</div>
      <div [ngStyle]="{'font-size':attr}">22px 字体</div>
      <div [ngStyle]="currentStyles">24px 加粗黑色斜体字</div>
   `,
  styles: ['.blackClassStyle {color: black} ' +
  '.blueClassStyle {color: blue} ' +
  '.largerFont {font-size: larger} ' +
  '.boldFont {font-weight:bold} ' +
  '.currentStyle  {font-style: italic; font-weight :  bold;font-size : 24px}'
  ]
})
export class DirectiveexamplesComponent {
  showName: boolean = true;
authors: any = [{name: '左丘明', book: '左传'}, {value: '陈寿', book: '三国志'}, {value: '范晔',
book: '后汉书'}];
  days = Days;
  day: Days = this.days.TUESDAY;
  flag = true ;
    objectStyleFormat = {'blackClassStyle': true, 'blueClassStyle': false, 'largerFont': true};
    arrayStyleFormat = ['blackClassStyle', 'largerFont', 'boldFont'];
    stringStyleFormat = "blackClassStyle boldFont ";
    currentStyles = { "font-style": "italic", "font-weight": "bold", "font-size":
"24px ", "color": "black"};
    attr = '22px'
}
export enum Days {
  MONDAY,
  TUESDAY,
  WEDNESDAY,
  THURSDAY,
  FRIDAY
}
```

3.7.2 模块

修改 src\examples 目录下的文件 examplesmodules1. module. ts,代码如例 3-9 所示。

【例 3-9】 修改文件 examplesmodules1. module. ts 的代码。

```
import {NgModule} from '@angular/core';
import{AppDirectiveexampleModule}from'./directiveexamples/app-directiveexample.module';
import {DirectiveexamplesComponent} from './directiveexamples/directiveexamples.component';
@NgModule({
```

```
    imports: [
      AppDirectiveexampleModule
    ],
    bootstrap: [DirectiveexamplesComponent]
})
export class ExamplesmodulesModule1 {}
```

3.7.3 运行结果

保持其他文件不变并成功运行程序后，在浏览器地址栏中输入 localhost:4200，结果如图 3-2 所示。

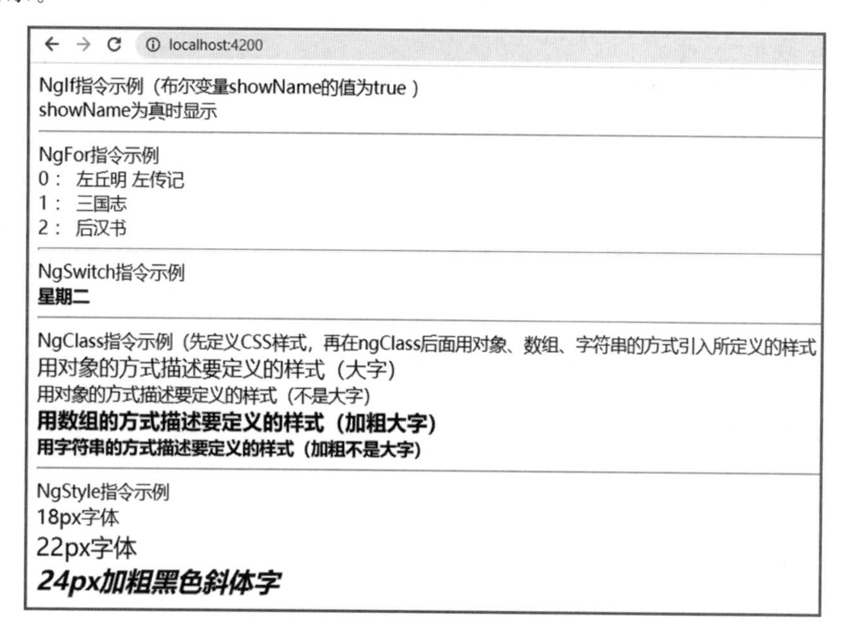

图 3-2 成功运行程序后在浏览器地址栏中输入 localhost:4200 的结果

习题 3

一、简答题

1. 简述对指令的理解。

2. 简述对各类指令的理解。

二、实验题

实现指令的应用开发。

第4章

组　件

4.1 组件概述

4.1.1 组件的实现

Angular 应用中包含一棵组件树，每个组件是定义行为的 Typescript 类。在默认情况下，组件文件中先添加 import 语句：import { Component } from '@angular/core';，在 import 语句之后添加@Component()装饰器；接着，在@Component()装饰器中添加一个 selector 语句来指定组件的 CSS 选择器(简称选择器)；再定义组件用于显示信息的 HTML 模板，也可以为组件的模板选择样式，还可以在单独的文件中定义组件模板的样式；最后，添加一个包含该组件的导出代码——class 语句。

4.1.2 组件的应用

每个组件都需要一个选择器，用于定义组件在模板中的使用方式。在模板 HTML 中出现一个组件选择器对应的标签时，选择器会触发 Angular，并把该组件实例化在对应位置。例如，一个组件的选择器被定义为 app-hello-world，当模板中出现< app-hello-world >标签时，这个选择器对应的标签就会让 Angular 实例化 app-hello-world 所在组件。

Angular 组件是一种特殊的指令，和其他指令不同，组件的模板中的每个元素只能具有一个组件实例。组件必须从属于某个 NgModule，才能被其他组件或应用程序使用。要想让组件成为某个 NgModule 中的一员，就要把它列在@NgModule()装饰器的元数据 declarations 字段中。除了这些用来对指令进行配置的选项之外，还可以通过实现生命周期钩子来控制组件的在运行时的行为。

4.1.3 组件和视图

组件用于控制用户看到的用户界面内容(称为视图)上的一小片区域；视图是由一个个组件定义和控制的。在组件类中通过定义组件的应用逻辑为视图提供支持。组件通过一些由属性和方法组成的 API 与视图交互。

组件可以通过配套模板来定义其视图，视图通常会分层次地进行组织，并以用户界面分区或页面为单位进行修改、显示或隐藏。与组件直接关联的外部模板会定义该组件的宿主视图。组件还可以在其内部定义一个带层次结构的视图，它可以包含一些内嵌的视图(即内部组件控制的视图)并作为它们的宿主。

带层次结构的视图可以包含同一模块中组件的视图，也可以(而且经常会)包含其他模块中定义的组件的视图。

4.1.4 元数据

用户访问 Angular 应用过程中，Angular 会创建、更新、销毁一些组件。Angular 应用可以通过一些可选的生命周期钩子(如 ngOnInit()方法)在每个特定的时机对组件等进行相关操作。

@Component()装饰器会指出紧随其后的组件并为其指定元数据。元数据告诉 Angular 到哪里获取它所需要的主要构造块，以创建和展示这个组件及其视图。它把一个模板和该组件关联起来，该组件及其模板共同描述了一个视图。除了包含或指向模板之外，@Component()装饰器的元数据还会配置如何在 HTML 中引用该组件以及该组件需要哪些服务等。

选择器会告诉 Angular，一旦在模板 HTML 中找到了选择器对应的标签，就创建并插入与选择器对应组件的一个实例。模板定义了该组件的宿主视图。providers 表明当前组件所需的服务提供者的一个数组，如 providers: [HeroService]告诉 Angular 该如何提供一个 HeroService 实例。

微课视频

4.2 组件样式及其应用

4.2.1 组件样式说明

Angular 应用中使用标准 CSS 来设置样式。这意味着 CSS 可以直接用于 Angular 应用中，如样式表、选择器、规则以及媒体查询等。另外，Angular 还能把组件样式捆绑在组件上，以实现比标准样式表更加模块化的设计。

可以使用 CSS 类名和选择器命名组件。CSS 类名和选择器局限于组件范围，不会与应用中其他地方的类名和选择器冲突。组件的样式不会因为别的地方修改了样式而被意外改变。可以让每个组件的 CSS 代码和它的 TypeScript、HTML 代码放在一起(即内部样式)，这将促成清爽整洁的项目结构(本书主要采用此种方式)。根据需要，可以修改或移除原组件的 CSS 代码构成一个独立文件(外部样式)。

可以通过设置 styles 或 styleUrls 元数据、内嵌在模板的 HTML 原组件中、通过 CSS 文件导入等方式把样式加入组件。具体来说，可以给@Component()装饰器添加 styles 数组型属性。这个数组中的每一个字符串(通常只有一个)定义一份 CSS。这些样式只对当前组件的模板生效，它们既不会作用于在模板中嵌入当前的任何组件，也不会作用于通过内容投影嵌入的组件。这种范围限制就是所谓的样式模块化特性。也可以通过把外部 CSS 文件添加到@Component()的 styleUrls 属性中以加载外部样式；又可以直接在组件的 HTML 模板中写< style >标签来内嵌 CSS 样式；还可以在组件的 HTML 模板中写< link >标签；可以利用标准的 CSS 的@import 规则来把其他 CSS 文件导入一个 CSS 文件中。

每个组件都会关联一个与其选择器相匹配的元素，这个元素称为宿主元素，模板会被渲染到其中。伪类选择器：host 可用于创建针对宿主元素自身的样式，而不是针对宿主内部的那些元素。其host把宿主元素作为样式目标的唯一方式，除此之外，没有别的办法指定它，因为宿主不是组件自身模板的一部分，而是父组件模板的一部分。host也可以与其他选择器组合使用，在：host后面添加选择器以选择子元素，如用：host h2定位组件视图内的< h2 >。注意不应该在：host前面添加除：host-context之外的选择器，因为此类选择器的作用域不会限于组件的视图，而会选择外部上下文，但这不符合组件样式的使用要求。

有时候，需要以某些来自宿主的祖先元素为条件来决定是否要应用某些样式，例如，在文档的< body >元素上可能有一个用于表示样式主题的CSS类。当基于CSS类来决定组件的样式时，可以使用伪类选择器：host-context，它的使用方法是以类似：host形式在当前组件宿主元素的祖先节点中查找CSS类，直到文档的根节点为止。注意，伪类选择器：host-context只能与其他选择器组合使用。

4.2.2 内部样式应用

在项目src\examples目录下创建子目录componentexamples，在src\examples\componentexamples目录下创建文件styleexamples. component. ts，代码如例4-1所示。

【例4-1】 创建文件styleexamples. component. ts的代码，演示内部样式的用法。

```
import {Component} from '@angular/core';
@Component({
  selector: 'style - root',
  template: `
    < p >注意样式的叠加及叠加的先后次序(优先级)</p >
    < div >内联 styles 属性接收一个包含 CSS 代码的字符串数组(通常一个字符串即可)</div >
    < h3 >子组件标签 h3,父组件中对其加粗,设置蓝色(:host 叠加样式到原来 h3 标签)</h3 >
    < link rel = "stylesheet" href = "./linkex.css">
    < div class = "added">用 link 引入 CSS 样式的方法</div >
    < div class = "linkandimport">用 link + import 一起引入 CSS 样式的方法</div >
  `,
  styles: ['div {font - family: "Times New Roman";font - size: larger;color: red}']
})
export class StyleexamplesComponent {
}
```

在src\examples\componentexamples目录下创建文件hostexamples. component. ts，代码如例4-2所示。

【例4-2】 创建文件hostexamples. component. ts的代码，演示对样式重新设置。

```
import {Component} from '@angular/core';
@Component({
  selector: 'host - root',
  template: `
    < hr >
    < div >对子组件的统一设置格式,被重新设置(为了对比)</div >
  `,
  styles: ['div {color: red}']
})
export class HostexamplesComponent {
}
```

4.2.3 内部样式和外部样式的综合应用

在 src\examples\componentexamples 目录下创建文件 outsideexamples. component. ts，代码如例 4-3 所示。

【例 4-3】 创建文件 outsideexamples. component. ts 的代码，演示组件内同时有内部样式和外部样式。

```
import {Component} from '@angular/core';
@Component({
  selector: 'outside - root',
  template: `
    < hr >
    < div class = "outside">使用外部样式文件的示例</div >
    < hr >
    < div class = "insideandoutside">使用外部样式文件(加粗)和内部样式(斜体)的示例</div >
  `,
  styleUrls: ['./outstyle.css'],
  styles: ['.insideandoutside {font - style: italic}']
})
export class OutsideexamplesComponent {}
```

在 src\examples\componentexamples 目录下创建文件 outstyle. css，代码如例 4-4 所示。

【例 4-4】 创建文件 outstyle. css 的代码，定义外部样式。

```
.outside {
  color: blueviolet;
  font - size: larger;
}
.insideandoutside {
  color:brown;
  font - weight: bold;
}
```

在 src\examples\componentexamples 目录下创建文件 added. css，代码如例 4-5 所示。

【例 4-5】 创建文件 added. css 的代码，定义基础 CSS。

```
.linkandimport {
  font - family: "Times New Roman";
  color: black;
  font - style: italic
}
```

在 src\examples\componentexamples 目录下创建文件 linkex. css，代码如例 4-6 所示。

【例 4-6】 创建文件 linkex. css 的代码，演示 CSS 文件中导入另外一个 CSS 文件。

```
@import './added.css';
.added {
  font - family: "Times New Roman";
  font - size: larger;
  color: green
}
```

4.2.4 :host 应用

在 src\examples\componentexamples 目录下创建文件 stylehome. component. ts，代码

如例 4-7 所示。

【例 4-7】 创建文件 stylehome. component. ts 的代码,演示:host 用法。

```
import {Component} from '@angular/core';
@Component({
  selector: 'root',
  template: `
    <style-root></style-root>
    <host-root></host-root>
    <outside-root></outside-root>
  `,
  styles: [
    ':host     {color: blue;font-weight: bold;}'
  ]
})
export class StylehomeComponent {}
```

4.2.5 模块和运行结果

在 src\examples\componentexamples 目录下创建文件 app-style. module. ts,代码如例 4-8 所示。

【例 4-8】 创建文件 app-style. module. ts 的代码,定义路由并声明组件。

```
import {NgModule} from '@angular/core';
import {StylehomeComponent} from "./stylehome.component";
import {BrowserModule} from "@angular/platform-browser";
import {RouterModule} from "@angular/router";
import {HostexamplesComponent} from "./hostexamples.component";
import {OutsideexamplesComponent} from "./outsideexamples.component";
import {StyleexamplesComponent} from "./styleexamples.component";
@NgModule({
  imports: [
    BrowserModule,
    RouterModule.forRoot([
      {path: 'style', component: StylehomeComponent},
    ]),
  ],
  declarations: [
    StylehomeComponent,
    HostexamplesComponent,
    OutsideexamplesComponent,
    StyleexamplesComponent
  ],
})
export class AppStyleModule { }
```

修改 src\examples 目录下的文件 examplesmodules1. module. ts,代码如例 4-9 所示。

【例 4-9】 修改文件 examplesmodules1. module. ts 的代码,设置启动组件。

```
import {NgModule} from '@angular/core';
import {AppBindexampleModule} from "./bindexamples/app-bindexample.module";
import {StylehomeComponent} from "./componentexamples/stylehome.component";
@NgModule({
  imports: [
    AppBindexampleModule
```

```
  ],
  bootstrap: [StylehomeComponent]
})
export class ExamplesmodulesModule1 {}
```

保持其他文件不变并成功运行程序后，在浏览器地址栏中输入 localhost:4200，结果如图 4-1 所示。

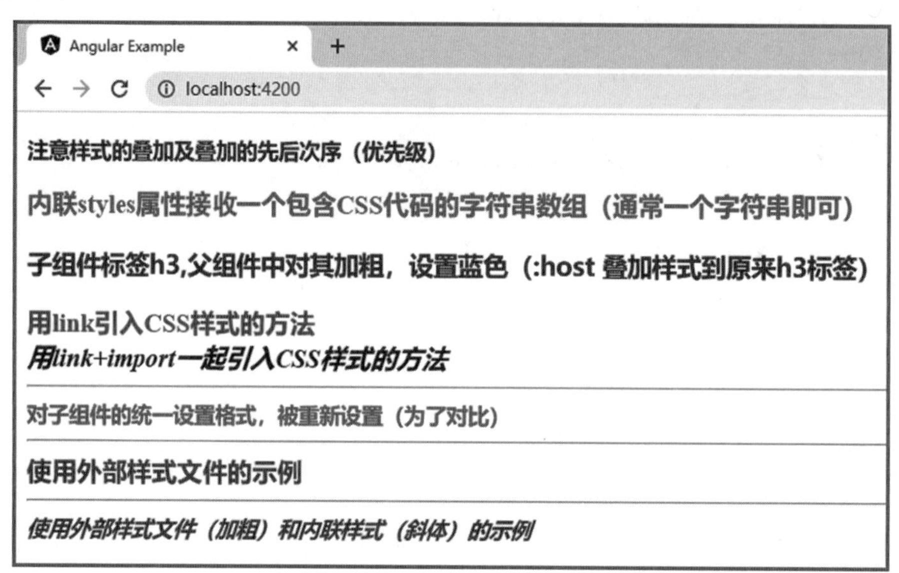

图 4-1 成功运行程序后在浏览器地址栏中输入 localhost:4200 的结果

4.3 组件生命周期

4.3.1 说明

生命周期是指 Angular 组件从开始渲染到销毁的全过程。Angular 实例化组件类(简称组件)并渲染组件视图及其子视图时，组件实例的生命周期就开始了。组件在整个生命周期一直伴随着变更检测，Angular 会检查数据绑定属性何时发生变化，并按需更新视图和组件实例。当 Angular 销毁组件实例并从 DOM 中移除它渲染的内容时，生命周期就结束了。Angular 在创建、更新和销毁指令实例时，指令有类似的生命周期。

可以用生命周期钩子方法来引发(tap)组件或指令生命周期中的关键事件。例如，初始化新实例，在需要时启动变更检测，在变更检测过程中响应更新，在删除实例之前进行清理。可以通过实现 Angular 的 core 库中定义的一个或多个生命周期钩子接口(即生命周期方法)来响应组件或指令生命周期中的事件。这些方法可以在适当的时候(如创建、更新或销毁实例时)对组件或指令实例进行操作。

组件或指令生命周期的每个接口都有一个钩子方法，它们的名字是由接口名加上 ng 前缀构成的。例如，接口 OnInit 的钩子方法叫作 ngOnInit()方法。如果在组件或指令类中实现了这个方法，Angular 就会在首次检查完组件或指令的输入属性后调用它。Angular 应用开发时，不必实现所有生命周期钩子方法，只要实现需要的方法就可以了。

4.3.2　生命周期方法

当应用程序通过调用构造函数来实例化一个组件或指令时，Angular 就会调用该实例在生命周期中适当位置实现了的钩子方法。Angular 可以用它们来执行不同的操作。

（1）**ngOnChanges**()方法。Angular 设置或重新设置数据绑定的输入属性时响应该方法。它接收当前和上一属性值的 SimpleChanges 对象。由于该方法调用频繁，该方法中执行的任何操作都会显著影响性能。如果组件绑定过输入属性，那么在 ngOnInit()方法之前以及所绑定的一个或多个输入属性的值发生变化时都会被调用。如果组件没有输入属性，或者使用它时没有提供任何输入属性，就不会调用 ngOnChanges()方法。

（2）**ngOnInit**()方法。Angular 第一次显示数据绑定和设置组件或指令的输入属性后，利用该方法初始化组件或指令。在第一轮 ngOnChanges()方法完成之后调用 ngOnInit()方法，且只调用一次，即使没有调用过 ngOnChanges()方法，也仍然会调用 ngOnInit()方法（如当模板中没有绑定任何输入属性时）。

（3）**ngDoCheck**()方法。检测并在发生 Angular 无法或不愿意自己检测组件或指令的变化时响应该方法；在每次执行变更检测时，需在调用 ngOnChanges()方法和首次执行变更检测时调用 ngOnInit()方法后，再调用该方法。

（4）**ngAfterContentInit**()方法。当 Angular 把外部内容投影到组件视图或指令所在的视图之后调用该方法；在第一次调用 ngDoCheck()方法后调用该方法，且只调用一次。

（5）**ngAfterContentChecked**()方法。当 Angular 检查完被投影到组件或指令中的内容之后调用该方法；在调用 ngAfterContentInit()方法和每次调用 ngDoCheck()方法之后调用该方法。

（6）**ngAfterViewInit**()方法。当 Angular 初始化完组件视图及其子视图或包含该指令的视图之后调用该方法；在第一次调用 ngAfterContentChecked()方法后调用该方法，且只调用一次。

（7）**ngAfterViewChecked**()方法。当 Angular 做完组件视图、子视图或包含该指令的视图等变更检测后调用该方法；在调用 ngAfterViewInit()方法和每次调用 ngAfterContentChecked()方法之后调用该方法。

（8）**ngOnDestroy**()方法。在 Angular 每次销毁指令或组件之前调用该方法并进行清扫工作。

4.4　组件生命周期的综合应用

微课视频

4.4.1　生命周期接口

在项目 src\examples 目录下创建子目录 lifecycleexamples，在 src\examples\lifecycleexamples 目录下创建文件 peekaboo.directive.ts，代码如例 4-10 所示。

【**例 4-10**】　创建文件 peekaboo.directive.ts 的代码，实现一个接口方法的指令。

```
import {Directive, OnInit} from "@angular/core";
import {LoggerService} from "./logger.service";
let nextId = 1;
```

```
@Directive({selector: '[appPeekABoo]'})
export class PeekABooDirective implements OnInit {
  constructor(private logger: LoggerService) {
  }
  //实现 ngOnInit()方法
  ngOnInit() {
    this.logIt('OnInit 接口方法');
  }
  logIt(msg: string) {
    this.logger.log(`# ${nextId++} ${msg}`);
  }
}
```

在 src\examples\lifecycleexamples 目录下创建文件 peekaboo. component. ts,代码如例 4-11 所示。

【例 4-11】 创建文件 peekaboo. component. ts 的代码,实现多个接口方法的组件。

```
import {LoggerService} from './logger.service';
import {PeekABooDirective} from './peekaboo.directive';
import {
  AfterContentChecked,
  AfterContentInit,
  AfterViewChecked,
  AfterViewInit,
  Component,
  DoCheck,
  Input,
  OnChanges,
  OnDestroy,
  OnInit,
  SimpleChanges
} from "@angular/core";  //接口
@Component({
  selector: 'peek-a-boo',
  template: '<div>您现在看到的历史人物是, {{name}}</div>'
})
export class PeekABooComponent extends PeekABooDirective implements OnChanges, OnInit, DoCheck,
  AfterContentInit, AfterContentChecked, AfterViewInit, AfterViewChecked, OnDestroy {
  @Input() name = '';
  private verb = '初始化';
  constructor(logger: LoggerService) {
    super(logger);
    const is = this.name ? '在' : '不在';
    this.logIt(`人名 name ${is}构造方法中`);
  }
  ngOnChanges(changes: SimpleChanges) {
    const changesMsgs: string[] = [];
    for (const propName in changes) {
      if (propName === 'name') {
        const name = changes['name'].currentValue;
        changesMsgs.push(`人名 name ${this.verb}成 "${name}"`);
      } else {
        changesMsgs.push(propName + ' ' + this.verb);
      }
```

```
    }
    this.logIt(`OnChanges 接口方法: ${changesMsgs.join('; ')}`);
    this.verb = 'changed 改变'; //next time it will be a change
  }
  ngDoCheck() {
    this.logIt('DoCheck 接口方法');
  }
  ngAfterContentInit() {
    this.logIt('AfterContentInit 接口方法');
  }
  ngAfterContentChecked() {
    this.logIt('AfterContentChecked 接口方法');
  }
  ngAfterViewInit() {
    this.logIt('AfterViewInit 接口方法');
  }
  ngAfterViewChecked() {
    this.logIt('AfterViewChecked 接口方法');
  }
  ngOnDestroy() {
    this.logIt('OnDestroy 接口方法');
  }
}
```

在 src\examples\lifecycleexamples 目录下创建文件 peekabooparent.component.ts，代码如例 4-12 所示。

【例 4-12】 创建文件 peekabooparent.component.ts 的代码，注册需要注入的依赖。

```
import {Component} from '@angular/core';
import {LoggerService} from './logger.service';
@Component({
  selector: 'peekaboo',
  template: `
    <div class = "parent">
      <button (click) = "toggleChild()">
        {{hasChild ? '销毁' : '创建'}} PeekABooComponent 组件
      </button>
      <button (click) = "updateHero()" [hidden] = "!hasChild">更新人物信息</button>
      <div class = "info">
        <peek - a - boo * ngIf = "hasChild" [name] = "heroName"></peek - a - boo>
        <div>生命周期日志</div>
        <div * ngFor = "let msg of hookLog" class = "log">{{msg}}</div>
      </div>
    </div>
  `,
  providers: [LoggerService]//注册需要注入的依赖
})
export class PeekabooparentComponent {
  hasChild = false;
  hookLog: string[] = [];
  heroName = '张三丰';
  private logger: LoggerService;
  constructor(logger: LoggerService) {
    this.logger = logger;
```

```
      this.hookLog = logger.logs;
    }
    toggleChild() {
      this.hasChild = !this.hasChild;
      if (this.hasChild) {
        this.heroName = '李斯';
        this.logger.clear();
      }
      this.hookLog = this.logger.logs;
      this.logger.tick();
    }
    updateHero() {
      this.heroName += '!';
      this.logger.tick();
    }
  }
```

在src\examples\lifecycleexamples目录下创建文件logger.service.ts,代码如例4-13所示。

【例4-13】 创建文件logger.service.ts的代码,使用注入器。

```
import {Injectable} from '@angular/core';
@Injectable()
export class LoggerService {
  logs: string[] = [];
  prevMsg = '';
  prevMsgCount = 1;
  log(msg: string) {
    if (msg === this.prevMsg) {
      this.logs[this.logs.length - 1] = msg + ` (${this.prevMsgCount += 1}x)`;
    } else {
      this.prevMsg = msg;
      this.prevMsgCount = 1;
      this.logs.push(msg);
    }
  }
  clear() {
    this.logs = [];
  }
  tick() {
    this.tick_then(() => {
    });
  }
  tick_then(fn: () => any) {
    setTimeout(fn, 0);
  }
}
```

注入器的主要工作是负责向服务与组件、指令等注入依赖。Injector主要通过Angular的provider(提供者)中的token查找需要注入的依赖。@Injectable()装饰器表明服务或者组件的构造函数的参数需要依赖注入器进行注入。所以如果构造函数中没有参数,可以无须@Injectable()装饰器。组件和指令为什么没有@Injectable()也能注入构造函数中的参数?

因为@Component()和@Directive()装饰器也具备告知框架构造函数参数需要依赖注入，所以无须在组件、指令中使用@Injectable()装饰器。

4.4.2　响应事件

在 src\examples\lifecycleexamples 目录下创建文件 spy. directive. ts，代码如例 4-14 所示。

【例 4-14】 创建文件 spy. directive. ts 的代码，定义指令。

```
import {Directive, OnDestroy, OnInit} from "@angular/core";
import {LoggerService} from "./logger.service";
let nextId = 1;
//监视应用它的任何元素
//指令 appSpy 的用法<div appSpy>...</div>
@Directive({selector: '[appSpy]'})
export class SpyDirective implements OnInit, OnDestroy {
  private id = nextId++;
  constructor(private logger: LoggerService) {
  }
  ngOnInit() {
    this.logger.log(`监视(Spy)组件 # ${this.id} onInit 接口方法`);
  }
  ngOnDestroy() {
    this.logger.log(`监视(Spy)组件 # ${this.id} onDestroy 接口方法`);
  }
}
```

在 src\examples\lifecycleexamples 目录下创建文件 spy. component. ts，代码如例 4-15 所示。

【例 4-15】 创建文件 spy. component. ts 的代码，响应事件。

```
import {Component} from '@angular/core';
import {LoggerService} from "./logger.service";
@Component({
  selector: 'spy-parent',
  template: `
    <div class="parent">
      <label for="hero-name">历史人物名称: </label>
      <input type="text" id="hero-name" [(ngModel)]="newName" (keyup.enter)="addHero()">
      <button (click)="addHero()">增加人物</button>
      <button (click)="reset()">重置人物</button>
      <div class="info">
        <div *ngFor="let hero of heroes" appSpy>
          {{hero}}
        </div>
        <div>监视(Spy)组件(每个组件对应一位历史人物)生命周期钩子(Hook)日志</div>
        <div *ngFor="let msg of logger.logs" class="log">{{msg}}</div>
      </div>
    </div>
  `,
  providers: [LoggerService],  //注册需要注入的依赖
  styles: ['div {font-family: "Times New Roman";color: red}']
})
```

```
export class SpyComponent {
  newName = '子路';
  heroes: string[] = ['颜回'];
  constructor(public logger: LoggerService) {
  }
  addHero() {
    if (this.newName.trim()) {
      this.heroes.push(this.newName.trim());
      this.newName = '';
      this.logger.tick();
    }
  }
  removeHero(hero: string) {
    this.heroes.splice(this.heroes.indexOf(hero), 1);
    this.logger.tick();
  }
  reset() {
    this.logger.log('重置');
    this.heroes = [];
    this.logger.tick();
  }
}
```

4.4.3 OnChanges 方法

在 src\examples\lifecycleexamples 目录下创建文件 hero.ts,代码如例 4-16 所示。

【例 4-16】 创建文件 hero.ts 的代码,定义类 Hero。

```
export class Hero {
  constructor(public name: string) {
  }
}
```

在 src\examples\lifecycleexamples 目录下创建文件 onchanges.component.ts,代码如例 4-17 所示。

【例 4-17】 创建文件 onchanges.component.ts 的代码,定义实现 OnChanges 接口的组件。

```
import {Component, Input, OnChanges, SimpleChanges} from "@angular/core";
import {Hero} from "./hero";
@Component({
  selector: 'on-changes',
  template: `
    <div class="info">
      <div>{{hero.name}} 会 {{power}}</div>
      <div>变更(历史人物及其能力)日志</div>
      <div *ngFor="let chg of changeLog" class="log">{{chg}}</div>
    </div>
  `

})
export class OnchangesComponent implements OnChanges {
  @Input() hero!: Hero;
  @Input() power = '';
```

```
  changeLog: string[] = [];
  ngOnChanges(changes: SimpleChanges) {
    for (const propName in changes) {
      const chng = changes[propName];
      const cur = JSON.stringify(chng.currentValue);
      const prev = JSON.stringify(chng.previousValue);
      this.changeLog.push(`${propName}: 现在值为 ${cur},原值为 ${prev}`);
    }
  }
  reset() {
    this.changeLog = [];
  }
}
```

在 src\examples\lifecycleexamples 目录下创建文件 onchangesparent.component.ts，代码如例 4-18 所示。该文件(父组件)把子组件的内容拿来用，利用@ViewChild()装饰器可以调用子组件的方法。

【例 4-18】 创建文件 onchangesparent.component.ts 的代码，调用例 4-17 定义的子组件。

```
import {Component, ViewChild} from "@angular/core";
import {Hero} from "./hero";
import {OnchangesComponent} from "./onchanges.component";
@Component({
  selector: 'on-changes-parent',
  template: `
  <label for="power-input">能力: </label>
  <!-- 双向绑定,就是说在组件中给 power 赋了值可以影响到 input 的值
  但 input 的值发生改变时会影响组件中 power 的值 -->
  <input type="text" id="power-input" [(ngModel)]="power">
  <label for="hero-name">历史人物名字: </label>
  <input type="text" id="hero-name" [(ngModel)]="hero.name">
  <button (click)="reset()">重置日志</button>
  <on-changes [hero]="hero" [power]="power"></on-changes>
  `
})
export class OnchangesparentComponent {
  hero!: Hero;
  power = '';
  @ViewChild(OnchangesComponent) childView!: OnchangesComponent;
  constructor() {
    this.reset();
  }
  reset() {
    this.hero = new Hero('张三丰');
    this.power = '太极拳';
    if (this.childView) {
      this.childView.reset();
    }
  }
}
```

4.4.4 AfterView 方法

在 src\examples\lifecycleexamples 目录下创建文件 childview.component.ts，代码如

例 4-19 所示。

【例 4-19】 创建文件 childview. component. ts 的代码，定义子组件。

```
import {Component} from '@angular/core';
@Component({
  selector: 'child-view',
  template: `
    <label for="hero-name">人名: </label>
    <input type="text" id="hero-name" [(ngModel)]="hero">
  `
})
export class ChildviewComponent {
  hero = '张三丰';
}
```

在 src\examples\lifecycleexamples 目录下创建文件 afterview. component. ts，代码如例 4-20 所示。

【例 4-20】 创建文件 afterview. component. ts 的代码，定义实现 AfterViewInit()等 AfterView 方法的组件。

```
import {AfterViewChecked, AfterViewInit, Component, ViewChild} from '@angular/core';
import {LoggerService} from './logger.service';
import {ChildviewComponent} from "./childview.component";
@Component({
  selector: 'after-view',
  template: `
    <child-view></child-view>
    <p *ngIf="comment" class="comment">
      {{comment}}
    </p>
  `
})
export class AfterviewComponent implements AfterViewChecked, AfterViewInit {
  comment = '';
  @ViewChild(ChildviewComponent) viewChild!: ChildviewComponent;
  private prevHero = '';
  constructor(private logger: LoggerService) {
    this.logIt('AfterView constructor');
  }
  ngAfterViewInit() {
    this.logIt('AfterViewInit');
    this.doSomething();
  }
  ngAfterViewChecked() {
    if (this.prevHero === this.viewChild.hero) {
      this.logIt('AfterViewChecked (no change)');
    } else {
      this.prevHero = this.viewChild.hero;
      this.logIt('AfterViewChecked');
      this.doSomething();
    }
  }
  private doSomething() {
```

```
    const c = this.viewChild.hero.length > 10 ? "That's a long name" : '';
    if (c !== this.comment) {
      this.logger.tick_then(() => this.comment = c);
    }
  }
  private logIt(method: string) {
    const child = this.viewChild;
    const message = `${method}: ${child ? child.hero : 'no'} child view`;
    this.logger.log(message);
  }
}
```

在 src\examples\lifecycleexamples 目录下创建文件 afterviewparent.component.ts，代码如例 4-21 所示。

【例 4-21】 创建文件 afterviewparent.component.ts 的代码，调用例 4-20 组件。

```
import {Component} from '@angular/core';
import {LoggerService} from './logger.service';
@Component({
  selector: 'after-view-parent',
  template: `
  <after-view  *ngIf="show"></after-view>
  <div class="info">
    <div>AfterView 变更日志</div>
    <button (click)="reset()">重置</button>
    <div *ngFor="let msg of logger.logs" class="log">{{msg}}</div>
  </div>
  `,
  providers: [LoggerService]
})
export class AfterviewparentComponent {
  show = true;
  constructor(public logger: LoggerService) {
  }
  reset() {
    this.logger.clear();
    this.show = false;
    this.logger.tick_then(() => this.show = true);
  }
}
```

4.4.5 AfterContent 方法

在 src\examples\lifecycleexamples 目录下创建文件 child.component.ts，代码如例 4-22 所示。

【例 4-22】 创建文件 child.component.ts 的代码，定义子组件。

```
import {Component} from '@angular/core';
@Component({
  selector: 'app-child',
  template: `
    <label for="hero-name">人名: </label>
    <input type="text" id="hero-name" [(ngModel)]="hero">
```

```
  `
})
export class ChildComponent {
  hero = '张三丰';
}
```

在 src\examples\lifecycleexamples 目录下创建文件 aftercontent. component. ts,代码如例 4-23 所示。

【例 4-23】 创建文件 aftercontent. component. ts 的代码,定义实现 AfterContentInit()等 AfterContent 方法的组件。

```
import {AfterContentChecked, AfterContentInit, Component, ContentChild} from '@angular/core';
import {ChildComponent} from './child.component';
import {LoggerService} from './logger.service';
@Component({
  selector: 'after-content',
  template: `
    <ng-content></ng-content>
    <div *ngIf="comment" class="comment">
      {{comment}}
    </div>
  `
})
export class AftercontentComponent implements AfterContentChecked, AfterContentInit {
  comment = '';
  @ContentChild(ChildComponent) contentChild!: ChildComponent;
  private prevHero = '';
  constructor(private logger: LoggerService) {
    this.logIt('AfterContent constructor');
  }
  ngAfterContentInit() {
    this.logIt('AfterContentInit');
    this.doSomething();
  }
  ngAfterContentChecked() {
    if (this.prevHero === this.contentChild.hero) {
      this.logIt('AfterContentChecked (no change)');
    } else {
      this.prevHero = this.contentChild.hero;
      this.logIt('AfterContentChecked');
      this.doSomething();
    }
  }
  private doSomething() {
    this.comment = this.contentChild.hero.length > 10 ? "That's a long name" : '';
  }
  private logIt(method: string) {
    const child = this.contentChild;
    const message = `${method}: ${child ? child.hero : 'no'} child content`;
    this.logger.log(message);
  }
}
```

在 src\examples\lifecycleexamples 目录下创建文件 aftercontentparent. component. ts,代码

如例 4-24 所示。

【例 4-24】 创建文件 aftercontentparent. component. ts 的代码，调用例 4-23 组件。

```
import {Component} from '@angular/core';
import {LoggerService} from './logger.service';
@Component({
  selector: 'after-content-parent',
  template: `
    <div class="parent">
      <div *ngIf="show">
        <after-content>
          <app-child></app-child>
        </after-content>
      </div>
      <div class="info">
        <div *ngFor="let msg of logger.logs" class="log">{{msg}}</div>
      </div>
    </div>
  `,
  providers: [LoggerService]
})
export class AftercontentparentComponent {
  show = true;
  constructor(public logger: LoggerService) {
  }
  reset() {
    this.logger.clear();
    this.show = false;
    this.logger.tick_then(() => this.show = true);
  }
}
```

4.4.6 DoCheck 方法

在 src\examples\lifecycleexamples 目录下创建文件 docheck. component. ts，代码如例 4-25 所示。

【例 4-25】 创建文件 docheck. component. ts 的代码，定义实现 DoCheck()方法的组件。

```
import {Component, DoCheck, Input} from '@angular/core';
import {Hero} from './hero';
@Component({
  selector: 'do-check',
  template: `
    <div class="info">
      <div>{{hero.name}} 会{{power}}</div>
    </div>
  `
})
export class DocheckComponent implements DoCheck {
  @Input() hero!: Hero;
  @Input() power = '';
  changeDetected = false;
  changeLog: string[] = [];
```

```
  oldHeroName = '';
  oldPower = '';
  noChangeCount = 0;
  ngDoCheck() {
    if (this.hero.name !== this.oldHeroName) {
      this.changeDetected = true;
 this.changeLog.push(`DoCheck: Hero name changed to "${this.hero.name}" from "${this.
oldHeroName}"`);
      this.oldHeroName = this.hero.name;
    }
    if (this.power !== this.oldPower) {
      this.changeDetected = true;
       this.changeLog.push(`DoCheck: Power changed to "${this.power}" from "${this.
oldPower}"`);
      this.oldPower = this.power;
    }
    if (this.changeDetected) {
      this.noChangeCount = 0;
    } else {
      const count = this.noChangeCount += 1;
      const noChangeMsg = `DoCheck called ${count}x when no change to hero or power`;
      if (count === 1) {
        this.changeLog.push(noChangeMsg);
      } else {
        this.changeLog[this.changeLog.length - 1] = noChangeMsg;
      }
    }
    this.changeDetected = false;
  }
  reset() {
    this.changeDetected = true;
    this.changeLog = [];
  }
}
```

在 src\examples\lifecycleexamples 目录下创建文件 docheckparent.component.ts，代码如例 4-26 所示。

【例 4-26】 创建文件 docheckparent.component.ts 的代码，调用例 4-25 组件。

```
import {Component, ViewChild} from '@angular/core';
import {DocheckComponent} from './docheck.component';
import {Hero} from './hero';
@Component({
  selector: 'do-check-parent',
  template: `
    <do-check [hero]="hero" [power]="power"></do-check>
  `
})
export class DocheckparentComponent {
  hero!: Hero;
  power = '';
  title = 'DoCheck';
  @ViewChild(DocheckComponent) childView!: DocheckComponent;
  constructor() {
```

```
    this.reset();
  }
  reset() {
    this.hero = new Hero('张三丰');
    this.power = '耍太极';
    if (this.childView) {
      this.childView.reset();
    }
  }
}
```

4.4.7 组件、模块和运行结果

在 src\examples\lifecycleexamples 目录下创建文件 lifecycleexample. component. ts，代码如例 4-27 所示。

【例 4-27】 创建文件 lifecycleexample. component. ts 的代码，定义组件。

```
import {Component} from '@angular/core';
@Component({
  selector: 'root',
  template: `
    < div >响应生命周期事件示例【PeekABoo】</div >
    < peekaboo ></peekaboo >
    < hr >
    < div >使用指令(directives)来监视(watch)DOM【Spy】</div >
    < spy - parent ></spy - parent >
    < hr >
    < div >使用变更检测钩子【onChanges】</div >
    < on - changes - parent ></on - changes - parent >
    < hr >
    < div >响应视图的变更【AfterView】</div >
    < after - view - parent ></after - view - parent >
    < hr >
    < div >响应被投影内容的变更【AfterContent】</div >
    < after - content - parent ></after - content - parent >
    < hr >
    < div >自定义变更检测逻辑【DoCheck】</div >
    < do - check - parent ></do - check - parent >
  `,
  styles: ['div {font - family: "Times New Roman";font - size: larger;color: red}']
})
export class LifecycleexampleComponent {}
```

在 src\examples\lifecycleexamples 目录下创建文件 app-lifecycle. module. ts，代码如例 4-28 所示。

【例 4-28】 创建文件 app-lifecycle. module. ts 的代码，定义路由并声明组件。

```
import {NgModule} from '@angular/core';
import {BrowserModule} from "@angular/platform - browser";
import {RouterModule} from "@angular/router";
import {LifecycleexampleComponent} from "./lifecycleexample.component";
import {PeekABooDirective} from "./peekaboo.directive";
import {PeekABooComponent} from "./peekaboo.component";
```

```
import {PeekabooparentComponent} from "./peekabooparent.component";
import {SpyDirective} from "./spy.directive";
import {SpyComponent} from "./spy.component";
import {OnchangesComponent} from "./onchanges.component";
import {OnchangesparentComponent} from "./onchangesparent.component";
import {ChildviewComponent} from "./childview.component";
import {AfterviewComponent} from "./afterview.component";
import {AfterviewparentComponent} from "./afterviewparent.component";
import {ChildComponent} from "./child.component";
import {AftercontentComponent} from "./aftercontent.component";
import {AftercontentparentComponent} from "./aftercontentparent.component";
import {DocheckComponent} from "./docheck.component";
import {DocheckparentComponent} from "./docheckparent.component";
@NgModule({
  imports: [
    BrowserModule,
    RouterModule.forRoot([
      {path: 'lifecycle', component: LifecycleexampleComponent},
    ]),
  ],
  declarations: [
    PeekABooDirective,
    PeekABooComponent,
    PeekabooparentComponent,
    SpyDirective,
    SpyComponent,
    OnchangesComponent,
    OnchangesparentComponent,
    ChildviewComponent,
    AfterviewComponent,
    AfterviewparentComponent,
    ChildComponent,
    AftercontentComponent,
    AftercontentparentComponent,
    DocheckComponent,
    DocheckparentComponent,
    LifecycleexampleComponent,
  ],
})
export class AppStyleModule { }
```

修改 src\examples 目录下的文件 examplesmodules1.module.ts，代码如例 4-29 所示。

【例 4-29】 修改文件 examplesmodules1.module.ts 的代码，设置启动组件。

```
import {NgModule} from '@angular/core';
import {AppLifecycleModule} from "./Lifecycleexamples/app-lifecycle.module";
import {LifecycleexampleComponent} from './lifecycleexamples/lifecycleexample.component';
@NgModule({
  imports: [
    AppLifecycleModule
  ],
  bootstrap: [LifecycleexampleComponent]
})
export class ExamplesmodulesModule1 {}
```

保持其他文件不变并成功运行程序后，在浏览器地址栏中输入 localhost:4200，结果如图 4-2 所示。

Angular Example
localhost:4200
响应生命周期事件示例【PeekABoo】
创建 PeekABooComponent组件
生命周期日志
使用指令（directives）来监视（watch）DOM【Spy】
历史人物名称：子路 增加人物 重置人物
颜回
监视（Spy）组件（每个组件对应一位历史人物）生命周期钩子（Hook）日志
监视（Spy）组件 #1 onInit接口方法
使用变更检测钩子【onChanges】
能力：太极拳 历史人物名字：张三丰 重置日志
张三丰 会 太极拳
变更（历史人物及其能力）日志
hero：现在值为 {"name":"张三丰"}，原值为 undefined
power：现在值为 "太极拳"，原值为 undefined
响应视图的变更【AfterView】
人名：张三丰
AfterView变更日志
重置
AfterView constructor: no child view
AfterViewInit: 张三丰 child view
AfterViewChecked: 张三丰 child view
响应被投影内容的变更【AfterContent】
人名：张三丰
AfterContent constructor: no child content
AfterContentInit: 张三丰 child content
AfterContentChecked: 张三丰 child content
自定义变更检测逻辑【DoCheck】
张三丰 会耍太极

图 4-2　成功运行程序后在浏览器地址栏中输入 localhost:4200 的结果

4.5　组件之间的交互及其应用

4.5.1　组件交互说明

两个或多个组件之间共享信息的方法包括以下 7 个。

（1）通过输入型绑定把数据从父组件传到子组件。

（2）通过 setter 截听（intercept）输入属性值的变化。

（3）用 ngOnChanges()方法来截听、监测输入属性值的变化并作出回应。

（4）父组件监听（listen）子组件的事件。子组件暴露一个 EventEmitter 属性，当事件发生时，子组件利用该属性 emit（向上弹射）事件，使父组件绑定到这个事件属性，并在事件发生时作出回应。子组件的 EventEmitter 属性是一个输出属性，通常带有@Output()装饰器。

（5）父组件与子组件通过本地变量互动。父组件不能使用数据绑定来读取子组件的属性或调用子组件的方法。但可以在父组件模板里新建一个本地变量来代表子组件，然后利用这个变量读取子组件的属性和调用子组件的方法。

（6）父组件调用@ViewChild()装饰器。因为父组件和子组件的连接必须全部在父组件的模板中进行，而子组件无法访问父组件代码。如果父组件的类需要依赖于子组件，就不能使用本地变量方法,因为父组件和子组件的实例彼此并不知道,因此父类也就不能访问子类中的属性和方法。组件之间的父子关系也不能通过在每个组件中各自定义本地变量来建立。当父组件需要访问子组件的属性和方法时,可以把子组件作为ViewChild注入父组件,要实现这一点,必须先导入对@ViewChild()装饰器以及ngAfterViewInit()方法的引用,再通过@ViewChild()装饰器将子组件注入私有属性。

(7) 父组件和子组件通过服务来通信。父组件和它的子组件共享同一个服务,利用该服务在组件家族内部实现双向通信。该服务实例的作用域被限制在父组件及其子组件内,这个组件子树之外的组件将无法访问该服务或者与它们通信。

4.5.2 父组件和子组件

在项目src\examples目录下创建子目录interactionexamples,在src\examples\interactionexamples目录下创建文件hero.ts,代码如例4-30所示。

【例4-30】 创建文件hero.ts的代码,定义接口和数组。

```
export interface Hero {
  id: number;
  name: string;
}
export const HEROES: Hero[] = [
  {id: 12, name: '李斯'},
  {id: 13, name: '王阳明'},
];
```

在src\examples\interactionexamples目录下创建文件childexample.component.ts,代码如例4-31所示。

【例4-31】 创建文件childexample.component.ts的代码,定义子组件。

```
import {Component, Input} from '@angular/core';
import {Hero} from './hero';
@Component({
  selector: 'childex',
  template: `
    <div>{{hero.name}}说: {{master}},您好.</div>
  `,
  styles: ['div {font-family: "Times New Roman";font-size: larger;color: red}']
})
export class ChildexampleComponent {
  //输入型属性,它们通常带@Input()装饰器
  @Input() hero!: Hero;
  @Input() master: string = '';
}
```

在src\examples\interactionexamples目录下创建文件parentexample.component.ts,代码如例4-32所示。

【例4-32】 创建文件parentexample.component.ts的代码,定义调用了例4-31组件的父组件。

```
import {Component, Input} from '@angular/core';
import {HEROES} from "./hero";
@Component({
```

```
  selector: 'parentex',
  template: `
    <div>通过 setter 截听(Intercept)输入属性(message)值的变化: {{message}}</div>
    <hr>
    <div>通过输入型绑定把数据从父组件传到子组件</div>
    <!-- 把子组件传到 *ngFor 循环器中,把自己的 master 属性绑定到子组件的 master 上,
    并把每个循环的 hero 实例绑定到子组件的 hero 属性 -->
    <div>{{heroes.length}}位历史人物问候{{master}} </div>
    <childex
      *ngFor = "let hero of heroes"
      [hero] = "hero"
      [master] = "master"
    >
    </childex>
  `
})
export class ParentexampleComponent {
  heroes = HEROES;
  master = 'Master';
  //注意要使用此内部变量,否则无法显示结果
  private _message: string = '';
  //属性_message
  @Input()
  get message(): string {
    return this._message;
  }
  set message(message: string) {
    this._message = (message && message.trim()) || '<no message set>';
  }
}
```

4.5.3　OnChanges 方法

在 src\examples\interactionexamples 目录下创建文件 onchangeexample.component.ts,代码如例 4-33 所示。

【例 4-33】　创建文件 onchangeexample.component.ts 的代码,定义实现了 OnChanges 方法的组件。

```
import {Component, Input, OnChanges, SimpleChanges} from '@angular/core';
@Component({
  selector: 'onchange-root',
  template: `
    <div>显示版本变化信息 Version {{major}}.{{minor}}</div>
    <div>版本更新日志</div>
    <ul>
      <li *ngFor = "let change of changeLog">{{change}}</li>
    </ul>
  `,
  styles: ['div {color: blue}']
})
//实现接口
export class OnchangeexampleComponent implements OnChanges {
  @Input() major = 0;
  @Input() minor = 0;
```

```
    changeLog: string[] = [];
    //实现方法
    ngOnChanges(changes: SimpleChanges) {
      const log: string[] = [];
      for (const propName in changes) {
        const changedProp = changes[propName];
        const to = JSON.stringify(changedProp.currentValue);
        if (changedProp.isFirstChange()) {
          log.push(`Initial value of ${propName} set to ${to}`);
        } else {
          const from = JSON.stringify(changedProp.previousValue);
          log.push(`${propName} changed from ${from} to ${to}`);
        }
      }
      this.changeLog.push(log.join(', '));
    }
}
```

4.5.4 事件

在 src\examples\interactionexamples 目录下创建文件 eventexample.component.ts，代码如例 4-34 所示。

【例 4-34】 创建文件 eventexample.component.ts 的代码，定义有事件的组件。

```
import {Component, EventEmitter, Input, Output} from '@angular/core';
@Component({
  selector: 'event-root',
  template: `
    <div>父组件监听子组件的示例</div>
    <div>
      今天的任务是: {{item}}
    </div>
    <label for="item-input">输入新任务</label>
    <input type="text" id="item-input" #newItem>
    <button (click)="addNewItem(newItem.value)">添加任务</button>
  `,
  styles: ['div {color: blue}']
})
export class EventexampleComponent {
  @Output() newItemEvent = new EventEmitter<string>();
  @Input() item = '';
  @Output() voted = new EventEmitter<boolean>();
  didVote = false;
  addNewItem(value: string) {
    this.newItemEvent.emit(value);
  }
  vote(agreed: boolean) {
    this.voted.emit(agreed);
    this.didVote = true;
  }
}
```

例 4-34 组件暴露 EventEmitter 属性(如 newItemEvent)，在事件发生时，它利用该属性

emit(向上弹射)事件。父组件例 4-39 绑定到这个事件属性，并在事件发生时作出回应。子组件的 EventEmitter 属性是一个输出属性，通常带有@Output()装饰器。

在 src\examples\interactionexamples 目录下创建文件 voteexample. component. ts，代码如例 4-35 所示。

【例 4-35】 创建文件 voteexample. component. ts 的代码，定义组件。

```
import {Component, EventEmitter, Input, Output} from '@angular/core';
@Component({
  selector: 'vote - root',
  template: `
    < div >投票人{{voter}}</div >
    < button (click) = "vote(true)" [disabled] = "didVote">赞成</button >
    < button (click) = "vote(false)" [disabled] = "didVote">反对</button >
  `,
  styles: ['div {color: blue}']
})
//实现接口
export class VoteexampleComponent {
  @Input() voter = '';
  @Output() voted = new EventEmitter < boolean >();
  didVote = false;
  vote(agreed: boolean) {
    this.voted.emit(agreed);
    this.didVote = true;
  }
}
```

在 src\examples\interactionexamples 目录下创建文件 timerchild. component. ts，代码如例 4-36 所示。

【例 4-36】 创建文件 timerchild. component. ts 的代码，定义组件。

```
import {Component, OnDestroy} from '@angular/core';
@Component({
  selector: 'timer - child',
  template: `< div >计数: {{message}}</div >
  `,
  styles: ['div {font - family: "Times New Roman";font - size: larger;color: red}']
})
export class TimerchildComponent implements OnDestroy {
  //进行倒计时
  //start()和 stop()方法负责控制时钟并在模板里显示倒计时的状态信息
  intervalId = 0;
  message = '';
  seconds = 11;
  ngOnDestroy() { this.clearTimer(); }
  start() { this.countDown(); }
  stop()  {
    this.clearTimer();
    this.message = `暂停在 T - ${this.seconds} 秒`;
  }
  private clearTimer() { clearInterval(this.intervalId); }
  private countDown() {
    this.clearTimer();
    this.intervalId = window.setInterval(() => {
```

```
        this.seconds -= 1;
        if (this.seconds === 0) {
          this.message = '归零!';
        } else {
          if (this.seconds < 0) { this.seconds = 10; } //reset
          this.message = `T- ${this.seconds} 秒并计数`;
        }
      }, 1000);
    }
}
```

4.5.5 本地变量

在 src\examples\interactionexamples 目录下创建文件 timerparent. component. ts，代码如例 4-37 所示。

【例 4-37】 创建文件 timerparent. component. ts 的代码，建本地变量代表例 4-36 组件。

```
import {Component} from '@angular/core';
@Component({
  selector: 'timer-root',
  template: `
    <hr>
    <div>父组件与子组件通过本地变量互动</div>
    <button (click)="timer.start()">开始倒计时</button>
    <button (click)="timer.stop()">暂停倒计时</button>
    <div class="seconds">{{timer.seconds}}</div>
    <timer-child #timer></timer-child>
  `,
  styles: ['div {font-family: "Times New Roman";font-size: larger;color: red}']
})
export class TimerparentComponent {
}
```

例 4-37 中组件不使用数据绑定来读取子组件的属性或调用子组件的方法。而是在模板里，新建一个本地变量(#timer)来代表子组件(即例 4-36)，然后利用这个变量来读取子组件的属性和调用子组件的方法。

4.5.6 @ViewChild()装饰器

在 src\examples\interactionexamples 目录下创建文件 timeview. component. ts，代码如例 4-38 所示。

【例 4-38】 创建文件 timeview. component. ts 的代码，注入例 4-37 组件。

```
import {AfterViewInit, Component, ViewChild} from '@angular/core';
import {TimerchildComponent} from "./timerchild.component";
@Component({
  selector: 'timer-view',
  template: `
    <hr>
    <div>父组件调用@ViewChild()</div>
    <button (click)="start()">开始倒计时</button>
```

```
    <button (click) = "stop()">暂停倒计时</button>
    <div class = "seconds">{{ seconds() }}</div>
    <timer - child></timer - child>
  `,
  styles: ['div {font - family: "Times New Roman";font - size: larger;color: red}']
})
export class TimeviewComponent implements AfterViewInit {
  @ViewChild(TimerchildComponent)
  private timerComponent! : TimerchildComponent;
  seconds() { return 0; }
  ngAfterViewInit() {
    setTimeout(
      () => this.seconds =  () => this.timerComponent.seconds,
      0);
  }
  start() { this.timerComponent.start(); }
  stop() { this.timerComponent.stop(); }
}
```

本地变量方法是个简单明了的方法。但是它也有局限性,因为父组件和子组件的连接必须全部在父组件的模板中进行。父组件本身对子组件没有访问权。如果父组件需要依赖于子组件,就不能使用本地变量方法。组件之间的父子关系不能通过在父子组件的类中定义本地变量来建立。这是因为这两个组件类的实例彼此并不知道,因此父类也就不能访问子类中的属性和方法。当父组件(如例 4-38)需要这种访问时,可以把子组件(如例 4-37)作为 ViewChild 注入父组件。

4.5.7 组件、模块和运行结果

在 src\examples\interactionexamples 目录下创建文件 interactionexample. component. ts,代码如例 4-39 所示。

【例 4-39】 创建文件 interactionexample. component. ts 的代码,定义组件。

```
import {Component} from '@angular/core';
@Component({
  selector: 'interaction - root',
  template: `
    <div>调用父组件</div>
    <parentex [message] = "message"></parentex>
    <hr>
    <div>使用生命周期钩子接口通过 ngOnChanges()来截听输入属性值的变化并作出回应</div>
    <button (click) = "newMinor()">新小版本</button>
    <button (click) = "newMajor()">新主版本(大版本)</button>
    <onchange - root [major] = "major" [minor] = "minor"></onchange - root>
    <hr>
    <event - root
      [item] = "currentItem"
      (newItemEvent) = "addItem( $ event)">
    </event - root>
    <ul>
      <li *ngFor = "let item of items">{{item}}</li>
    </ul>
    <h2>投票事项: Angular 学习曲线较陡?</h2>
```

```
    <vote-root *ngFor="let voter of voters"
               [voter]="voter"
               (voted)="onVoted($event)">
    </vote-root>
    <div>赞成总票数：{{agreed}}，反对总票数：{{disagreed}}</div>
  `,
  styles: ['div {font-family: "Times New Roman";font-size: larger;color: red}']
})
export class InteractionexampleComponent {
  message = 'hello'
  major = 1;
  minor = 23;
  currentItem = '学习 Angular';
  items = ['学习 TypeScript', '学习 Vue', '学习 JavaScript'];
  agreed = 0;
  disagreed = 0;
  voters = ['颜回', '子贡', '子路'];
  newMinor() {
    this.minor++;
  }
  newMajor() {
    this.major++;
    this.minor = 0;
  }
  addItem(newItem: string) {
    this.items.push(newItem);
  }
  onVoted(agreed: boolean) {
    if (agreed) {
      this.agreed++;
    } else {
      this.disagreed++;
    }
  }
}
```

在 src\examples\interactionexamples 目录下创建文件 interactionhome. component. ts，代码如例 4-40 所示。

【例 4-40】 创建文件 interactionhome. component. ts 的代码，定义组件。

```
import {Component} from '@angular/core';
@Component({
  selector: 'root',
  template: `
    <interaction-root></interaction-root>
    <timer-root></timer-root>
    <timer-view></timer-view>
  `,
  styles: ['div {font-family: "Times New Roman";font-size: larger;color: red}']
})
export class InteractionhomeComponent {
}
```

在 src\examples\interactionexamples 目录下创建文件 app-interaction. module. ts，代

码如例 4-41 所示。

【例 4-41】 创建文件 app-interaction. module. ts 的代码，定义路由并声明组件。

```
import {NgModule} from '@angular/core';
import {BrowserModule} from '@angular/platform-browser';
import {RouterModule} from "@angular/router";
import {ChildexampleComponent} from "./childexample.component";
import {ParentexampleComponent} from "./parentexample.component";
import {OnchangeexampleComponent} from "./onchangeexample.component";
import {EventexampleComponent} from "./eventexample.component";
import {VoteexampleComponent} from "./voteexample.component";
import {TimerchildComponent} from "./timerchild.component";
import {TimerparentComponent} from "./timerparent.component";
import {TimeviewComponent} from "./timeview.component";
import {InteractionhomeComponent} from "./interactionhome.component";
import {InteractionexampleComponent} from "./interactionexample.component";
@NgModule({
  imports: [
    BrowserModule,
    RouterModule.forRoot([
      {path: 'interaction', component: InteractionhomeComponent},
    ]),
  ],
  declarations: [
    ChildexampleComponent,
    ParentexampleComponent,
    OnchangeexampleComponent,
    EventexampleComponent,
    VoteexampleComponent,
    TimerchildComponent,
    TimerparentComponent,
    TimeviewComponent,
    InteractionhomeComponent,
    InteractionexampleComponent,
  ],
})
export class AppInteractionModule { }
```

修改 src\examples 目录下的文件 examplesmodules1. module. ts，代码如例 4-42 所示。

【例 4-42】 修改文件 examplesmodules1. module. ts 的代码。

```
import {NgModule} from '@angular/core';
import {InteractionhomeComponent} from './interactionexamples/interactionhome.component';
import {InteractionhomeComponent} from './interactionexamples/interactionhome.component';
@NgModule({
  imports: [
    AppInteractionModule
  ],
  bootstrap: [InteractionhomeComponent]
})
export class ExamplesmodulesModule1 {}
```

保持其他文件不变并成功运行程序后，在浏览器地址栏中输入 localhost:4200，结果如图 4-3 所示。

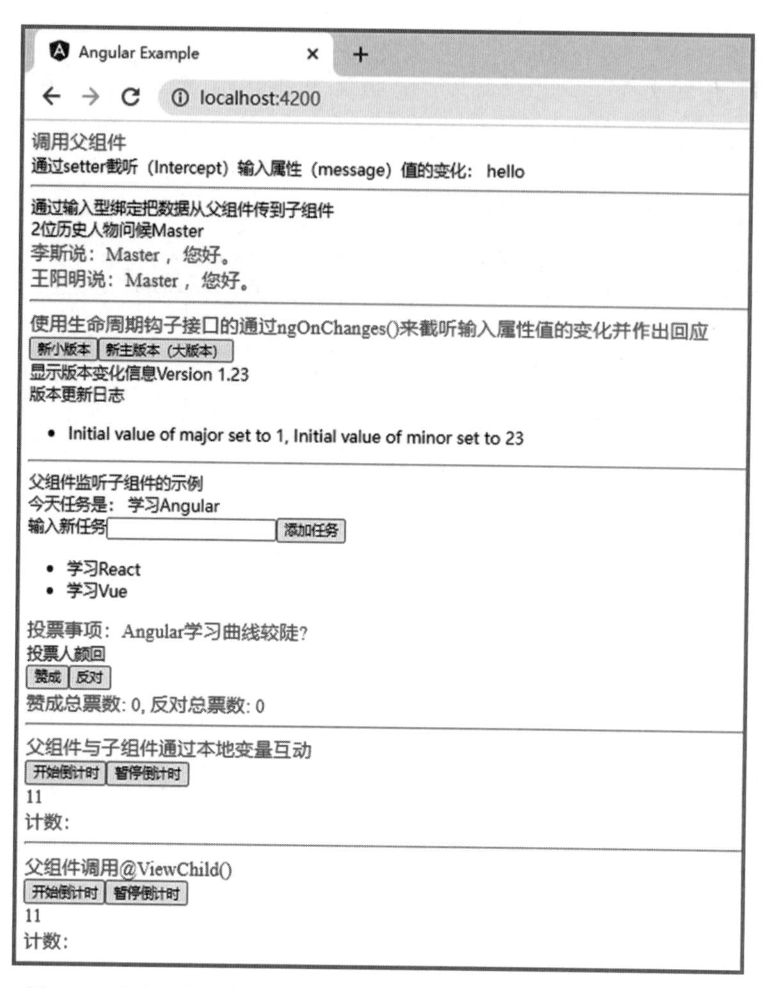

图 4-3　成功运行程序后在浏览器地址栏中输入 localhost:4200 的结果

4.6 Angular 元素及其应用

4.6.1 Angular 元素含义及其原理

Angular 元素(Element)是打包成自定义元素的 Angular 组件。所谓自定义元素,就是一套与具体框架无关的用于定义新 HTML 元素的 Web 标准。自定义元素扩展了 HTML,它允许定义一个由 JavaScript 或 TypeScript 代码创建和控制的标签。浏览器会维护一个自定义元素的注册表 CustomElementRegistry,把一个可实例化的 JavaScript 类映射到 HTML 标签上。自定义元素目前在一些浏览器中实现了原生支持,而一些浏览器或者尚未决定支持,或者已经制订了支持计划。

如果要在运行期间把一个组件添加到应用中,就得将其定义成动态组件,然后加载它并把它附加到 DOM 中的元素上,装配所有的依赖、变更检测和事件处理。若用 Angular 自定义组件(元素),则会让这个过程更简单、更透明。把组件转换成自定义元素,就为在 Angular 应用中创建动态 HTML 内容提供了一种简单的方式。Angular 会为自定义元素自动提供所有基础设施和框架，而开发人员要做的就是定义所需的各种事件处理逻辑。如果不准备在应用程序中用它，就要在编译时根据条件设置将相关组件排除出去。

把组件转换成自定义元素，可以让所有所需的 Angular 基础设施都在浏览器中可用。创建自定义元素的方式简单直观，它会自动把组件定义的视图连同变更检测与数据绑定等 Angular 功能映射为相应的内置 HTML 等价物。组件的属性和逻辑会直接映射到 HTML 属性和浏览器的事件系统中。自定义元素功能使 Angular 元素可以运用在由其他框架所构建的 Web 应用中。Angular 框架的一个小型的、自包含版本将会作为服务被注入，以提供组件的变更检测和数据绑定功能，从而简化对 Angular 元素的处理。

自定义元素会在添加到 DOM 中时自行启动，在从 DOM 中移除时自行销毁。自定义元素是一种特殊的组件，这个过程与组件生命周期对应。一旦自定义元素添加到了任何页面的 DOM 中，它的外观和行为就和其他的 HTML 元素一样。在 Angular 应用中，直接添加到 DOM 中的 HTML 内容无须 Angular 处理，除非使用动态组件来借助代码把 HTML 标签与应用数据关联起来并参与变更检测。使用自定义元素，所有这些装配工作都是自动完成的。当自定义元素放进页面中时，浏览器会创建一个已注册类的实例。其内容由组件模板提供，使用 Angular 模板语法，并且使用组件和 DOM 数据进行渲染。组件的输入属性对应于该元素的输入特性。

4.6.2　Angular 元素相关 API

@angular/elements 包导出一个 createCustomElement()方法 API 以支持把 Angular 组件及其依赖转换成自定义元素。它在 Angular 组件接口与变更检测功能和内置 DOM 的 API 之间建立了一个桥梁。使用 createCustomElement()方法来把组件转换成一个可注册成浏览器中自定义元素的类。注册完这个配置好的类之后，就可以像内置 HTML 元素一样使用这个新元素(类)了，如直接把它加到 DOM 中。createCustomElement()方法会收集该组件的 Observable 型属性，提供浏览器创建和销毁实例时所需的 Angular 功能，还会对变更进行检测并做出响应。这个转换过程实现了 NgElementConstructor 接口，并创建了一个构造器类，用于生成该组件的一个自举型实例。

使用内置的 customElements.define()方法把配置好的构造器和相关的自定义元素标签注册到浏览器的 CustomElementRegistry 中。当浏览器遇到这个已注册元素的标签时，就会使用自定义元素的构造器来创建一个自定义元素的实例。用于创建组件(和自定义元素对应)的 API 会解析该组件以查找输入属性，并在这个自定义元素上定义相应的特性。API 把属性名转换成与自定义元素兼容的形式(自定义元素不区分大小写)，生成的属性名会使用中线分隔的小写形式。例如，带有@Input('myInputProp')inputProp 属性的组件，其对应的自定义元素会带有一个 my-input-prop 属性。组件的输出属性会用 HTML 自定义事件的形式进行分发，自定义事件的名字就是这个输出属性的名字。例如，带有@Output()valueChanged=new EventEmitter()属性的组件，其相应的自定义元素将会分发名叫 valueChanged 的事件，事件中所携带的数据存储在该事件对象的 detail 属性中。如果提供了别名(@Output()装饰器中的参数指定的名字)，就分发改用之后的别名。例如，语句@Output('myClick')clicks=new EventEmitter<string>()；会导致分发名为 myClick 事件。

一般的 DOM API，如 document.createElement()或 document.querySelector()方法，

会返回一个与指定的参数相匹配的元素类型。例如,调用 document.createElement('a')会返回 HTMLAnchorElement,TypeScript 就会知道它有一个 href 属性。当调用未知元素(如自定义的元素名 popup-element)时,该方法会返回泛化类型(即基类),如 HTMLELement;TypeScript 就无法推断出所返回元素的正确类型。用 Angular 创建的自定义元素会扩展(继承)NgElement 类型(而它扩展了 HTMLElement)。除此之外,自定义元素还拥有相应组件的每个输入属性。如果要让自定义元素获得正确的类型,还可使用一些选项。要获得精确类型,最直接的方式是把相关 DOM 方法的返回值转换成正确的类型。要做到这一点,可以使用 NgElement 和 WithProperties 类型。这是一种让自定义元素快速获得 TypeScript 特性(如类型检查和自动完成支持)的好办法。不过,如果要在多个地方使用它,可能会有点啰唆,因为不得不在每个地方对返回类型做转换。另一种方式是对每个自定义元素的类型只声明一次。可以扩展 HTMLElementTagNameMap,TypeScript 会在 DOM 方法(如 document.createElement()方法、document.querySelector()方法等)中用它来根据标签名推断返回元素的类型。

4.6.3 Angular 元素应用示例

在项目 src\examples 目录下创建子目录 elementexamples,在 src\examples\elementexamples 目录下创建文件 popup.service.ts,代码如例 4-43 所示。

【例 4-43】 创建文件 popup.service.ts 的代码,定义指令。

```
import {ApplicationRef, ComponentFactoryResolver, Injectable, Injector} from '@angular/core';
import {NgElement, WithProperties} from '@angular/elements';
import {PopupComponent} from './popup.component';
@Injectable()
export class PopupService {
  constructor(private injector: Injector,
              private applicationRef: ApplicationRef,
              private componentFactoryResolver: ComponentFactoryResolver) {}
  //预先动态加载方法(要求设置基础结构),在将弹出窗口添加到 DOM 之前
  showAsComponent(message: string) {
    //创建元素
    const popup = document.createElement('popup-component');
    //创建组件
    const factory = this.componentFactoryResolver.resolveComponentFactory(PopupComponent);
    const popupComponentRef = factory.create(this.injector, [], popup);
    this.applicationRef.attachView(popupComponentRef.hostView);
    //监听 closed 事件
    popupComponentRef.instance.closed.subscribe(() => {
      document.body.removeChild(popup);
      this.applicationRef.detachView(popupComponentRef.hostView);
    });
    popupComponentRef.instance.message = message;
    //添加到 DOM
    document.body.appendChild(popup);
  }
  //使用新的自定义元素方法将弹出窗口添加到 DOM 中
  showAsElement(message: string) {
const popupEl: NgElement & WithProperties<PopupComponent> = document.createElement('popup-element') as any;
```

```
    popupEl.addEventListener('closed', () => document.body.removeChild(popupEl));
    popupEl.message = message;
    document.body.appendChild(popupEl);
  }
}
```

在 src\examples\elementexamples 目录下创建文件 popup.component.ts，代码如例 4-44 所示。

【例 4-44】 创建文件 popup.component.ts 的代码，定义组件。

```
import {Component, EventEmitter, HostBinding, Input, Output} from '@angular/core';
import {animate, state, style, transition, trigger} from '@angular/animations';
@Component({
  selector: 'my-popup',
  template: `
    <span>Popup: {{message}}</span>
    <button (click)="closed.next()">&#x2716;</button>
  `,
  animations: [
    trigger('state', [
      state('opened', style({transform: 'translateY(0%)'})),
      state('void, closed', style({transform: 'translateY(100%)', opacity: 0})),
      transition('* => *', animate('100ms ease-in')),
    ])
  ],
  styles: [`
    :host {
      position: absolute;
      bottom: 0;
      left: 0;
      right: 0;
      background: #009cff;
      height: 48px;
      padding: 16px;
      display: flex;
      justify-content: space-between;
      align-items: center;
      border-top: 1px solid black;
      font-size: 24px;
    }
    button {
      border-radius: 50%;
    }
  `]
})
export class PopupComponent {
  @HostBinding('@state')
  state: 'opened' | 'closed' = 'closed';
  @Input()
  get message(): string {return this._message;}
  set message(message: string) {
    this._message = message;
    this.state = 'opened';
  }
```

```
    private _message = '';
    @Output()
    closed = new EventEmitter<void>();
}
```

在 src\examples\elementexamples 目录下创建文件 elementexamples. component. ts，代码如例 4-45 所示。

【例 4-45】 创建文件 elementexamples. component. ts 的代码，定义自定义元素。

```
import {Component, Injector} from "@angular/core";
import {PopupService} from "./popup.service";
import {PopupComponent} from "./popup.component";
import {createCustomElement} from "@angular/elements";
@Component({
  selector: 'root',
  template: `
    <input #input value="对话框内容(组件或元素)">
    <button (click)="popup.showAsComponent(input.value)">作为组件显示</button>
    <button (click)="popup.showAsElement(input.value)">作为元素显示</button>
  `,
  styles: ['div {font-family: "Times New Roman";font-size: larger;color: red}']
})
export class ElementexamplesComponent {
  constructor(injector: Injector, public popup: PopupService) {
    //将组件 PopupComponent 定义成一个自定义元素
    const PopupElement = createCustomElement(PopupComponent, {injector});
    //将自定义元素注册到浏览器
    customElements.define('popup-element', PopupElement);
  }
}
```

在 src\examples\elementexamples 目录下创建文件 app-element. module. ts，代码如例 4-46 所示。

【例 4-46】 创建文件 app-element. module. ts 的代码，定义路由并声明组件。

```
import {NgModule} from '@angular/core';
import {BrowserModule} from '@angular/platform-browser';
import {RouterModule} from "@angular/router";
import {ElementexamplesComponent} from "./elementexamples.component";
import {PopupComponent} from "./popup.component";
@NgModule({
  imports: [
    BrowserModule,
    RouterModule.forRoot([
      {path: 'element', component: ElementexamplesComponent},
    ]),
  ],
  declarations: [
    PopupComponent,
    ElementexamplesComponent,
  ],
})
export class AppElementModule { }
```

修改 src\examples 目录下的文件 examplesmodules1. module. ts,代码如例 4-47 所示。

【例 4-47】 修改文件 examplesmodules1. module. ts 的代码,设置启动组件。

```
import {NgModule} from '@angular/core';
import {AppBindexampleModule} from "./bindexamples/app-bindexample.module";
import {ElementexamplesComponent} from './elementexamples/elementexamples.component';
import {PopupService} from "./elementexamples/popup.service";
import {BrowserAnimationsModule} from "@angular/platform-browser/animations";
@NgModule({
  imports: [
    AppBindexampleModule,
    BrowserAnimationsModule,//新增
  ],
  providers: [PopupService]//新增
  bootstrap: [ElementexamplesComponent]
})
export class ExamplesmodulesModule1 {}
```

保持其他文件不变后运行程序,成功运行程序后,在浏览器地址栏中输入 localhost:4200,结果如图 4-4 所示。

图 4-4 成功运行程序后在浏览器地址栏中输入 localhost:4200 的结果

习题 4

一、简答题

1. 简述对组件的理解。
2. 简述对组件样式的理解。
3. 简述对组件生命周期的理解。
4. 简述对组件之间交互的理解。
5. 简述对 Angular 元素的理解。

二、实验题

1. 实现组件样式的应用开发。
2. 实现组件生命周期的应用开发。
3. 实现组件之间交互的应用开发。
4. 实现 Angular 元素的应用开发。

第 5 章

组件的组合、分解及其应用

微课视频

5.1 内容投影及其应用

5.1.1 常见的内容投影

Angular 内容投影是一种组件组合模式,可以在一个组件中插入或投影另一个组件,并使用另一个组件的内容。Angular 中内容投影的常见实现方法包括单插槽内容投影、多插槽内容投影和有条件的内容投影。在使用单插槽内容投影场景下,组件可以从单一来源(如组件)接收内容;而在多插槽内容投影场景下,组件可以从多个来源接收内容。换言之,一个组件可以插入一个或多个其他组件,使用条件内容投影的组件仅在满足特定条件时才渲染内容。

1. 单插槽内容投影

内容投影的最基本形式之一是单插槽内容投影。单插槽内容投影是指创建一个组件且可以在其中投影另一个组件。创建使用单插槽内容投影的组件的步骤包括创建一个组件;在创建的组件模板中添加< ng-content >元素并将希望投影的内容写入其中;有了< ng-content >元素,组件的用户(调用者)就可以将自己的消息另一个组件的内容投影到该组件中。< ng-content >元素是一个占位符,它不会创建真正的 DOM 元素,但会被替换成调用时的消息(内容),其自定义属性将被忽略。

2. 多插槽内容投影

一个组件可以具有多个插槽,每个插槽可以指定一个 CSS 选择器,该选择器会决定将哪些内容写入该插槽。此种模式被称为多插槽内容投影。使用此模式,可以用< ng-content >的 select 属性指定希望投影内容出现的位置。创建使用多插槽内容投影的组件的步骤包括创建一个组件;在创建的组件模板中添加< ng-content >元素并将希望投影的内容写入其中;将 select 属性添加到< ng-content >元素。Angular 使用的选择器支持标签名、属性、CSS 类和伪类:not 的任意组合。使用 question 属性的内容将投影到带有 select=[question]属性的< ng-content >元素。如果组件中包含不带 select 属性的< ng-content >元素,则该实例将接收所有与其他< ng-content >元素都不匹配的投影组件。

3. 条件内容投影

如果组件需要有条件地渲染内容或多次渲染内容,则应配置该组件以接收一个包含有条件渲染的内容的< ng-template >元素。在这种情况下,不建议使用< ng-content >元素。

因为只要组件的使用者(provider)提供了内容,即使该组件从未定义< ng-content >元素或该< ng-content >元素位于 ngIf 语句的内部,该内容也总会被初始化。使用< ng-template >元素可以让组件根据条件显式渲染内容,并可以进行多次渲染。在显式渲染< ng-template >元素之前,Angular 不会初始化该元素的内容。

< ng-template >进行条件内容投影的步骤包括创建一个组件;在接收< ng-template >元素创建的组件中使用< ng-container >元素渲染该模板;将< ng-container >元素包装在另一个元素(例如 div 元素)中后应用条件逻辑;在要投影内容的模板中将投影的内容写入< ng-template >元素中。组件可以使用@ContentChild()装饰器或@ContentChildren()装饰器获得对此模板内容的引用(即 TemplateRef)。借助于 TemplateRef,组件可以使用 ngTemplateOutlet 指令或 ViewContainerRef. createEmbeddedView()方法来渲染所引用的内容。如果是多插槽内容投影,则可以使用@ContentChildren()装饰器获取投影元素的查询列表。在某些情况下,可能希望将内容投影为其他元素,例如,如果要投影的内容可能是另一个元素的子元素,那么可以用 ngProjectAs 属性来完成此操作。< ng-container >元素是一个逻辑结构,可用于对其他 DOM 元素进行分组;但< ng-container > 本身不会在 DOM 树中渲染。

5.1.2 内容投影的应用

在项目 src\examples 目录下创建子目录 contentprojectexamples,在 src\examples\contentprojectexamples 目录下创建文件 zippybasic. component. ts,代码如例 5-1 所示。

【例 5-1】 创建文件 zippybasic. component. ts 的代码,定义带单插槽内容投影功能的组件。

```
import {Component} from '@angular/core';
@Component({
  selector: 'app - zippy - basic',
  template: `
    < div >单插槽内容投影(Single - slot content projection)</div >
    < ng - content ></ng - content >
  `
})
export class ZippybasicComponent {}
```

在 src\examples\contentprojectexamples 目录下创建文件 zippymultislot. component. ts,代码如例 5-2 所示。

【例 5-2】 创建文件 zippymultislot. component. ts 的代码,定义带多插槽内容投影功能的组件。

```
import {Component} from '@angular/core';
@Component({
  selector: 'app - zippy - multislot',
  template: `
    < div >多插槽内容投影(Multi - slot content projection)</div >
    默认:
    < ng - content ></ng - content >
    问题:
    < ng - content select = "[question]"></ng - content >
  `
})
export class ZippymultislotComponent {}
```

在 src\examples\contentprojectexamples 目录下创建文件 zippyngprojectas. component. ts，代码如例 5-3 所示。

【例 5-3】 创建文件 zippyngprojectas. component. ts 的代码，定义带 select 属性的组件。

```
import {Component} from '@angular/core';
@Component({
  selector: 'app-zippy-ngprojectas',
  template: `
    <div>使用 ngProjectAs 属性的内容投影(Content projection with ngProjectAs)</div>
    默认:
    <ng-content></ng-content>
    问题:
    <ng-content select="[question]"></ng-content>
  `
})
export class ZippyngprojectasComponent {}
```

在 src\examples\contentprojectexamples 目录下创建文件 zippy. component. ts，代码如例 5-4 所示。

【例 5-4】 创建文件 zippy. component. ts 的代码，定义带有条件内容投影功能的组件。

```
import {Component, ContentChild, Input} from "@angular/core";
import {ZippycontentDirective} from "./zippycontent.directive";
let nextId = 0;
@Component({
  selector: 'app-example-zippy',
  template: `
    <ng-content></ng-content>
    <div *ngIf="expanded" [id]="contentId">
      <ng-container [ngTemplateOutlet]="content.templateRef"></ng-container>
    </div>
  `,
})
export class ZippyComponent {
  contentId = `zippy-${nextId++}`;
  @Input() expanded = false;
  @ContentChild(ZippycontentDirective) content!: ZippycontentDirective;
}
```

在 src\examples\contentprojectexamples 目录下创建文件 zippytoggle. directive. ts，代码如例 5-5 所示。

【例 5-5】 创建文件 zippytoggle. directive. ts 的代码，定义指令。

```
import {Directive, HostBinding, HostListener} from "@angular/core";
import {ZippyComponent} from "./zippy.component";
@Directive({
  selector: 'button[appExampleZippyToggle]',
})
export class ZippytoggleDirective {
  @HostBinding('attr.aria-expanded') ariaExpanded = this.zippy.expanded;
  @HostBinding('attr.aria-controls') ariaControls = this.zippy.contentId;
  @HostListener('click') toggleZippy() {
```

```
    this.zippy.expanded = !this.zippy.expanded;
  }
  constructor(public zippy: ZippyComponent) {}
}
```

在 src\examples\contentprojectexamples 目录下创建文件 zippycontent.directive.ts，代码如例 5-6 所示。

【例 5-6】 创建文件 zippycontent.directive.ts 的代码，定义指令。

```
import {Directive, TemplateRef} from "@angular/core";
@Directive({
  selector: '[appExampleZippyContent]'
})
export class ZippycontentDirective {
  constructor(public templateRef: TemplateRef<unknown>) {}
}
```

在 src\examples\contentprojectexamples 目录下创建文件 contentprojectexample.component.ts，代码如例 5-7 所示。

【例 5-7】 创建文件 contentprojectexample.component.ts 的代码，内容投影的综合应用。

```
import {Component} from "@angular/core";
@Component({
  selector: 'root',
  template: `
    <div>内容投影(Content Projection)</div>
    <app-zippy-basic>
      <p>内容投影很酷吧?</p>
    </app-zippy-basic>
    <hr />
    <app-zippy-multislot>
      <p question>
        内容投影很酷吧?
      </p>
      <p>开始学习内容投影吧!</p>
    </app-zippy-multislot>
    <hr />
    <div>zippy 示例</div>
    <app-example-zippy>
      <button appExampleZippyToggle>内容投影很酷吧?</button>
      <ng-template appExampleZippyContent>
        这取决于用它做什么
      </ng-template>
    </app-example-zippy>
    <hr />
    <app-zippy-ngprojectas>
      <p>开始学习内容投影吧!</p>
      <ng-container ngProjectAs="[question]">
        <p>内容投影很酷吧?</p>
      </ng-container>
    </app-zippy-ngprojectas>
  `,
  styles: ['p {font-family: Lato}']
```

```
})
export class ContentprojectexampleComponent {
}
```

在 src\examples\contentprojectexamples 目录下创建文件 app-content. module. ts，代码如例 5-8 所示。

【例 5-8】 创建文件 app-content. module. ts 的代码，定义路由并声明组件。

```
import {NgModule} from '@angular/core';
import {BrowserModule} from '@angular/platform-browser';
import {RouterModule} from "@angular/router";
import {ZippybasicComponent} from "./zippybasic.component";
import {ZippymultislotComponent} from "./zippymultislot.component";
import {ZippyngprojectasComponent} from "./zippyngprojectas.component";
import {ZippytoggleDirective} from "./zippytoggle.directive";
import {ZippycontentDirective} from "./zippycontent.directive";
import {ZippyComponent} from "./zippy.component";
import {ContentprojectexampleComponent} from "./contentprojectexample.component";
@NgModule({
  imports: [
    BrowserModule,
    RouterModule.forRoot([
      {path: 'content', component: ContentprojectexampleComponent},
    ]),
  ],
  declarations: [
    ZippybasicComponent,
    ZippymultislotComponent,
    ZippyngprojectasComponent,
    ZippytoggleDirective,
    ZippycontentDirective,
    ZippyComponent,
    ContentprojectexampleComponent,
  ],
})
export class AppEcontentModule { }
```

修改 src\examples 目录下的文件 examplesmodules1. module. ts，代码如例 5-9 所示。

【例 5-9】 修改文件 examplesmodules1. module. ts 的代码，设置启动组件。

```
import {NgModule} from '@angular/core';
import {AppEcontentModule} from "./contentprojectexamples/app-econtent.module";
import { ContentprojectexampleComponent } from './contentprojectexamples/contentprojectexample.
component';
@NgModule({
  imports: [
    AppEcontentModule,
  ],
  bootstrap: [ContentprojectexampleComponent]
})
export class ExamplesmodulesModule1 {}
```

保持其他文件不变并成功运行程序后，在浏览器地址栏中输入 localhost:4200，结果如图 5-1 所示。

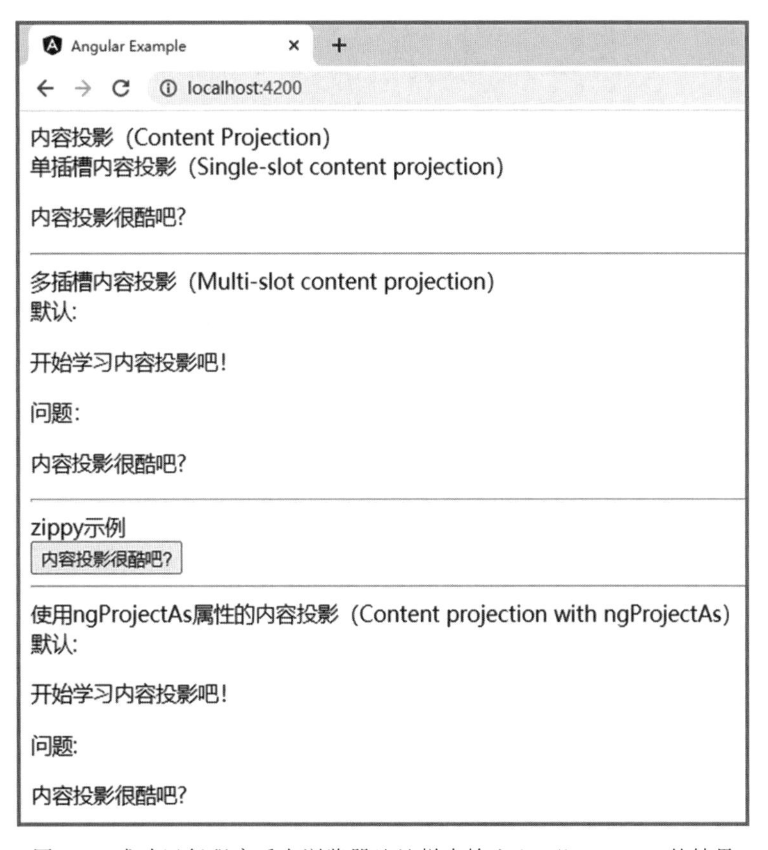

图 5-1　成功运行程序后在浏览器地址栏中输入 localhost:4200 的结果

5.2　视图封装及其应用

5.2.1　视图封装模式

Angular 应用中组件的样式可以封装在组件的宿主元素中，这样它们就不会影响应用的其余部分的样式。组件的装饰器提供了 encapsulation 选项用来控制如何基于模式对每个组件的应用视图进行封装。encapsulation 选项包括 Shadow Dom、Emulated、None 等模式。

1. Shadow Dom 模式

Shadow Dom 模式下 Angular 使用浏览器内置的 Shadow DOM API 将组件的视图包含在 Shadow Root（用作组件的宿主元素）中，并以隔离的方式应用所提供的样式。Shadow Dom 模式仅适用于内置支持阴影 DOM 的浏览器。并非所有浏览器都支持它。

2. Emulated 模式

Emulated 模式下 Angular 会修改组件 CSS 选择器，使它们只应用于组件的视图而不影响应用中的其他元素（模拟 Shadow DOM 行为）。此模式也是默认模式和推荐模式。使用 Emulated 模式时，Angular 会预处理所有组件的样式，以便它们仅应用于组件的视图。在运行的 Angular 应用的 DOM 中，使用 Emulated 的组件所在的元素附加了一些额外的属性(如_nghost 属性、_ngcontent 属性)。_nghost 属性被添加到包裹组件视图的元素(即宿主元素)中，这个元素是 Shadow DOM 封装中的 ShadowRoot；组件的宿主元素通常就是这种情况。_ngcontent 属性被添加到组件视图中的子元素上，这些属性用于将元素与其各自模拟的 ShadowRoot(具有匹配_nghost 属性的宿主元素)相匹配。Angular 独自实现这些属性的具体取值且隐藏实现细节。这些取值是自动生成的，不应在应用代码中引用；它们常用于生成组件的样式，这些组件样式会被注入 DOM 的< head >部分；并经过后期处理以便每个 CSS 选择器都能使用适当的_nghost 或_ngcontent 属性对样式进行扩充。这些修改后的选择器可以确保样式以相互独立且有针对性的方式应用于组件的视图。

3. None 模式

None 模式表示 Angular 不使用任何形式的视图封装，这意味着为组件指定的任何样式都是全局的，可以影响应用中存在的任何 HTML 元素。这种模式在本质上与将样式包含在 HTML 中是一样的。

可以在组件的装饰器中针对每个组件指定封装模式，这意味着应用程序中不同的组件可以使用不同的封装策略，但不建议这样做。Shadow Dom 组件的样式仅添加到 Shadow DOM 宿主中，确保它们仅影响各自组件视图中的元素。Emulated 组件的样式会添加到文档的< head >中，以使它们在整个应用中可用，但它们的选择器只会影响它们各自组件模板中的元素。None 组件的样式会添加到文档的< head >中，使它们在整个应用中可用，会影响文档中的任何匹配元素(因为是全局的)。Emulated 和 None 组件的样式会添加到每个 Shadow Dom 组件的 Shadow DOM 宿主中。None 组件的样式将影响 Shadow DOM 中的匹配元素。

5.2.2 视图封装的应用

在项目 src\examples 目录下创建子目录 viewencapsulationexamples，在 src\examples\viewencapsulationexamples 目录下创建文件 noencapsulation. component. ts，代码如例 5-10 所示。

【例 5-10】 创建文件 noencapsulation. component. ts 的代码，定义使用 None 模式的组件。

```
import {Component, ViewEncapsulation} from "@angular/core";
@Component({
  selector: 'app - no - encapsulation',
  template: `
    < div class = "inside">
    < div > None </div >
    < div class = "none - message">没有任何视图封装</div >
    </div >
  `,
  styles: ['div, .none - message { color: red}   ' +
  '.inside{border: 1px solid black;width: 200px}'],
  //使用全局样式,没有任何视图封装
  encapsulation: ViewEncapsulation.None,
})
export class NoencapsulationComponent { }
```

在 src\examples\viewencapsulationexamples 目录下创建文件 emulatedencapsulation.component.ts，代码如例 5-11 所示。

【例 5-11】 创建文件 emulatedencapsulation.component.ts 的代码，定义 Emulated 模式组件。

```
import {Component, ViewEncapsulation} from "@angular/core";
@Component({
  selector: 'app-emulated-encapsulation',
  template: `
    <div class="middle">
    <div>Emulated(默认编译器)</div>
    <div class="emulated-message">使用垫片(shimmed) CSS 来模拟原生行为</div>
    <app-no-encapsulation></app-no-encapsulation>
    </div>
  `,
  styles: ['h2, .emulated-message {color: black;} ' +
  '.middle{border: 2px solid blue; width: 400px}'],
  encapsulation: ViewEncapsulation.Emulated,
})
export class EmulatedencapsulationComponent { }
```

在 src\examples\viewencapsulationexamples 目录下创建文件 shadowdomencapsulation.component.ts，代码如例 5-12 所示。

【例 5-12】 创建文件 shadowdomencapsulation.component.ts 的代码，定义 ShadowDom 模式组件。

```
import {Component, ViewEncapsulation} from "@angular/core";
@Component({
  //用 ViewEncapsulation.ShadowDom 时，selector 的命名须用连接符，如 app-root
  selector: 'app-root',
  template: `
    <div class="outside">
      <div class="shadow-message"> ShadowDOM 封装</div>
      <app-emulated-encapsulation></app-emulated-encapsulation>
      <app-no-encapsulation></app-no-encapsulation>
    </div>
  `,
  styles: ['h2, .shadow-message {color: blue;}' +
  '.outside{border: 1px solid red;width: 600px}'],
  encapsulation: ViewEncapsulation.ShadowDom,
})
export class ShadowdomencapsulationComponent { }
```

在 src\examples\viewencapsulationexamples 目录下创建文件 viewencapsulationexample.component.ts，代码如例 5-13 所示。

【例 5-13】 创建文件 viewencapsulationexample.component.ts 的代码，定义组件。

```
import {Component} from "@angular/core";
@Component({
  selector: 'root',
  template: `
```

```
    <app-root></app-root>
  `
})
export class ViewencapsulationexampleComponent { }
```

5.2.3 模块和运行结果

在 src\examples\viewencapsulationexamples 目录下创建文件 app-encapsulation.module.ts，代码如例 5-14 所示。

【例 5-14】 创建文件 app-encapsulation.module.ts 的代码，定义路由并声明组件。

```
import {NgModule} from '@angular/core';
import {BrowserModule} from '@angular/platform-browser';
import {RouterModule} from "@angular/router";
import {EmulatedencapsulationComponent} from "./emulatedencapsulation.component";
import {NoencapsulationComponent} from "./noencapsulation.component";
import {ShadowdomencapsulationComponent} from "./shadowdomencapsulation.component";
import {ViewencapsulationexampleComponent} from "./viewencapsulationexample.component";
@NgModule({
  imports: [
    BrowserModule,
    RouterModule.forRoot([
      {path: 'encapsulation', component: ViewencapsulationexampleComponent},
    ]),
  ],
  declarations: [
    EmulatedencapsulationComponent,
    NoencapsulationComponent,
    ShadowdomencapsulationComponent,
    ViewencapsulationexampleComponent,
  ],
})
export class AppEncapsulationModule { }
```

修改 src\examples 目录下的文件 examplesmodules1.module.ts，代码如例 5-15 所示。

【例 5-15】 修改文件 examplesmodules1.module.ts 的代码，设置启动组件。

```
import {NgModule} from '@angular/core';
import {AppEncapsulationModule} from"./viewencapsulationexamples/app-encapsulation.module";
import {ViewencapsulationexampleComponent} from './viewencapsulationexamples/viewencapsul-
ationexample.component';
@NgModule({
  imports: [
    AppEncapsulationModule,
  ],
  bootstrap: [ViewencapsulationexampleComponent]
})
export class ExamplesmodulesModule1 {}
```

保持其他文件不变并成功运行程序后，在浏览器地址栏中输入 localhost:4200，结果如图 5-2 所示。

Angular Example
localhost:4200
Shadow DOM封装
Emulated（默认编译器）
使用垫片（shimmed）CSS 来模拟原生行为
None
没有任何视图封装
None
没有任何视图封装

图 5-2 成功运行程序后在浏览器地址栏中输入 localhost:4200 的结果

5.3 依赖注入及其应用

微课视频

5.3.1 依赖注入概述

依赖注入(DI)是一种设计模式，在这种设计模式中，类(如组件、模块、服务等)会从外部源请求依赖项而不是创建它们。依赖项是指某个类执行其功能所需的服务或对象。Angular 的 DI 框架会在实例化某个类时为其提供依赖；可以使用 DI 来提高应用的灵活性和模块化程度。@Injectable()装饰器会指定 Angular 在 DI 体系中使用所定义的类。

注入某些服务会使它们对组件可见。要将依赖项注入组件的 constructor()方法中，提供具有此依赖项类型的构造函数参数。当创建一个带有参数的 constructor()方法的类时，还需要指定参数类型和关于这些参数的元数据，以便 Angular 可以注入正确的服务。

通过配置提供者，可以把服务提供给那些需要它们的应用部件。依赖提供者会使用 DI 令牌来配置注入器，会将提供者与依赖项注入令牌(或叫 DI 令牌)并关联起来。注入器会用它来提供这个依赖值的具体的运行时版本，允许 Angular 创建任何内部依赖项的映射。DI 令牌会充当该映射的键名，如果把服务类指定为提供者令牌，那么注入器的默认行为是用 new 来实例化该类。尽管许多依赖项的值是通过类提供的，但扩展的 provide 对象可以将不同种类的提供者与 DI 令牌相关联；也可以用一个替代提供者来配置注入器，以指定另一些同样能提供日志功能的对象。因此，可以使用服务类来配置注入器，以提供一个替代类、一个对象或一个工厂函数。

5.3.2 依赖注入的实现方法

类提供者的语法实际上是一种简写形式，它会扩展成一个由 Provider 接口定义的提供者配置对象。不同的类可以提供相同的服务。如果替代类提供者有自己的依赖，就在父模块或组件的元数据属性 providers 中指定那些依赖。要为类提供者设置别名，在 providers 数组中使用 useExisting 属性指定别名和类提供者。通常，编写同一个父组件别名提供者的变体时会使用 forwardRef；若要为多个父类型指定别名(每个类型都有自己的类接口令牌)，则要配置 provideParent()方法以接收更多的参数。

要注入一个对象，可以用 useValue 选项来配置注入器。常用的对象字面量是配置对象。若要提供并注入配置对象，则要在@NgModule()装饰器的 providers 数组中指定该对象，可以定义和使用一个 InjectionToken 对象来为非类的依赖选择一个提供者令牌。借助@Inject()参数装饰器，可以把这个配置对象注入构造函数中。

虽然 TypeScript 的 AppConfig 接口可以在类中提供类型支持，但它在依赖注入时却没有任何作用。在 TypeScript 中，接口是一项设计时的部件，它没有可供 DI 框架使用的运行时表示形式或令牌。当转译器把 TypeScript 转换成 JavaScript 时，接口就会消失，因为 JavaScript 没有接口。由于 Angular 在运行时没有接口，所以接口不能作为令牌，也不能注入它。如果想在运行前，根据尚不可用的信息创建可变的依赖值，则可以使用工厂提供者(即采用工厂模式)。

5.3.3 服务类

在项目 src\examples 根目录下创建 injectexamples 子目录，在 src\examples\injectexamples 目录下创建文件 hero.ts，代码如例 5-16 所示。

【例 5-16】 创建文件 hero.ts 的代码，定义接口和数组。

```
export interface Hero {
  id: number;
  name: string;
  isSecret: boolean;
}
export const HEROES: Hero[] = [
  {id: 1, isSecret: false, name: '张三丰'},
  {id: 2, isSecret: false, name: '李斯'},
  {id: 3, isSecret: false, name: '王阳明'},
  {id: 4, isSecret: false, name: '朱熹'},
];
```

在 src\examples\injectexamples 目录下创建文件 hero.service.ts，代码如例 5-17 所示。

【例 5-17】 创建文件 hero.service.ts 的代码，定义类 HeroService。

```
import {Injectable} from '@angular/core';
import {Logger} from "./logger.service";
import {UserService} from "./user.service";
import {HEROES} from "./hero";
@Injectable({
  providedIn: 'root',
  useFactory: (logger: Logger, userService: UserService) =>
    new HeroService(logger, userService.user.isAuthorized),
  deps: [Logger, UserService],      //要用到类 Logger 和 UserService
})
export class HeroService {
  constructor(
    private logger: Logger,
    private isAuthorized: boolean) {
  }
  getHeroes() {
    const auth = this.isAuthorized ? 'authorized ' : 'unauthorized';
    this.logger.log(`Getting heroes for ${auth} user.`);
```

```
    return HEROES.filter(hero => this.isAuthorized || !hero.isSecret);
  }
}
```

在 src\examples\injectexamples 目录下创建文件 user.service.ts，代码如例 5-18 所示。

【例 5-18】 创建文件 user.service.ts 的代码，定义类 User 和 UserService。

```
import {Injectable} from '@angular/core';
export class User {
  constructor(
    public name: string,
    public isAuthorized = false) {
  }
}
const bob = new User('Bob', false);
@Injectable({
  providedIn: 'root'
})
export class UserService {
  user = bob;
}
```

在 src\examples\injectexamples 目录下创建文件 logger.service.ts，代码如例 5-19 所示。

【例 5-19】 创建文件 logger.service.ts 的代码，定义类 Logger。

```
import {Injectable} from '@angular/core';
@Injectable({
  providedIn: 'root'
})
export class Logger {
  logs: string[] = [];
  log(message: string) {
    this.logs.push(message);
    console.log(message);
  }
}
```

5.3.4 组件

在 src\examples\injectexamples 目录下创建文件 herolist.component.ts，代码如例 5-20 所示。

【例 5-20】 创建文件 herolist.component.ts 的代码，定义组件 app-hero-list。

```
import {Component} from '@angular/core';
import {Hero} from './hero';
import {HeroService} from './hero.service';
@Component({
  selector: 'app-hero-list',
  template: `
    <div *ngFor="let hero of heroes">
      {{hero.id}} - {{hero.name}}
      ({{hero.isSecret ? 'secret' : 'public'}})
```

```
    </div>
  `,
})
export class HeroListComponent {
  heroes: Hero[];
  constructor(heroService: HeroService) {
    this.heroes = heroService.getHeroes();
  }
}
```

在 src\examples\injectexamples 目录下创建文件 heroes.component.ts,代码如例 5-21 所示。

【例 5-21】 创建文件 heroes.component.ts 的代码,定义组件 app-heroes。

```
import {Component} from '@angular/core';
@Component({
  selector: 'app-heroes',
  template: `
    <div>人名列表</div>
    <app-hero-list></app-hero-list>
  `
})
export class HeroesComponent {
}
```

在 src\examples\injectexamples 目录下创建文件 heroestsp.component.ts,代码如例 5-22 所示。

【例 5-22】 创建文件 heroestsp.component.ts 的代码,定义组件 app-heroes-tsp。

```
import {Component} from '@angular/core';
@Component({
  selector: 'app-heroes-tsp',
  template: `
    <div>人名列表</div>
    <app-hero-list></app-hero-list>
  `
})
export class HeroestspComponent {
}
```

在 src\examples\injectexamples 目录下创建文件 injectexamples.component.ts,代码如例 5-23 所示。

【例 5-23】 创建文件 injectexamples.component.ts 的代码,调用三个组件以形成对比。

```
import {Component} from '@angular/core';
@Component({
  selector: 'root',
  template: `
    <app-heroes id="authorized" *ngIf="isAuthorized"></app-heroes>
    <app-heroes id="unauthorized" *ngIf="!isAuthorized"></app-heroes>
    <hr>
    <div>人名列表</div>
    <app-hero-list></app-hero-list>
```

```
    <hr>
    <app-heroes-tsp id="tspAuthorized" *ngIf="isAuthorized"></app-heroes-tsp>
  `
})
export class InjectexamplesComponent {
  isAuthorized = true;
}
```

5.3.5 模块和运行结果

在 src\examples\injectexamples 目录下创建文件 app-inject. module. ts,代码如例 5-24 所示。

【例 5-24】 创建文件 app-inject. module. ts 的代码,定义路由并声明组件。

```
import {NgModule} from '@angular/core';
import {BrowserModule} from '@angular/platform-browser';
import {RouterModule} from "@angular/router";
import {InjectexamplesComponent} from "./injectexamples.component";
import {HeroesComponent} from "./heroes.component";
import {HeroListComponent} from "./herolist.component";
import {HeroestspComponent} from "./heroestsp.component";
@NgModule({
  imports: [
    BrowserModule,
    RouterModule.forRoot([
      {path: 'inject', component: InjectexamplesComponent},
    ]),
  ],
  declarations: [
    HeroesComponent,
    HeroListComponent,
    HeroestspComponent,
    InjectexamplesComponent
  ],
})
export class AppInjectModule { }
```

修改 src\examples 目录下的文件 examplesmodules1. module. ts,代码如例 5-25 所示。

【例 5-25】 修改文件 examplesmodules1. module. ts 的代码,设置启动组件。

```
import {NgModule} from '@angular/core';
import {AppInjectModule} from "./injectexamples/app-inject.module";
import {InjectexamplesComponent} from './injectexamples/injectexamples.component';
@NgModule({
  imports: [
    AppInjectModule,
  ],
  bootstrap: [InjectexamplesComponent]
})
export class ExamplesmodulesModule1 {}
```

保持其他文件不变并成功运行程序后,在浏览器地址栏中输入 localhost:4200,结果如图 5-3 所示。

图 5-3 成功运行程序后在浏览器地址栏中输入 localhost:4200 的结果

习题 5

一、简答题

1. 简述对内容投影的理解。
2. 简述对视图封装的理解。
3. 简述对依赖注入的理解。

二、实验题

1. 实现内容投影的应用开发。
2. 实现视图封装的应用开发。
3. 实现依赖注入的应用开发。

第 6 章

路由及其应用

6.1 路由概述

6.1.1 路由的含义、实现和规则

1. 含义

在单页面应用中，可以通过显示或隐藏与特定组件相对应的部分视图来更改用户界面（应用视图），而不用去服务器获取新页面。运行应用程序时，需要在定义好的不同视图之间转换。可以用路由（Router）实现从一个视图到另一个视图的导航。路由会通过将浏览器 URL 解释为更改视图的操作指令来启用导航。用于处理路由的程序称为路由器。

2. 实现

创建路由的基本步骤包括创建路由相关的组件；设置路由数组（Routers）并为 @NgModule()装饰器配置 imports 和 exports 数组；把 RouterModule 和 Routes 导入路由所在的模块（简称路由模块）中；把路由模块导入应用主模块并把它添加到 imports 数组中；在使用组件之前把它们导入。

Routes 定义一个路由数组，每一个条目都会把一个 URL 路径映射到组件。每个路由都是包含 path 和 component 两个属性的对象。其中，path 属性定义路由的 URL 路径；component 属性定义要让 Angular 用作相应路径的组件。如果定义了路由，就可以把它们添加到应用程序中。首先，添加转换到路由相关新组件的链接，即把要添加路由的链接赋值给 routerLink 属性（即将该属性的值设置为新组件），以便在用户单击各个链接时显示这个值。接下来，修改组件模板已包含的< router-outlet >标签。该元素会通知 Angular 可以用所选路由的组件更新应用的视图。

3. 规则

路由的顺序很重要，因为 Router 在匹配路由时使用先到先得策略，所以要先放置更具体的路由。首先列出静态路径的路由，然后可以设置一个与默认路由匹配的空路径路由。通配符路由可以是最后一个路由，因为它匹配每一个 URL。把通配符路由作为最后一个路由的好处是当所请求的 URL 与任何路由器路径都不匹配时，Router 就会选择通配符路由。

通常，当用户使用导航访问应用时，如果希望把信息从一个组件传递到另一个组件，就可以用一个路由把这种类型的信息传给应用组件。要做到这一点，可以使用 ActivatedRoute 接口。若要从路由中获取信息，则可以把 ActivatedRoute 和 ParamMap 导

入组件。这些 import 语句添加了组件所需的几个重要元素。通过把 ActivatedRoute 的一个实例(即参数)添加到应用的构造函数(有参构造函数)中来注入它并使用 ngOnInit()方法来访问 ActivatedRoute 并跟踪 name 参数。

为了处理应用中用户试图导航到不存在的视图时,可以设置通配符路由。两个星号 ** 表明定义的是通配符路由。例如,要显示 404 页面,可以设置一个通配符路由,并将组件属性设置为要用于 404 页面的组件。如果请求的 URL 与前面列出的路径不匹配,那么路由器会选择这个路由,并跳转到 PageNotFoundComponent 组件对应的页面(视图)。

要设置页面(视图)重定向,可以使用重定向源的 path、要重定向目标的组件和一个 pathMatch 值来配置路由,以告诉路由器该如何匹配 URL。随着应用变得越来越复杂,可能要创建一些根组件之外的相对路径路由。这些嵌套路由类型称为子路由。子路由和其他路由一样,同时需要 path 和 component。唯一的区别是要把子路由放在父路由的 children 数组中。

可以配置路由定义来实现惰性加载模块(只会在需要时才加载这些模块,而不是在应用启动时就加载全部)。可以在后台预加载一些应用部件来改善用户体验。使用路由守卫来防止用户未经授权就导航到应用的某些部分。要想使用路由守卫,可以考虑使用无组件路由,因为这对于保护子路由很方便。

路由中链接参数数组是指一个由路由器将其解释为路由指南的数组。可以将该数组绑定到 RouterLink 属性或将该数组作为参数传给 Router. navigate()方法。链接参数数组保存路由导航时所需的路径、必备路由参数和可选路由参数。它们将进入该路由的 URL,可以把 RouterLink 指令绑定到一个数组;可以用一级、两级或多级路由来实现应用。链接参数数组提供了用来表示任意深度路由的链接参数数组以及任意合法的路由参数序列、必须的路由器参数以及可选的路由参数对象。当路由器导航到一个新的组件视图时,它会用该视图的 URL 来更新浏览器的当前地址以及历史。

必须在开发项目的早期就选择一种路由策略,因为一旦该应用进入了生产阶段,用户就会使用并依赖应用的 URL 引用。几乎所有的 Angular 项目都会使用默认的 HTML 5 风格。使用该风格生成的 URL 更易于被用户理解,也为将来做服务端渲染预留了空间。在服务器端渲染指定的页面是一项可以在应用首次加载时大幅提升响应速度的技术。路由器使用浏览器的 history. pushState 进行视图导航。

6.1.2 路由的工作步骤

在单页面应用中,应用程序的所有功能都存在于同一个 HTML 页面中。当用户访问应用的不同特性时,浏览器只需渲染那些用户需要关心的部分,而不用重新加载整个页面。这种模式可以显著改善应用的用户体验。

路由工作的一般步骤包括定义如何导航到组件;使用参数把信息传给组件;通过嵌套多个路由来构造路由体系;检查用户是否可以访问路由;控制应用是否可以放弃未保存的更改;通过预先获取路由数据和惰性加载特性模块来提高性能;需要特定的条件来加载组件。

RouterOutlet 是一个来自路由器库的指令,虽然它的用法像组件一样。它充当占位符,用于在模板中标记出路由器应该把该组件显示在哪个出口的位置。

（1）导入。

Angular 的 Router 是一项可选服务，它为指定的 URL 提供特定的组件视图，它位于包 @angular/router 中，可以从任何其他的 Angular 包中导入需要的功能代码。带路由的 Angular 应用中有一个 Router 服务的单例实例，当浏览器的 URL 发生变化时，路由器会查找相应的 Route，以便根据它确定要显示的组件。在配置之前，路由器没有任何可用的路由。

由于路径中没有前导斜杠，路由器会解析并构建最终的 URL，这样就可以在应用视图导航时使用相对路径和绝对路径了。相对路径允许定义相对于当前 URL 段的路径，可以使用./或者不带前导斜杠(/)来指定当前级别；可以用../符号将路径上升一个级别(表示父级)；还可以使用 NavigationExtras 中的 relativeTo 属性指定父级。有时，应用程序中的某个特性需要访问路由的部件，例如查询参数或片段；例如，/hero/42 的 URL 中，42 是 id 参数的值。路由中的 data 属性是存放与特定路由关联的任意数据的地方，且每个被激活的路由都可以访问 data 属性。可以用它来存储页面标题、面包屑文本和其他只读静态数据等内容。可以用路由器守卫来检索动态数据。

路由中的空路径表示应用程序的默认路径。如果需要查看导航生命周期(从导航开始到导航结束的过程)中发生了什么事件，可以把 enableTracing 选项作为路由器默认配置的一部分。这会把导航生命周期中发生的每个路由器事件都输出到浏览器控制台中。enableTracing 通常用于调试，可以把 enableTracing：true 选项作为第二个参数传给 RouterModule.forRoot()方法。

（2）链接。

如果想通过某些用户操作(如单击< a >标签)进行导航，则可以使用 RouterLink。RouterLink 指令可以让路由器控制元素。导航路径是固定的，所以可以给 RouterLink 赋值一个字符串(即一次性绑定)。如果导航路径更加动态，可以给它绑定到一个模板表达式，该表达式要返回一个链接参数数组。路由器会把该数组解析成一个完整的 URL。

RouterLinkActive 指令会根据当前的 RouterState 切换到处于活动状态的 RouterLink 上所绑定的 CSS 类。当这个链接处于活动状态时，路由器就会加上 RouterLink 所绑定的字符串(并在非活动状态时删除)。可以把 RouterLinkActive 指令设置成字符串，也可以把它绑定到一个返回字符串的组件属性上。RouterLinkActive 指令会在 HTML 元素上或元素内包含的相关 RouterLink 属性处于活动、非活动状态时，从 HTML 元素上添加、移除类。ActivatedRoute 是一个提供给每个路由组件的服务，其中包含当前路由专属的信息，例如路由参数、静态数据、解析数据、全局查询参数和全局片段等。活动路由链接会级联到路由树的每个级别，这样父路由和子路由链接就可以同时处于活动状态。

（3）状态。

在每个成功的导航生命周期结束后，路由器都会构建一个 ActivatedRoute 对象树，它构成了路由器的当前状态。可以从任何地方使用应用程序的 Router 服务和 RouterState 属性来访问当前的 RouterState。RouterState 是路由器的当前状态，包括一棵当前活动的路由的树以及遍历这棵路由树的便捷方法。RouterState 中的每个 ActivatedRoute 都提供了向上或向下遍历路由树的方法，用于从父路由、子路由和兄弟路由中获取信息。

可以通过注入名为 ActivatedRoute 的路由服务获得路由的路径和参数。它提供了路由器事件、路由器部件、路由 RouterModule、NgModule 等内容。Router 在每次导航过程中都会通过 Router. events 属性发出导航事件。这些事件的范围贯穿从导航开始和结束之间(即导航生命周期)的多个时间点。NgModule 提供了一些必要的服务提供者和一些用于在应用视图间导航的路由相关指令。

微课视频

6.2 路由的应用开发

6.2.1 基础组件

在项目 src\examples 根目录下创建 routerexamples 子目录，在 src\examples\routerexamples 目录下创建文件 books. ts，代码如例 6-1 所示。

【例 6-1】 创建文件 books. ts 的代码，定义接口和数组。

```
export interface Book {
  author : string;
  name : string;
  bookclass : string;
}
export const BOOKS: Book[] = [
  {author: '孔子', name: '论语', bookclass: '儒家'},
  {author: '董仲舒', name: '春秋繁露', bookclass: '儒家'},
  {author: '王阳明', name: '传习录', bookclass: '儒家'},
  {author: '六祖', name: '坛经', bookclass: '释家'},
  {author: '老子', name: '道德经', bookclass: '道家'},
];
```

在 src\examples\routerexamples 目录下创建文件 booknamelist. component. ts，代码如例 6-2 所示。

【例 6-2】 创建文件 booknamelist. component. ts 的代码，定义视图组件。

```
import {Component} from '@angular/core';
import {BOOKS} from "./books";
@Component({
  selector: 'app-name-list',
  template: `
    <!-- 书名 -->
    <div *ngFor="let book of books;let i = index">
      <div>{{i+1}}:<<{{book.name}}>></div>
    </div>
  `,
})
export class BooknamelistComponent {
  books = BOOKS
}
```

在 src\examples\routerexamples 目录下创建文件 bookauthorlist. component. ts，代码如例 6-3 所示。

【例 6-3】 创建文件 bookauthorlist. component. ts 的代码，定义视图组件。

```
import {Component} from '@angular/core';
import {BOOKS} from "./books";
```

```
@Component({
  selector: 'app-author-list',
  template: `
    <!-- 作者 -->
    <div *ngFor="let book of books;let i = index ">
      <div>{{i+1}}:{{book.author}} </div>
    </div>
  `,
})
export class BookauthorlistComponent {
  books = BOOKS
}
```

在 src\examples\routerexamples 目录下创建文件 bookclasslist. component. ts,代码如例 6-4 所示。

【例 6-4】 创建文件 bookclasslist. component. ts 的代码,定义视图组件。

```
import {Component} from '@angular/core';
import {BOOKS} from "./books";
@Component({
  selector: 'app-class-list',
  template: `
    <!-- 经典类型 -->
    <div *ngFor="let book of books;let i = index">
      <div>{{i+1}}:{{book.bookclass}}</div>
    </div>
  `,
})
export class BookclasslistComponent {
  books = BOOKS
}
```

在 src\examples\routerexamples 目录下创建文件 pagenotfound. component. ts,代码如例 6-5 所示。

【例 6-5】 创建文件 pagenotfound. component. ts 的代码,定义视图组件。

```
import {Component} from '@angular/core';
@Component({
  selector: 'app-page-not-found',
  template: `
    <div>找不到页面</div>
  `,
})
export class PagenotfoundComponent{
}
```

在 src\examples\routerexamples 目录下创建文件 error. component. ts,代码如例 6-6 所示。

【例 6-6】 创建文件 error. component. ts 的代码,定义视图组件。

```
import {Component} from '@angular/core';
@Component({
  selector: 'app-error',
  template: `
```

```
    <div>出错了</div>
  `,
})
export class ErrorComponent{
}
```

6.2.2 路由设置

在 src\examples\routerexamples 目录下创建文件 first. component. ts,代码如例 6-7 所示。

【例 6-7】 创建文件 first. component. ts 的代码,定义视图组件。

```
import {Component} from '@angular/core';
@Component({
  selector: 'app-first',
  template: `
    <div>
      <div>将路由信息独立成一个模块(文件)示例 first work</div>
    </div>
  `,
})
export class FirstComponent {}
```

在 src\examples\routerexamples 目录下创建文件 second. component. ts,代码如例 6-8 所示。

【例 6-8】 创建文件 second. component. ts 的代码,定义视图组件。

```
import {Component} from '@angular/core';
@Component({
  selector: 'app-second',
  template: `
    <div>
      <div>将路由信息独立成一个模块(文件)示例 second work</div>
    </div>
  `,
})
export class SecondComponent {}
```

在 src\examples\routerexamples 目录下创建文件 app-routing1. module. ts,代码如例 6-9 所示。

【例 6-9】 创建文件 app-routing. module1. ts 的代码,定义简单路由。

```
import {NgModule} from '@angular/core';
import {RouterModule, Routes} from '@angular/router';
import {FirstComponent} from "./first.component";
import {SecondComponent} from "./second.component";
const routes: Routes = [
  {path: 'first', component: FirstComponent},
  {path: 'second', component: SecondComponent},
];
@NgModule({
  imports: [RouterModule.forRoot(routes)],
  exports: [RouterModule]
})
export class AppRoutingModule {}
```

6.2.3 路由链接

在 src\examples\routerexamples 目录下创建文件 firstoutside.component.ts，代码如例 6-10 所示。

【例 6-10】 创建文件 firstoutside.component.ts 的代码，定义路由链接。

```
import {Component} from '@angular/core';
@Component({
  selector: 'app-first-outside',
  template: `
    <div>
      <div>嵌套路由信息 first outside</div>
      <a class="button" routerLink="/child-a" routerLinkActive="activebutton">a</a> |
      <a class="button" routerLink="/child-b" routerLinkActive="activebutton">b</a>
    </div>
  `,
})
export class FirstoutsideComponent {}
```

在 src\examples\routerexamples 目录下创建文件 childa.component.ts，代码如例 6-11 所示。

【例 6-11】 创建文件 childa.component.ts 的代码，定义视图组件。

```
import {Component} from '@angular/core';
@Component({
  selector: 'app-child-a',
  template: `
    <div>
      <div>内部路由信息 child a</div>
    </div>
  `,
})
export class ChildaComponent {}
```

在 src\examples\routerexamples 目录下创建文件 childb.component.ts，代码如例 6-12 所示。

【例 6-12】 创建文件 childb.component.ts 的代码，定义视图组件。

```
import {Component} from '@angular/core';
@Component({
  selector: 'app-child-b',
  template: `
    <div>
      <div>内部路由信息 child b</div>
    </div>
  `,
})
export class ChildbComponent {}
```

6.2.4 多级路由

在 src\examples\routerexamples 目录下创建文件 app-routing2.module.ts，代码如

例 6-13 所示。

【例 6-13】 创建文件 app-routing2. module. ts 的代码,定义多级路由。

```
import {NgModule} from '@angular/core';
import {RouterModule, Routes} from '@angular/router';
import {FirstoutsideComponent} from "./firstoutside.component";
import {ChildaComponent} from "./childa.component";
import {ChildbComponent} from "./childb.component";
const routes: Routes = [
  {
    path: 'first-outside',
    component: FirstoutsideComponent,
    children: [
      {
        path: 'child-a',
        component: ChildaComponent,
      },
      {
        path: 'child-b',
        component: ChildbComponent,
      },
    ],
  },
];
@NgModule({
  imports: [RouterModule.forRoot(routes)],
  exports: [RouterModule]
})
export class AppRouting2Module {}
```

6.2.5 带参数的路由

在 src\examples\routerexamples 目录下创建文件 profile. component. ts,代码如例 6-14 所示。

【例 6-14】 创建文件 profile. component. ts 的代码,定义视图组件。

```
import {Component, OnInit} from '@angular/core';
import {ActivatedRoute, ParamMap} from '@angular/router';
import {map} from 'rxjs/operators';
@Component({
  selector: 'app-profile',
  template: `
    <p>
      Hello {{ username$ | async }}!
    </p>
  `,
})
export class ProfileComponent implements OnInit {
  username$ = this.route.paramMap
    .pipe(
      map((params: ParamMap) => params.get('username'))
    );
  constructor(private route: ActivatedRoute) { }
  ngOnInit() { }
}
```

在 src\examples\routerexamples 目录下创建文件 app-routing3. module. ts，代码如例 6-15 所示。

【例 6-15】 创建文件 app-routing3. module. ts 的代码，定义带参数的路由。

```
import {NgModule} from '@angular/core';
import {BrowserModule} from '@angular/platform-browser';
import {FormsModule} from '@angular/forms';
import {RouterModule, UrlSegment} from '@angular/router';
import {ProfileComponent} from "./profile.component";
@NgModule({
  imports: [
    BrowserModule,
    FormsModule,
    RouterModule.forRoot([
      {
        matcher: (url) => {
          if (url.length === 1 && url[0].path.match(/^@[\w]+$/gm)) {
            return {
              consumed: url,
              posParams: {
                username: new UrlSegment(url[0].path.substr(1), {})
              }
            };
          }
          return null;
        },
        component: ProfileComponent
      }
    ])],
  declarations: [ProfileComponent],
})
export class AppRouting3Module {}
```

6.2.6 组件、模块和运行结果

在 src\examples\routerexamples 目录下创建文件 routerexamples. component. ts，代码如例 6-16 所示。

【例 6-16】 创建文件 routerexamples. component. ts 的代码，定义综合视图组件。

```
import {Component} from '@angular/core';
@Component({
  selector: 'root',
  template: `
    <div>路由示例</div>
    <hr>
    <nav>
      <a class="button" routerLink="/name-list" routerLinkActive="activebutton">书名</a> |
      <a class="button" routerLink="/author-list" routerLinkActive="activebutton">作者
</a> |
      <a class="button" routerLink="/class-list" routerLinkActive="activebutton">经典
类型</a> |
    <a class="button" routerLink="/page-not-found" routerLinkActive="activebutton">没
有找到页面</a> |
```

```
    <a class = "button" routerLink = "/error - list" routerLinkActive = "activebutton">出错
了</a>
    </nav>
    <hr>
    <div class = "result">输出结果</div>
    <router - outlet></router - outlet>
    <hr>
    <div>路由信息独立成一个模块(文件)</div>
    <a class = "button" routerLink = "/first" routerLinkActive = "activebutton"> first </a> |
    <a class = "button" routerLink = "/second" routerLinkActive = "activebutton"> second </a>
    <hr>
    <div>嵌套路由信息</div>
    <a class = "button" routerLink = "/first - outside" routerLinkActive = "activebutton">
first - outside </a> |
    < a class = " button" routerLink = "/first - outside/child - a" routerLinkActive =
"activebutton"> a </a>
    < a class = " button" routerLink = "/first - outside/child - b" routerLinkActive =
"activebutton"> b </a>
    <hr>
    <div>自定义路由匹配器</div>
    问候<a routerLink = "/@Zhangsanfeng">张三丰</a>
  `,
  styles: ['.result {color:red}' +
  '.button {\n' +
  '    box - shadow: inset 0 1px 0 0 #ffffff;\n' +
  '    background: #ffffff linear - gradient(to bottom, #ffffff 5 % , #f6f6f6 100 % );\n' +
  '    border - radius: 6px;\n' +
  '    border: 1px solid #dcdcdc;\n' +
  '    display: inline - block;\n' +
  '    cursor: pointer;\n' +
  '    color: #666666;\n' +
  '    font - family: Arial, sans - serif;\n' +
  '    font - size: 15px;\n' +
  '    font - weight: bold;\n' +
  '    padding: 6px 24px;\n' +
  '    text - decoration: none;\n' +
  '    text - shadow: 0 1px 0 #ffffff;\n' +
  '    outline: 0;\n' +
  '}\n' +
  '.activebutton {\n' +
  '    box - shadow: inset 0 1px 0 0 #dcecfb;\n' +
  '    background: #bddbfa linear - gradient(to bottom, #bddbfa 5 % , #80b5ea 100 % );\n' +
  '    border - radius: 6px;\n' +
  '    border: 1px solid #84bbf3;\n' +
  '    display: inline - block;\n' +
  '    cursor: pointer;\n' +
  '    color: #ffffff;\n' +
  '    font - family: Arial, sans - serif;\n' +
  '    font - size: 15px;\n' +
  '    font - weight: bold;\n' +
  '    padding: 6px 24px;\n' +
  '    text - decoration: none;\n' +
  '    text - shadow: 0 1px 0 #528ecc;\n' +
  '    outline: 0;\n' +
```

```
    '}\n '
  ]
})
export class RouterexamplesComponent {}
```

在 src\examples\routerexamples 目录下创建文件 app-router. module. ts，代码如例 6-17 所示。

【例 6-17】 创建文件 app-router. module. ts 的代码，定义路由并声明组件。

```
import {NgModule} from '@angular/core';
import {BrowserModule} from '@angular/platform-browser';
import {RouterModule} from "@angular/router";
import {RouterexamplesComponent} from "./routerexamples.component";
import {BooknamelistComponent} from "./booknamelist.component";
import {BookauthorlistComponent} from "./bookauthorlist.component";
import {BookclasslistComponent} from "./bookclasslist.component";
import {PagenotfoundComponent} from "./pagenotfound.component";
import {ErrorComponent} from "./error.component";
import {FirstComponent} from "./first.component";
import {SecondComponent} from "./second.component";
import {FirstoutsideComponent} from "./firstoutside.component";
import {ChildaComponent} from "./childa.component";
import {ChildbComponent} from "./childb.component";
import {AppRoutingModule} from "./app-routing.module";
import {AppRouting2Module} from "./app-routing2.module";
import {AppRouting3Module} from "./app-routing3.module";
@NgModule({
  imports: [
    BrowserModule,
    RouterModule.forRoot([
      {path: 'home', component: RouterexamplesComponent},
      {path: 'name-list', component: BooknamelistComponent},
      {path: 'author-list', component: BookauthorlistComponent},
      {path: 'class-list', component: BookclasslistComponent},
      {path: 'page-not-found', component: PagenotfoundComponent},
      {path: '', redirectTo: '/name-list', pathMatch: 'full'},
      {path: '**', component: ErrorComponent}
    ]),
    AppRoutingModule     //新增路由模块
    AppRouting2Module,
    AppRouting3Module,
  ],
  declarations: [
    BooknamelistComponent,
    BookauthorlistComponent,
    PagenotfoundComponent,
    BookclasslistComponent,
    ErrorComponent,
    FirstComponent,
    SecondComponent,
    FirstoutsideComponent,
    ChildaComponent,
    ChildbComponent,
    RouterexamplesComponent,
  ],
```

```
})
export class AppRouterModule { }
```

修改 src\examples 目录下的文件 examplesmodules1. module. ts，代码如例 6-18 所示。

【例 6-18】 修改文件 examplesmodules1. module. ts 的代码，设置启动组件。

```
import {NgModule} from '@angular/core';
import {RouterexamplesComponent} from './routerexamples/routerexamples.component';
import {AppRouterModule} from "./routerexamples/app-router.module";
@NgModule({
  imports: [
    AppRouterModule
  ],
  bootstrap: [RouterexamplesComponent]
})
export class ExamplesmodulesModule1 {}
```

保持其他文件不变并成功运行程序后，在浏览器地址栏中输入 localhost:4200 后将自动跳转到 localhost:4200/name-list，结果如图 6-1 所示。

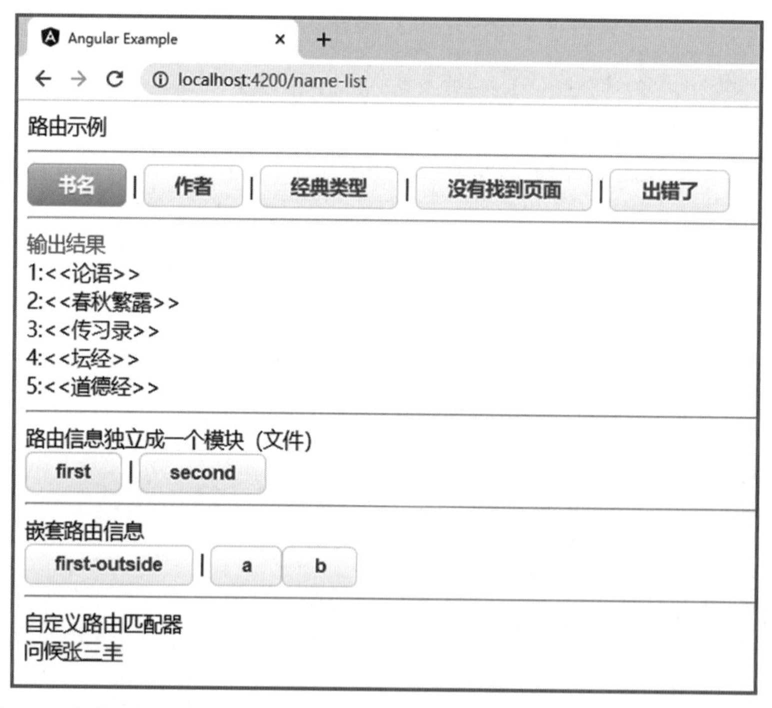

图 6-1　成功运行程序后在浏览器地址栏中输入 localhost:4200/name-list 的结果

习题 6

一、简答题

简述对路由的理解。

二、实验题

完成路由的应用开发。

第 7 章

表单及其应用

7.1 表单概述

7.1.1 表单的含义、分类和实现

1. 含义

用表单处理用户输入是许多应用程序常见的基础功能。例如，可以通过表单来让用户登录、修改个人档案、输入敏感信息以及执行各种数据输入任务。Angular 提供了响应式表单和模板驱动表单两种表单处理用户输入。两者都从应用视图(用户界面)中捕获用户输入事件、验证用户输入、创建表单模型、修改数据模型，并提供跟踪这些更改的途径。

2. 分类

表单可分为响应式表单和模板驱动表单，它们以不同的方式处理和管理表单数据。

(1) 响应式表单可以直接在组件中定义表单模型，提供对底层表单对象模型直接、显式的访问，更加健壮，可扩展、可复用和可测试。如果表单是应用程序的关键部分或者已经使用了响应式表单，可以优先使用响应式表单。如果表单是应用的核心部分，那么可伸缩性就非常重要。此时，能够跨组件复用表单的模型变得至关重要。此种场景下，也应优先使用响应式表单。在应用视图和数据模型之间使用同步数据流，从而可以更轻松地创建大型表单。响应式表单需要较少的测试设置，测试时不需要深入理解变更检测，就能正确测试表单的更新和验证。

(2) 模板驱动表单中，表单模型是隐式的，依赖模板中的指令来创建和操作底层的对象模型。它们对于向应用程序添加一个简单的表单非常有用，比如电子邮件列表注册表单。模板驱动表单专注于简单的场景，它们很容易被添加到应用中，但在可扩展性、可复用性方面不如响应式表单。它们抽象出了底层表单 API，并且在应用视图和数据模型之间使用异步数据流。测试程序依赖于手动触发变更检测才能正常运行，并且需要进行更多的设置操作。

响应式表单和模板驱动表单的一些关键差异如表 7-1 所示。

表 7-1 响应式表单和模板驱动表单的区别

对 比 项 目	响应式表单	模板驱动表单
建立表单模型	显式的，在组件类中创建	隐式的，由指令创建
数据模型	结构化和不可变的	非结构化和可变的
数据流	同步	异步
表单验证	函数	指令

3. 实现

响应式表单和模板驱动表单都会跟踪与用户交互的表单输入元素以及组件模型中的表单数据发生的值变更。这两种方法共享同一套底层构建块，只在如何创建和管理常用表单控件实例方面有所不同。响应式表单和模板驱动表单都是建立在FormControl、FormGroup、FormArray、ControlValueAccessor等基础类之上的。其中，FormControl用于追踪单个表单控件的值和验证状态。FormGroup用于追踪一个表单控件组的值和状态。FormArray用于追踪表单控件数组的值和状态。ControlValueAccessor用于在Angular的FormControl实例和内置DOM元素之间创建一个桥梁。

当应用程序包含一个表单时，Angular必须让视图与组件模型之间互相保持同步；当用户通过视图更改数据取值时，新值必须反映到数据模型中；当程序逻辑改变数据模型中的数据取值时，这些改变后的值也必须更新到视图中。

对来自用户或程序化变更时的数据处理方式，响应式表单和模板驱动表单也有所不同。响应式表单中，从视图到模型的修改以及从模型到视图的修改都是同步的，而且不依赖于用户界面的渲染方式。视图中数值发生变化时，新值会传给FormControl实例；FormControl实例会通过valueChanges这个可观察对象发出一个新值(值变更的信息)；valueChanges的任何一个订阅者都会收到这个新值。而模型到视图的一般步骤包括favoriteColorControl.setValue()方法被调用，它会更新FormControl的值；FormControl实例会通过valueChanges发出新值；valueChanges的任何订阅者都会收到这个新值；视图对应部分更新为新值。

在模板驱动表单中，每一个表单元素都是和一个负责管理内部表单模型的指令关联起来的。

(1) 视图到模型的流动步骤包括视图中数值发生变化时，触发(trigger) FormControl实例上的setValue()方法；FormControl实例通过valueChanges这个可观察对象发出新值；valueChanges的任何订阅者都会收到新值；控件值访问器ControlValueAccessor还会调用NgModel.viewToModelUpdate()方法，它会发出一个ngModelChange事件；组件中的属性修改成ngModelChange事件所发出的值。

(2) 模型到视图的流动步骤包括组件中修改值；变更检测开始，Angular调用NgModel指令上的ngOnChanges()方法；ngOnChanges()方法把一个异步任务排入队列以设置内部FormControl实例的值；变更检测完成；在下一个检测周期用来为FormControl实例赋值的任务就会执行；FormControl实例通过valueChanges发出最新值；valueChanges的任何订阅者都会收到这个新值；控件值访问器ControlValueAccessor会使用最新值来修改视图。

响应式表单通过以不可变的数据结构提供数据模型来保持数据模型的纯粹性。每当在数据模型更改时，FormControl实例都会返回一个新的数据模型，而不会更新现有的数据模型。这使得能够通过该控件的可观察对象跟踪对数据模型的更改。此种更改往往只在可观察对象中发生，不会在其他地方发生(即更改发生地的单一性)，这让变更检测更有效率，更不容易出错，因为它只需在发生单一性更改(即对象引用发生变化)时进行更新。此种设计用到了观察者模式。由于数据更新遵循响应式模式，因此可以把它和可观察对象的各种运算符集成起来以转换数据。

模板驱动的表单依赖于可变性和双向数据绑定，可以在模板中作出更改时更新组件中的数据模型，属性被修改为新值。由于使用双向数据绑定时无法跟踪对数据模型进行的单一性更改,因此变更检测在需要确定何时更新时效率较低。

7.1.2 表单的验证和测试

验证是管理任何表单时必备的一部分工作。无论是检查必填项,还是查询外部 API 来进行验证,Angular 都提供了一组内置的验证器以及创建自定义验证器所需的能力。响应式表单把自定义验证器定义成函数,它以要验证的控件作为参数。模板驱动表单和模板指令紧密相关,并且必须提供包装了验证函数的自定义验证器指令。

响应式表单提供了相对简单的测试策略,因为它们能提供对表单和数据模型的同步访问,而且不必渲染用户界面就能测试它们。在这些测试中,控件和数据是通过控件进行查询和操纵的,不需要进行变更检测。使用模板驱动表单编写测试就需要详细了解变更检测过程,以及指令在每个变更检测周期中如何运行,以确保在正确的时间查询、测试或更改元素。

7.2 响应式表单

7.2.1 表单控件

响应式表单提供了一种模型驱动的方式来处理表单输入,其中的值会随时间而变化。对表单状态的每一次变更都会返回一个新的状态,这样可以在变化时维护模型的整体性。响应式表单采用观察者设计模式,它是围绕 Observable 流构建的,它可以同步访问。因为在请求时可以确信这些数据是一致的、可预料的。这个流的任何一个消费者都可以安全地操纵这些数据。使用 Observable 的操作符提供了不可变性,并且通过 Observable 流提供了变化追踪功能。而模板驱动表单允许直接在模板中修改数据,因为它们依赖嵌入模板中的指令,并借助可变数据来异步跟踪变化。使用表单控件有以下 4 个步骤。

(1) 在应用程序中注册响应式表单模块。该模块声明了一些要用在响应式表单中的指令。要从@angular/forms 包中导入 ReactiveFormsModule,并把它添加到 NgModule 的 imports 数组中。

(2) 导入 FormControl 类,生成一个新的 FormControl 实例,并把它保存在组件中,将其保存为类的属性。可以用 FormControl 的构造函数设置初始值,通过在组件中创建这些控件,可以直接对表单控件的状态进行监听、修改和校验。

(3) 在模板中注册 FormControl。在组件中创建了控件之后,还要把它和模板中的一个表单控件关联起来。

(4) 把组件添加到模板中来显示表单。通过可观察对象 valueChanges,可以在模板中使用 AsyncPipe 或在组件中使用 subscribe()方法来监听表单值的变化。使用 value 属性能获得当前值的一份快照。一旦修改了表单控件所关联的元素,显示的值就跟着变化。

响应式表单还能通过每个实例的属性和方法提供关于特定控件的更多信息。AbstractControl 的属性和方法用于控制表单状态,并在处理表单校验时决定何时显示信息。

响应式表单还可以用编程的方式灵活地修改控件的值,而不需要借助于用户的交互。FormControl 提供了一个 setValue()方法,它会修改表单控件的值,并且验证与控件结构相对应的值的结构。例如,当从后端 API 或服务接收到了表单数据时,可以通过 setValue()

方法把原来的值替换为新的值。

7.2.2 表单组

表单中通常会包含几个相互关联的控件。响应式表单提供了两种把多个相关控件分组到同一个输入表单中的方法。表单组定义了一个带有一组控件的表单，可以把它们放在一起管理。表单数组定义了一个动态表单，可以在运行时动态地添加和删除控件；也可以通过嵌套表单数组来创建更复杂的表单。FormGroup的实例能跟踪一组FormControl实例（对应一个表单控件）的表单状态。创建FormGroup时，其中的每个控件都会根据其名字进行跟踪。如果某个控件的状态或值变化了，父控件（FormGroup实例）会发出一次新的状态变更或值变更事件。

将表单组添加到组件中的步骤包括创建一个FormGroup实例；把FormGroup模型关联到视图；保存表单数据。要初始化FormGroup，要为构造函数提供一个控件组对象，对象中的每个名字都要和表单控件的名字一一对应。独立的表单控件被收集到了一个控件组中。FormGroup实例用对象的形式提供了它的模型值，这个值来自组中每个控件的值。由于FormGroup实例是一组FormControl实例的集合，FormGroup实例拥有和FormControl实例相同的属性（如value、untouched）和方法（如setValue()方法）。

表单组可以同时接收单个表单控件实例和其他表单组实例作为其子控件。这可以让复杂的表单模型更容易维护，并在逻辑上对它们进行分组。通过构建复杂的表单，能在更小的分区中管理不同类别的信息。使用嵌套的FormGroup可以把大型表单组织成一些稍小的、易管理的分组（部件）。制作复杂的表单的步骤包括创建一个嵌套的表单组；在模板中对这个嵌套表单分组；某些类型的信息属于同一个组。

要创建一个嵌套组，就要把一个嵌套的元素添加到表单组的实例中。来自内嵌控件组的状态和值的变更将会冒泡到它的父控件组（事件冒泡机制），以维护整体模型的一致性。在修改了组件中的模型之后，还要通过修改模板的方式把FormGroup实例对接到它的输入元素。当修改包含多个FormGroup实例的值时，可能只希望更新模型中的一部分数值，而不是完全替换掉内容。此时，有两种更新模型值的方式：一是使用setValue()方法来为单个控件设置新值。setValue()方法会严格遵循表单组的结构，并整体地替换控件的值；二是使用patchValue()方法可以用对象中所定义的任何属性为表单模型进行替换。setValue()方法的严格检查可以帮助捕获复杂表单嵌套中的错误，而patchValue()方法在遇到错误时可能只会默默地失败。patchValue()方法要针对模型的结构进行更新，且只会更新表单模型中所定义的属性。

7.2.3 多个表单控件的创建

1. FormBuilder

当需要与多个表单打交道时，手动创建多个表单控件实例会非常烦琐。而FormBuilder服务就提供了一些便捷方法来生成表单控件。FormBuilder在幕后使用与手动创建多个表单控件同样的方式来创建和返回这些实例，只是用起来更简单。此种设计称为建造者模式。通过导入FormBuilder类、注入FormBuilder服务、生成表单内容可以利用这项服务。FormBuilder是一个可注入的服务提供者，它是由ReactiveFormModule提供的，只要把它添加到组件的构造函数中就可以注入这个依赖。FormBuilder 服务有 control()、group()和 array()3 个方法。这些方法都是工厂方法，用于在组件中分别生成 FormControl、FormGroup 和 FormArray。

2. FormArray

FormArray 是除 FormGroup 之外实现多个表单控制的另一个选择，可用于管理任意数量的匿名控件。像 FormGroup 实例一样，也可以往 FormArray 中动态插入和移除控件，并且 FormArray 实例的值和验证状态也是根据它的子控件计算得来的。不需要为每个控件定义一个名字作为 key。因此，如果事先不知道子控件的数量，FormArray 就是一个很好的选择。

要定义一个动态表单，执行的步骤如下。

(1) 从@angular/form 中导入 FormArray 类，以使用它的类型信息。FormBuilder 服务用于创建 FormArray 实例。

(2) 定义一个 FormArray 控件。可以通过把一组(从零个到多个)控件定义在一个数组中来初始化一个 FormArray。使用 FormBuilder. array()方法来定义该数组，并用 FormBuilder. control()方法往该数组中添加初始控件。

(3) 使用 getter()方法访问 FormArray 控件。表单数组 FormArray 实例用一个数组来代表不确定数量的控件。通过 getter()方法来访问控件很方便，这种方法还能很容易地重复处理更多控件。因为返回的控件的类型是 AbstractControl，所以要为该方法提供一个显式的类型转换声明(将 AbstractControl 类强制转换成 FormArray 类)来访问 FormArray 特有的语法。用 FormArray. push()方法把该控件添加为数组中的新条目。在这个模板中，这些控件会被迭代，把每个控件都显示为一个独立的输入框。

(4) 在模板中显示这个表单数组。

3. AbstractControl

AbstractControl 是所有三种表单控件类(FormControl、FormGroup 和 FormArray)的抽象基类。它提供了一些公共的行为和属性。其中，FormControl 管理单个表单控件的值和有效性状态，它对应于 HTML 的表单控件(如< input >或< select >)。FormGroup 管理一组 AbstractControl 实例的值和有效性状态，该组的属性中包括了它的子控件，组件中的顶层表单就是 FormGroup。FormArray 管理一些 AbstractControl 实例数组的值和有效性状态，一个可注入的服务，提供一些用于创建控件实例的工厂方法。

7.3 表单验证及实现

7.3.1 表单验证含义和验证器函数

表单验证用于确保用户的输入是完整和正确的，可以提高整体的数据质量。为了往模板驱动表单中添加验证机制，要添加一些验证属性，就像原生的 HTML 表单验证器一样。Angular 会用指令来实现验证功能。每当表单控件中的值发生变化时，Angular 就会进行验证，并生成一个验证错误的列表(对应着 INVALID 状态)或者 null(对应着 VALID 状态)。

表单验证的实现可以通过在表单组件中导入一个验证器函数，再把这个验证器函数添加到表单中的相应字段，并添加逻辑来处理验证状态等步骤添加表单验证。最常见的表单验证是做一个必填字段。响应式表单包含了一组开箱即用的常用验证器函数，这些函数接收一个控件，用以验证并根据验证结果返回一个错误对象或空值(Null)。当往表单控件上添加了一个必填字段时，它的初始值是无效的。这种无效状态(INVALID)会传播到其父 FormGroup 元素中，也使父 FormGroup 元素的状态变为无效。可以通过该 FormGroup 实例的 status 属性来访问其当前状态。

在响应式表单中，组件是必备的。因此，不能通过模板上的属性来添加验证器函数，而应该在组件中直接把验证器函数添加到表单控件模型上(FormControl)。然后，一旦控件发生了变化，Angular 就会调用这些验证器函数。

验证器函数可以是同步函数，也可以是异步函数。同步验证器函数接收一个控件实例，然后返回一组验证错误或 null。可以在实例化一个 FormControl 时把它作为构造函数的第二个参数传进去。异步验证器函数接收一个控件实例并返回一个 promise 或 Observable，它稍后会发出一组验证错误或 null。在实例化 FormControl 时，可以把它们作为第三个参数传入。出于性能方面的考虑，只有在所有同步验证器函数都通过之后，Angular 才会运行异步验证，并在每一个异步验证器函数都执行完之后，才会设置这些验证错误。

可以选择编写自己的验证器函数，也可以使用 Angular 的内置验证器函数。在模板驱动表单中用作属性的那些内置验证器函数(如 required 和 minlength)也都可以作为 Validators 类中的函数使用(如 Validators. required)。所有这些验证器函数都是同步的，所以它们可作为第二个参数传递，通过把这些验证器函数放到一个数组中传入来支持多项验证。内置的验证器函数并不是总能精确匹配应用中的用例，因此有时需要创建一个自定义验证器函数。

7.3.2 不同类型表单的验证

1. 响应式表单验证

在响应式表单中，通常会通过它所属的控件组(FormGroup)的 getter()方法来访问表单控件，但有时候也会为模板定义 getter()方法作为简短形式。这个表单与模板驱动的版本不同，它不再导出任何指令。

自定义异步验证器函数和同步验证器函数很像，只是前者必须返回一个稍后会输出 null 或验证错误对象的承诺(promise)或可观察对象。如果是可观察对象，那么它必须在某个时间点被完成(complete)，表单就会使用它输出的最后一个值作为验证结果。HTTP 服务是自动完成的，但是某些自定义的可观察对象可能需要手动调用 complete()方法。在响应式表单中，通过直接把该函数传给 FormControl 来添加自定义验证器函数。

2. 模板驱动表单验证

在模板驱动表单中，要为模板添加一个指令，该指令包含了 validator()函数。

Angular 会自动把很多控件属性作为 CSS 类映射到控件所在的元素上。可以使用.ng-valid、. ng-invalid、. ng-pending、. ng-pristine、. ng-dirty、. ng-untouched、. ng-touched、. ng-submitted(只对 form 元素添加)等类来根据表单状态给表单控件元素添加样式。

跨字段交叉验证器是一种自定义验证器，可以对表单中不同字段的值进行比较，并针对它们的组合进行接受或拒绝。例如，可能有一个提供互不兼容选项的表单，以便让用户选择

A 或 B,而不能两者都选。某些字段值也可能依赖于其他值；用户可能只有选择了 A 之后才能选择 B。

对于模板驱动表单,必须创建一个指令来包装验证器函数。为了提供更好的用户体验,当表单无效时,要显示一个恰当的错误信息。这在模板驱动表单和响应式表单中都是一样的。

异步验证器函数实现了 AsyncValidatorFn 和 AsyncValidator 接口。它们与其同步版本非常相似,但是它们之间有不同之处。validate()函数必须返回一个 promise 或可观察对象,返回的可观察对象必须是有尽的,这意味着它必须在某个时刻完成(complete)。要把无尽的可观察对象转换成有尽的,可以在管道中加入过滤操作符,比如 first、last、take 或 takeUntil。异步验证在同步验证完成后才会发生,并且只有在同步验证成功时才会执行。如果更基本的验证方法已经发现了无效输入,就无须进行后面的验证,这种检查顺序可以让表单避免使用昂贵的异步验证流程(例如 HTTP 请求)。异步验证开始之后,表单控件就会进入 pending 状态。可以检查控件的 pending 属性,并用它来给出对验证中的视觉反馈。

3. 其他

在默认情况下,所有验证程序在每次表单值更改后都会运行。对于同步验证器函数,这通常不会对应用性能产生明显的影响。但是,异步验证器函数通常会执行某种 HTTP 请求来验证控件。每次按键后调度一次 HTTP 请求都会给后端 API 带来压力,应该尽可能避免。

在默认情况下,Angular 通过在< form >元素上添加 novalidate 属性来禁用原生 HTML 表单验证,并使用指令将这些属性与框架中的验证器函数相匹配。如果想重新将原生验证与基于 Angular 的验证器函数结合使用,可以使用 ngNativeValidate 指令来重新启用它。

7.4 动态表单及其构建

为了更快、更轻松地生成表单的不同版本,可以根据描述业务对象模型的元数据来创建动态表单模板；然后根据数据模型中的变化,使用该模板自动生成新的表单。

如果一个表单内容必须经常更改以满足快速变化的业务需求和监管需求,动态表单技术就特别有用。例如,问卷调查需要在不同的上下文中获取用户的意见,且用户看到的表单格式和样式应该保持不变,而在调查问卷中提的实际问题则会因上下文的不同而有差异。

在构建动态表单时,由于动态表单是基于响应式表单的,为了让应用访问响应式表达式指令,根模块会从@angular/forms 库中导入 ReactiveFormsModule。动态表单需要一个对象模型来描述此表单功能所需的全部场景,会使用一个服务来根据表单模型创建输入控件的分组集合。动态表单本身是一个容器组件。表单模板使用元数据的动态数据绑定来渲染表单,而不用做任何与具体问题有关的硬编码。它能动态添加控件元数据和验证标准。

7.5 表单的综合应用开发

微课视频

7.5.1 表单基础

在项目 src\examples 根目录下创建 formexamples 子目录,在 src\examples\formexamples

目录下创建文件 heroes. service. ts，代码如例 7-1 所示。

【例 7-1】 创建文件 heroes. service. ts 的代码，定义类。

```
import {Injectable} from '@angular/core';
import {Observable, of} from 'rxjs';
import {delay} from 'rxjs/operators';
const ALTER_EGOS = ['Eric'];
@Injectable({providedIn: 'root'})
export class HeroesService {
  isAlterEgoTaken(alterEgo: string): Observable<boolean> {
    const isTaken = ALTER_EGOS.includes(alterEgo);
    return of(isTaken).pipe(delay(400));
  }
}
```

在 src\examples\formexamples 目录下创建文件 nameeditor. component. ts，代码如例 7-2 所示。

【例 7-2】 创建文件 nameeditor. component. ts 的代码，定义组件。

```
import {Component} from '@angular/core';
import {FormControl} from '@angular/forms';
@Component({
  selector: 'app-name-editor',
  template: `
    <label for="name">请输入人名:</label>
    <input id="name" type="text" [formControl]="name">
    <label>输入的人名是: {{ name.value }}</label>
    <button (click)="updateName()">更新默认的名字</button>
  `,
  styles: ['label {\n' +
  '  font-weight: bold;\n' +
  '  padding-bottom: .5rem;\n' +
  '  padding-top: 1rem;\n' +
  '  display: inline-block;\n' +
  '}\n' +
  '\n' +
  'button {\n' +
  '  max-width: 300px;\n' +
  '}']
})
export class NameeditorComponent {
  name = new FormControl('');
  updateName() {
    this.name.setValue('张三丰');
  }
}
```

7.5.2 表单组

在 src\examples\formexamples 目录下创建文件 profileeditor. component. ts，代码如例 7-3 所示。

【例 7-3】 创建文件 profileeditor. component. ts 的代码，定义组件。

```
import {Component} from '@angular/core';
import {FormArray, FormBuilder, Validators} from '@angular/forms';
@Component({
  selector: 'app-profile-editor',
  template: `
    <form [formGroup] = "profileForm" (ngSubmit) = "onSubmit()">
<div for = "first-name">
名字: <input id = "first-name" type = "text" formControlName = "firstName" required>
</div>
      <label formGroupName = "address">
        <div>地址: <input id = "street" type = "text" formControlName = "street"></div>
      </label>
      <div formArrayName = "aliases">
        <button (click) = "addAlias()" type = "button">添加别名</button>
        <div *ngFor = "let alias of aliases.controls; let i = index">
          <label for = "alias-{{ i }}">别名:
<input id = "alias-{{ i }}" type = "text" [formControlName] = "i">
</label>
        </div>
      </div>
      <button type = "submit" [disabled] = "!profileForm.valid">提交</button>
    </form>
    <div>表单值: {{ profileForm.value | json }} 表单状态: {{ profileForm.status }}
<button (click) = "updateProfile()">更新</button>
</div>
  `,
  styles: ['form { padding-top: 1rem;} label {display: block;margin: .5em 0; font-weight:
bold; }']
})
export class ProfileeditorComponent {
  profileForm  =  this.fb.group({
    firstName: ['', Validators.required],
    address: this.fb.group({
      street: [''],
    }),
    aliases: this.fb.array([
      this.fb.control('')
    ])
  });
  constructor(private fb: FormBuilder) {
  }
  get aliases() {
    return this.profileForm.get('aliases') as FormArray;
  }
  updateProfile() {
    this.profileForm.patchValue({
      firstName: '张三丰',
      address: {
        street: '123 Drew Street'
      }
    });
  }
  addAlias() {
    this.aliases.push(this.fb.control(''));
```

```
  }
  onSubmit() {
    console.warn(this.profileForm.value);
  }
}
```

7.5.3 验证器函数

在 src\examples\formexamples 目录下创建文件 alter-ego. directive. ts，代码如例 7-4 所示。

【例 7-4】 创建文件 alter-ego. directive. ts 的代码，定义指令。

```
import {Directive, forwardRef, Injectable} from '@angular/core';
import {
  AsyncValidator,
  AbstractControl,
  NG_ASYNC_VALIDATORS,
  ValidationErrors
} from '@angular/forms';
import {catchError, map} from 'rxjs/operators';
import {HeroesService} from './heroes.service';
import {Observable, of} from 'rxjs';
@Injectable({providedIn: 'root'})
export class UniqueAlterEgoValidator implements AsyncValidator {
  constructor(private heroesService: HeroesService) {}
  validate(
    ctrl: AbstractControl
  ): Promise<ValidationErrors | null> | Observable<ValidationErrors | null> {
    return this.heroesService.isAlterEgoTaken(ctrl.value).pipe(
      map(isTaken => (isTaken ? { uniqueAlterEgo: true } : null)),
      catchError(() => of(null))
    );
  }
}
@Directive({
  selector: '[appUniqueAlterEgo]',
  providers: [
    {
      provide: NG_ASYNC_VALIDATORS,
      useExisting: forwardRef(() => UniqueAlterEgoValidator),
      multi: true
    }
  ]
})
export class UniqueAlterEgoValidatorDirective {
  constructor(private validator: UniqueAlterEgoValidator) {}
  validate(control: AbstractControl) {
    this.validator.validate(control);
  }
}
```

在 src\examples\formexamples 目录下创建文件 forbidden-name. directive. ts，代码如例 7-5 所示。

【例 7-5】　创建文件 forbidden-name. directive. ts 的代码，定义指令。

```
import {Directive, Input} from '@angular/core';
import {AbstractControl, NG_VALIDATORS, ValidationErrors, Validator, ValidatorFn} from '@
angular/forms';
export function forbiddenNameValidator(nameRe: RegExp): ValidatorFn {
  return (control: AbstractControl): ValidationErrors | null => {
    const forbidden = nameRe.test(control.value);
    return forbidden ? {forbiddenName: {value: control.value}} : null;
  };
}
@Directive({
  selector: '[appForbiddenName]',
  providers: [{provide: NG_VALIDATORS, useExisting: ForbiddenValidatorDirective, multi:
true}]
})
export class ForbiddenValidatorDirective implements Validator {
  @Input('appForbiddenName') forbiddenName = '';
  validate(control: AbstractControl): ValidationErrors | null {
    return this.forbiddenName ? forbiddenNameValidator(new RegExp(this.forbiddenName, 'i'))
(control)
        : null;
  }
}
```

在 src\examples\formexamples 目录下创建文件 identity-revealed. directive. ts，代码如例 7-6 所示。

【例 7-6】　创建文件 identity-revealed. directive. ts 的代码，定义指令。

```
import {Directive} from '@angular/core';
import {AbstractControl, NG_VALIDATORS, ValidationErrors, Validator, ValidatorFn} from '@
angular/forms';
export const identityRevealedValidator: ValidatorFn = (control: AbstractControl): ValidationErrors
| null => {
  const name = control.get('name');
  const alterEgo = control.get('alterEgo');
  return name && alterEgo && name.value === alterEgo.value ? {identityRevealed: true} : null;
};
@Directive({
  selector: '[appIdentityRevealed]',
  providers: [{provide: NG_VALIDATORS, useExisting: IdentityRevealedValidatorDirective,
multi: true}]
})
export class IdentityRevealedValidatorDirective implements Validator {
  validate(control: AbstractControl): ValidationErrors | null {
    return identityRevealedValidator(control);
  }
}
```

7.5.4　动态表单

在 src\examples\formexamples 目录下创建文件 questionbase. ts，代码如例 7-7 所示。

【例 7-7】　创建文件 questionbase. ts 的代码，定义类。

```
export class QuestionBase<T> {
  value: T|undefined;
  key: string;
  label: string;
  required: boolean;
  order: number;
  controlType: string;
  type: string;
  options: {key: string, value: string}[];
  constructor(options: {
      value?: T;
      key?: string;
      label?: string;
      required?: boolean;
      order?: number;
      controlType?: string;
      type?: string;
      options?: {key: string, value: string}[];
    } = {}) {
    this.value = options.value;
    this.key = options.key || '';
    this.label = options.label || '';
    this.required = !!options.required;
    this.order = options.order === undefined ? 1 : options.order;
    this.controlType = options.controlType || '';
    this.type = options.type || '';
    this.options = options.options || [];
  }
}
```

在 src\examples\formexamples 目录下创建文件 questiondropdown. ts，代码如例 7-8 所示。

【例 7-8】 创建文件 questiondropdown. ts 的代码，定义类。

```
import {QuestionBase} from "./questionbase";
export class TextboxQuestion extends QuestionBase<string> {
  override controlType = 'textbox';
}
```

在 src\examples\formexamples 目录下创建文件 questiontextbox. ts，代码如例 7-9 所示。

【例 7-9】 创建文件 questiontextbox. ts 的代码，定义类。

```
import {QuestionBase} from "./questionbase";
export class TextboxQuestion extends QuestionBase<string> {
  override controlType = 'textbox';
}
```

在 src\examples\formexamples 目录下创建文件 questioncontrol. service. ts，代码如例 7-10 所示。

【例 7-10】 创建文件 questioncontrol. service. ts 的代码，定义类。

```
import {Injectable} from '@angular/core';
import {FormControl, FormGroup, Validators} from '@angular/forms';
```

```
import {QuestionBase} from "./questionbase";
@Injectable()
export class QuestionControlService {
  constructor() { }
  toFormGroup(questions: QuestionBase<string>[] ) {
    const group: any = {};
    questions.forEach(question => {
group[question.key] = question.required ? new FormControl(question.value || '', Validators.
required)
                                        : new FormControl(question.value || '');
    });
    return new FormGroup(group);
  }
}
```

在 src\examples\formexamples 目录下创建文件 question.service.ts，代码如例 7-11 所示。

【例 7-11】 创建文件 question.service.ts 的代码，定义类。

```
import {Injectable} from '@angular/core';
import {of} from 'rxjs';
import {QuestionBase} from "./questionbase";
import {DropdownQuestion} from "./questiondropdown";
import {TextboxQuestion} from "./questiontextbox";
@Injectable()
export class QuestionService {
  getQuestions() {
    const questions: QuestionBase<string>[] = [
      new  DropdownQuestion({
        key: 'brave',
        label: '勇气等级',
        options: [
          {key: 'solid',   value: '牢固'},
          {key: 'great',   value: '巨大'},
          {key: 'good',    value: '好'},
          {key: 'unproven', value: '未经证实'}
        ],
        order: 3
      }),
      new TextboxQuestion({
        key: 'firstName',
        label: '姓名',
        value: 'Bombasto',
        required: true,
        order: 1
      }),
      new TextboxQuestion({
        key: 'emailAddress',
        label: 'Email',
        type: 'email',
        order: 2
      })
    ];
    return of(questions.sort((a, b) => a.order - b.order));
```

```
    }
}
```

在 src\examples\formexamples 目录下创建文件 dynamicformquestion. component. ts,代码如例 7-12 所示。

【例 7-12】 创建文件 dynamicformquestion. component. ts 的代码,定义组件。

```
import {Component, Input} from '@angular/core';
import {FormGroup} from '@angular/forms';
import {QuestionBase} from "./questionbase";
@Component({
  selector: 'app-question',
  template: `
    <div [formGroup] = "form">
      <label [attr.for] = "question.key">{{question.label}}</label>
      <div [ngSwitch] = "question.controlType">
        <input *ngSwitchCase = "'textbox'" [formControlName] = "question.key"
               [id] = "question.key" [type] = "question.type">
         <select [id] = "question.key" *ngSwitchCase = "'dropdown'" [formControlName] =
"question.key">
          <option *ngFor = "let opt of question.options" [value] = "opt.key">{{opt.value}}
</option>
        </select>
      </div>
      <div class = "errorMessage" *ngIf = "!isValid">{{question.label}} is required</div>
    </div>
  `
})
export class DynamicformquestionComponent {
  @Input() question!: QuestionBase<string>;
  @Input() form!: FormGroup;
  get isValid() {return this.form.controls[this.question.key].valid;}
}
```

7.5.5 其他组件

在 src\examples\formexamples 目录下创建文件 dynamicform. component. ts,代码如例 7-13 所示。

【例 7-13】 创建文件 dynamicform. component. ts 的代码,定义组件。

```
import {Component, Input, OnInit} from '@angular/core';
import {FormGroup} from '@angular/forms';
import {QuestionBase} from "./questionbase";
import {QuestionControlService} from "./questioncontrol.service";
@Component({
  selector: 'app-dynamic-form',
  template: `
    <div>
      <form (ngSubmit) = "onSubmit()" [formGroup] = "form">
        <div *ngFor = "let question of questions" class = "form-row">
          <app-question [question] = "question" [form] = "form"></app-question>
        </div>
        <div class = "form-row">
```

```
          <button type = "submit" [disabled] = "!form.valid">保存</button>
        </div>
      </form>
      <div *ngIf = "payLoad" class = "form-row">
        <strong>Saved the following values</strong><br>{{payLoad}}
      </div>
    </div>
  `,
  providers: [ QuestionControlService ]
})
export class DynamicformComponent implements OnInit {
  @Input() questions: QuestionBase<string>[] | null = [];
  form!: FormGroup;
  payLoad = '';
  constructor(private qcs: QuestionControlService) {}
  ngOnInit() {
    this.form = this.qcs.toFormGroup(this.questions as QuestionBase<string>[]);
  }
  onSubmit() {
    this.payLoad = JSON.stringify(this.form.getRawValue());
  }
}
```

在 src\examples\formexamples 目录下创建文件 heroformreactive.component.ts，代码如例 7-14 所示。

【例 7-14】 创建文件 heroformreactive.component.ts 的代码，定义组件。

```
import {Component, OnInit} from '@angular/core';
import {FormControl, FormGroup, Validators} from '@angular/forms';
import {UniqueAlterEgoValidator} from "./alter-ego.directive";
import {forbiddenNameValidator} from "./forbidden-name.directive";
import {identityRevealedValidator} from "./identity-revealed.directive";
@Component({
  selector: 'app-hero-form-reactive',
  template: `
    <div class = "container">
      <hr>
      <form [formGroup] = "heroForm" #formDir = "ngForm">
        <div [hidden] = "formDir.submitted">
          <div class = "cross-validation"
 [class.cross-validation-error] = "heroForm.errors?.['identityRevealed'] && (heroForm.
touched || heroForm.dirty)">
            <div class = "form-group">
              <label for = "name">名字</label>
              <input type = "text" id = "name" class = "form-control"
                     formControlName = "name" required>
              <div *ngIf = "name.invalid && (name.dirty || name.touched)"
                   class = "alert alert-danger">
                <div *ngIf = "name.errors?.['required']">
                  名字不能为空，请填写名字
                </div>
                <div *ngIf = "name.errors?.['minlength']">
                  名字不能少于 4 个字符
                </div>
```

```
                <div *ngIf="name.errors?.['forbiddenName']">
                  名字不能是 Bob
                </div>
              </div>
            </div>
            <div class="form-group">
              <label for="alterEgo">曾用名</label>
              <input type="text" id="alterEgo" class="form-control"
                     formControlName="alterEgo">
              <div *ngIf="alterEgo.pending">验证...</div>
              <div *ngIf="alterEgo.invalid" class="alert alert-danger alter-ego-errors">
                <div *ngIf="alterEgo.errors?.['uniqueAlterEgo']">
                  Alter ego is already taken.
                </div>
              </div>
            </div>
            <div *ngIf="heroForm.errors?.['identityRevealed'] && (heroForm.touched ||
heroForm.dirty)"
                 class="cross-validation-error-message alert alert-danger">
              Name cannot match alter ego.
            </div>
          </div>
          <div class="form-group">
            <label for="power">能力</label>
            <select id="power" class="form-control"
                    formControlName="power" required>
              <option *ngFor="let p of powers" [value]="p">{{p}}</option>
            </select>
            <div *ngIf="power.invalid && power.touched" class="alert alert-danger">
              <div *ngIf="power.errors?.['required']">Power is required.</div>
            </div>
          </div>
          <button type="submit"
                  class="btn btn-default"
                  [disabled]="heroForm.invalid">提交
          </button>
          <button type="button" class="btn btn-default"
                  (click)="formDir.resetForm({})">重置
          </button>
        </div>
      </form>
      <div class="submitted-message" *ngIf="formDir.submitted">
        <p>You've submitted your hero, {{ heroForm.value.name }}!</p>
        <button (click)="formDir.resetForm({})">Add new hero</button>
      </div>
    </div>
  `,
  styles: ['.cross-validation-error input {\n' +
  '  border-left: 5px solid red;\n' +
  '}'],
})
export class HeroFormReactiveComponent implements OnInit {
  powers = ['写作', '唱歌', '跳舞'];
  hero = {name: '', alterEgo: '', power: this.powers[0]};
```

```
  heroForm!: FormGroup;
  constructor(private alterEgoValidator: UniqueAlterEgoValidator) {
  }
  get name() {
    return this.heroForm.get('name')!;
  }
  get power() {
    return this.heroForm.get('power')!;
  }
  get alterEgo() {
    return this.heroForm.get('alterEgo')!;
  }
  ngOnInit(): void {
    this.heroForm = new FormGroup({
      name: new FormControl(this.hero.name, [
        Validators.required,
        Validators.minLength(4),
        forbiddenNameValidator(/bob/i)
      ]),
      alterEgo: new FormControl(this.hero.alterEgo, {
        asyncValidators: [this.alterEgoValidator.validate.bind(this.alterEgoValidator)],
        updateOn: 'blur'
      }),
      power: new FormControl(this.hero.power, Validators.required)
    }, {validators: identityRevealedValidator});
  }
}
```

在 src\examples\formexamples 目录下创建文件 formhome.component.ts，代码如例 7-15 所示。

【例 7-15】 创建文件 formhome.component.ts 的代码，定义组件。

```
import {Component} from '@angular/core';
import {QuestionService} from "./question.service";
import {Observable} from "rxjs";
import {QuestionBase} from "./questionbase";
@Component({
  selector: 'root',
  template: `
    <div>表单示例</div>
    <app-name-editor></app-name-editor>
    <hr>
    <app-profile-editor></app-profile-editor>
    <app-hero-form-reactive></app-hero-form-reactive>
    <hr>
    <app-dynamic-form [questions]="questions$ | async"></app-dynamic-form>
  `,
  providers: [QuestionService]
})
export class FormhomeComponent {
  questions$: Observable<QuestionBase<any>[]>;
  constructor(service: QuestionService) {
    this.questions$ = service.getQuestions();
  }
}
```

7.5.6 模块和运行结果

在 src\examples\formexamples 目录下创建文件 app-form-example. module. ts，代码如例 7-16 所示。

【例 7-16】 创建文件 app-form-example. module. ts 的代码，声明组件并导入模块。

```
import {BrowserModule} from '@angular/platform-browser';
import {NgModule} from '@angular/core';
import {ReactiveFormsModule} from '@angular/forms';
import {NameeditorComponent} from "./nameeditor.component";
import {ProfileeditorComponent} from "./profileeditor.component";
import {FormhomeComponent} from "./formhome.component";
import {HeroFormReactiveComponent} from "./heroformreactive.component";
import {IdentityRevealedValidatorDirective} from "./identity-revealed.directive";
import {UniqueAlterEgoValidatorDirective} from "./alter-ego.directive";
import {ForbiddenValidatorDirective} from "./forbidden-name.directive";
import {DynamicformComponent} from "./dynamicform.component";
import {DynamicformquestionComponent} from "./dynamicformquestion.component";
@NgModule({
  declarations: [
    NameeditorComponent,
    ProfileeditorComponent,
    HeroFormReactiveComponent,
    ForbiddenValidatorDirective,
    IdentityRevealedValidatorDirective,
    UniqueAlterEgoValidatorDirective,
    DynamicformComponent,
    DynamicformquestionComponent,
    FormhomeComponent
  ],
  imports: [
    BrowserModule,
    ReactiveFormsModule
  ],
})
export class AppFormExampleModule {
}
```

修改 src\examples 目录下的文件 examplesmodules1. module. ts，代码如例 7-17 所示。

【例 7-17】 修改文件 examplesmodules1. module. ts 的代码，设置启动组件。

```
import {NgModule} from '@angular/core';
import {AppRouterModule} from "./routerexamples/app-router.module";
import {FormhomeComponent} from './formexamples/formhome.component';
import {AppFormExampleModule} from "./formexamples/app-form-example.module";
@NgModule({
  imports: [
    AppFormExampleModule,
  ],
  bootstrap: [FormhomeComponent]
})
export class ExamplesmodulesModule1 {}
```

保持其他文件不变并成功运行程序后，在浏览器地址栏中输入 localhost:4200 后将自动跳转到 localhost:4200/name-list，也可以在浏览器中输入 localhost:4200/form，结果如图 7-1 所示。

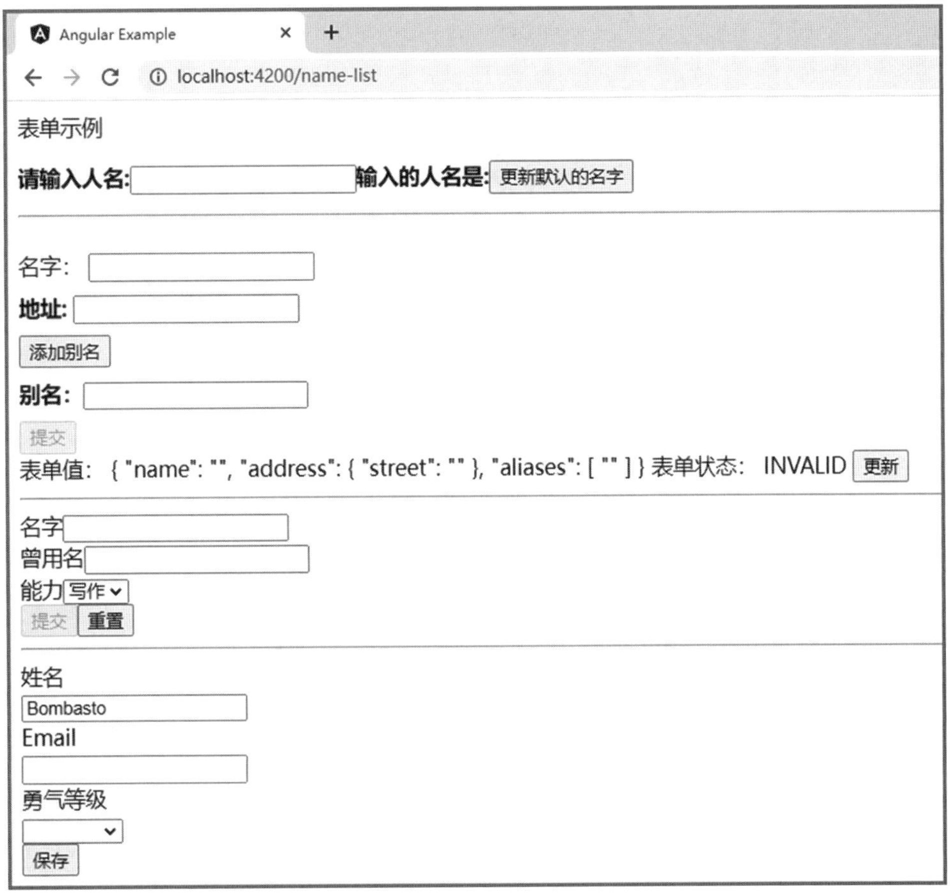

图 7-1　成功运行程序后在浏览器地址栏中输入 localhost:4200 的结果

习题 7

一、简答题

简述对表单的理解。

二、实验题

完成表单的应用开发。

第 8 章

HTTP客户端服务及其应用

8.1 HTTP 客户端服务

1. 含义

大多数前端应用都要通过 HTTP 协议与服务器通信，才能下载或上传数据并访问其他后端服务。Angular 给应用程序提供了一个 HTTP 客户端 API（即服务类 HttpClient）。HTTP 客户端服务提供了请求类型化响应对象、简单的错误处理、各种特性的可测试性、请求和响应的拦截机制、Angular 的应用设计基础等主要功能。要想使用 HttpClient，就要先从@angular/common/http 中导入 HttpClientModule。大多数应用都会在根模块中导入它。HttpClient 服务为基于它的所有任务都提供了可观察对象，并且导入范例代码片段中的 RxJS 可观察对象和操作符。对于 RxJS 的语法规则可以参考官方文档。

替代了 HttpClient 模块中的 HttpBackend 服务会模拟 REST 式的后端行为。使用 HttpClient. get()方法从服务器获取的数据有两个参数，一个是要获取的端点 URL，另一个是可以用来配置请求的选项对象。该方法会发送一个 HTTP 请求，并返回一个 Observable（观察对象）它会在收到观察者响应时发出所请求到的数据。返回的类型取决于调用时传入的 observe 和 responseType 属性参数。其中，observe 用于指定要返回的响应内容；responseType 指定返回数据的格式。例如，应用程序经常会从服务器请求 JSON 数据，调用 HttpClient. get()方法时需要设置选项的默认值{observe：'body'，responseType：'json'}获取 JSON 数据；可以用 options 对象来配置传出请求的各个方面；使用 params、属性可以配置带 HTTP URL 参数的请求；reportProgress 选项可以在传输大量数据时监听进度事件。

声明响应对象的类型可以自定义 HttpClient 请求的方式，以便让输出更容易、更明确。所指定的响应类型会在编译时充当类型断言。指定响应类型是在向 TypeScript 声明它应该把响应对象当做给定类型来使用。这是一种构建时检查，它并不能保证服务器会实际给出这种类型的响应对象。该服务器需要自己确保返回服务器 API 中指定的类型。要指定响应对象类型，首先要定义一个具有必需属性的接口。因为响应对象是普通对象，无法自动转换成类的实例。当把接口作为类型参数传给 HttpClient. get()方法时，可以使用 RxJS 的 map 操作符来根据用户界面(UI)的需求转换响应数据。然后，把转换后的数据传给异步管道。要访问接口中定义的属性，必须将从 JSON 普通对象显式转换为所需的响应类型。不是所有的 API 都会返回 JSON 数据。

当服务器不支持 CORS(Cross-Origin Resource Sharing,跨域资源共享)协议时,应用程序可以使用 HttpClient 跨域发出 JSONP 请求。通过在 NgModule 的 imports 中包含 HttpClientJsonpModule 来使用 JSONP。Angular 的 JSONP 请求会返回一个 Observable。遵循订阅可观察对象变量的模式,并在使用 async 管道管理结束之前,使用 RxJS 的 map 操作符转换响应。

2. 服务的异常、错误和方法

如果请求在服务器上失败了,那么 HttpClient 就会返回一个错误对象而不是一个成功的响应对象。执行服务器请求的同一个服务中也应该执行错误检查、解释和解析。发生错误时,可以获取失败的详细信息,以便通知用户。在某些情况下,也可以自动重试该请求。当数据访问失败时,应用会给用户提供有用的反馈。原始的错误对象作为反馈并不是特别有用。除了检测到错误已经发生之外,还需要获取错误详细信息并使用这些细节来撰写用户友好的响应。

服务器端可能会拒绝该请求,并返回状态码为 404 或 500 的 HTTP 响应对象;客户端也可能出现问题。例如,网络错误会让请求无法成功完成。RxJS 操作符也可能会抛出异常,这些错误会产生 ErrorEvent 对象。这些错误的 status 为 0,并且其 error 属性包含一个 ProgressEvent 对象,此对象的 type 属性可以提供更详细的信息。HttpClient 在其 HttpErrorResponse 中会捕获两种错误。可以检查这个响应对象是否存在错误。

有时候,错误只是临时性的,只要重试就可能会自动消失。例如,在移动端场景中可能会遇到网络中断的情况,只要重试一下就能拿到正确的结果。RxJS 库提供了几个重试操作符(或称为方法)。例如,retry 操作符会自动重新订阅一个失败的 Observable。重新订阅 HttpClient 方法会导致它重新发出 HTTP 请求。除了从服务器获取数据外,HttpClient 还支持其他一些 HTTP 方法,如 PUT、POST 和 DELETE,可以用它们来修改远程数据。应用程序经常在提交表单时通过 POST 请求向服务器发送数据。HttpClient.post()方法的类型参数可以用它来指出期望服务器返回特定类型的数据。该方法需要一个资源 URL 和两个额外的参数:body(要在请求体中 POST 的数据)和 options(一个包含方法选项的对象,用来指定必要的请求头)。

DELETE(及 PUT)操作时必须调用 subscribe()方法(订阅方法),否则什么都不会发生。在调用方法返回的可观察对象的 subscribe()方法之前,HttpClient 方法不会发起 HTTP 请求。这适用于 HttpClient 的所有方法,如 DELETE、PUT。

AsyncPipe 会自动订阅(以及取消订阅)。HttpClient 的所有方法返回的可观察对象都被设计为不能自动更新的。HTTP 请求的执行都是延期执行的,使得可以用 tap 和 catchError 这样的操作符来在实际执行 HTTP 请求之前,先对这个可观察对象进行扩展。调用 subscribe()方法会引发这个可观察对象的执行,导致 HttpClient 会组合并把 HTTP 请求发给服务器。可以把这些可观察对象看做实际 HTTP 请求的蓝图。实际上,每个 subscribe()方法都会完成对初始化可观察对象的一次单独的、独立的执行。订阅两次的操作就会导致发起两个 HTTP 请求。

很多服务器都需要额外的头来执行保存操作。例如,服务器可能需要一个授权令牌,或者需要 Content-Type 头来显式声明请求体的 MIME 类型。可以使用 HttpParams 类和

params 选项在 HttpRequest 中添加 URL 查询字符串。其中,HttpParams 是不可变对象。如果需要更新选项,需要保留 setter()方法的返回值。

8.2 拦截机制

8.2.1 拦截器的含义和原理

可以声明一些拦截器来实现拦截机制,这些拦截器可以检查并转换从应用中发给服务器的 HTTP 请求。这些拦截器还可以在返回应用的途中检查和转换来自服务器的响应。多个拦截器(拦截链)构成了请求、响应处理器的双向链表。拦截器可以用一种常规的、标准的方式完成对每一次 HTTP 的请求、响应时执行从认证到记日志等很多项隐式任务。如果没有拦截机制,那么开发人员将不得不对每次 HttpClient 调用都显式地实现这些任务。要实现拦截器,就要创建一个实现了 HttpInterceptor 接口 intercept()方法的类。intercept()方法会把请求转换成一个最终返回 HTTP 响应体的 Observable 观察对象。

拦截器的设计采用了职责链模式。大多数拦截器都会在传入时检查请求,然后把(可能被修改过的)请求转发给 next 对象(下一个对象)的 handle()方法来处理,而 next 对象实现了 HttpHandler 接口。next 对象的 handle()方法也能把 HTTP 请求转换成 HttpEvents 组成的 Observable 观察对象,它最终包含的是来自服务器的响应。intercept()方法可以检查这个可观察对象,并在把它返回给调用者之前修改它。这个无操作的拦截器会使用原始请求调用 next.handle()方法,并返回 next.handle()方法返回的可观察对象,而不做任何后续处理。

next 对象表示拦截器链表中的下一个拦截器。这个链表中的最后一个 next 对象就是 HttpClient 的后端处理器,它会把请求发给服务器,并接收服务器的响应。大多数的拦截器都会调用 next.handle()方法,以便这个请求流能传送到下一个拦截器,并最终传给后端处理器。拦截器也可以不调用 next.handle()方法,使这个链路短路,并返回一个带有人工构造出来的内容响应的自己的 Observable。这是一种常见的中间件模式,在像 Express.js 这样的框架中会找到这种实现模式。

由于拦截器是 HttpClient 服务的可选依赖,所以必须在提供 HttpClient 的同一个(或其各级父注入器)注入器中提供拦截器。依赖注入机制创建完 HttpClient 之后再提供的拦截器将会被忽略。

8.2.2 拦截器的处理方法

Angular 会按拦截器的提供顺序来使用它们(即依次调用拦截器)。例如,想处理 HTTP 请求的身份验证并记录它们,然后再将它们发送到服务器。要完成此任务,可以先提供身份验证的 AuthInterceptor 服务,再提供记录的 LoggingInterceptor 服务。发出的请求将依次从 AuthInterceptor 到 LoggingInterceptor 得到处理。对请求的响应过程则沿相反的方向流动,即从 LoggingInterceptor 回到 AuthInterceptor。拦截过程中的最后一个拦截器始终是处理与服务器通信的 HttpBackend 服务。

大多数 HttpClient 方法都会返回 HttpResponse<any>类型的可观察对象。HttpResponse 类本身是一个事件,它的类型是 HttpEventType.Response。单个 HTTP 请求可以生成其他类型的多个事件,如包括报告上传和下载进度的事件。HttpInterceptor.intercept()方法和 HttpHandler. handle()方法会返回 HttpEvent < any >型的可观察对象。

很多拦截器只关心发出的请求，而对 next. handle()方法返回的事件流不会做任何修改。有些拦截器需要检查并修改 next. handle()方法的响应。虽然拦截器有能力改变请求和响应，但 HttpRequest 和 HttpResponse 实例的属性却是只读(readonly)的。拦截器应该在没有任何修改的情况下返回每一个事件，除非它有令人信服的理由去做修改。

Angular 应用程序可能会在发送多次请求之后才能成功完成请求任务，这就意味着拦截器可能会多次重复处理同一个请求。如果拦截器可以修改原始的请求对象，那么重试阶段的操作(再次处理请求)就会从原始请求开始，而不是修改过的请求。这种不可变性可以确保这些拦截器在每次重试时看到的都是同样的原始请求。

如果必须修改请求体对象，需要的步骤包括复制请求体对象并在副本中进行修改；使用 clone()方法克隆这个请求体对象；用修改过的副本替换被克隆的请求体；克隆时清除请求体。有时，需要清除请求体而不是替换它。为此，需要将克隆后的请求体设置为 null。如果把克隆后的请求体设为 undefined，那么 Angular 会认为想让请求体保持原样。

8.2.3 拦截器的作用

Angular 应用程序通常会使用拦截器来设置外发请求的默认请求头。修改头的拦截器可以用于很多不同的操作，如认证和授权控制、控制缓存行为、跨站请求伪造(Cross-Site Request Forgery，CSRF)防护。

因为拦截器可以同时处理请求和响应，所以它们也可以对整个 HTTP 操作执行计时和记录日志等任务。拦截器可被用来自定义实现替换内置的 JSON 解析；还可以自行处理请求，而不用转发给 next. handle()方法。例如，缓存某些请求和响应以便提升性能，可以把这种缓存操作委托给某个拦截器，而不破坏现有的各个数据服务。

HttpClient. get()方法通常会返回一个可观察对象，它会发出一个值(数据或错误)。拦截器可以把它改成一个可以发出多个值的可观察对象。Angular 应用程序有时会传输大量数据，而这些传输(如文件上传)可能要花很长时间。此时可以通过提供关于此类传输的进度反馈，为用户提供更好的体验。每个进度事件都会引发变更检测，所以只有当需要在用户界面上报告进度时，才应该开启它们。

如果需要发一个 HTTP 请求来响应用户的输入，那么每次按键就发送一个请求的处理方式效率显然不高，最好是等用户停止输入后再发送请求。这种技术叫作防抖。

跨站请求伪造(XSRF 或 CSRF)是一个攻击技术，它能让攻击者假冒一个已认证的用户在网站上执行未知的操作。HttpClient 使用一种通用的机制来防范 XSRF 攻击。当执行 HTTP 请求时，一个拦截器会从 cookie 中读取 XSRF 令牌(默认名字为 XSRF-TOKEN)，并且把它设置为一个 HTTP 头 X-XSRF-TOKEN，由于只有运行在自己的域名下的合法代码才能读取这个 cookie，因此后端可以确认这个 HTTP 请求来自合法的客户端应用程序，而不是攻击者的应用程序。默认情况下，拦截器会在所有的修改型请求中(如 POST 等)把请求头发送给使用相对 URL 的请求。但不会在 GET 方法的 HEAD 请求中发送，也不会发送给使用绝对 URL 的请求。

要获得这种优点，服务器需要在页面加载或首个 GET 请求时把 XSRF-TOKEN 令牌写入可被 JavaScript 读到的会话 cookie 中。在后续的请求中，服务器可以验证这个 cookie

是否与 X-XSRF-TOKEN 的值一致,以确保只有运行在自己域名下的合法代码才能发起这个请求。这个令牌必须对每个用户都是唯一的,并且必须能被服务器验证,因此不能由客户端自己生成令牌。把这个令牌设置成服务器的站点认证信息并且加了盐(salt)的摘要,以提升安全性。密码学中,可以在散列(hash)前将指定的字符串插入散列内容中的任意固定位置,此种插入字符串的操作称为加盐。为了防止多个 Angular 应用程序共享同一个域名或子域时出现冲突,要给每个 Angular 应用程序分配一个唯一的 cookie 名称。HttpClient 支持的只是 XSRF 防护方案的客户端。后端服务必须配置为给页面设置 cookie 并且要验证请求头,以确保全都是合法的请求。否则,就会导致 Angular 的默认防护措施失效。如果后端服务中对 XSRF 令牌的 cookie 或头使用了不一样的名字,就要使用 HttpClientXsrfModule.withConfig()覆盖掉默认值。

8.2.4 拦截器的测试

如同所有的外部依赖一样,必须把 HTTP 后端也进行 Mock(模拟)处理,以便测试时可以模拟这种与后端的互动。@angular/common/http/testing 库能让这种 Mock 工作变得直截了当。Angular 的 HTTP 测试库是专为其中的测试模式而设计的。在这种模式下,会首先在应用中执行代码并发起请求。然后,这个测试会期待发起(或未发起)过某个请求,并针对这些请求进行断言,最终通过对每个所预期的请求进行刷新(flush)来响应这些请求。最终,测试可能会验证这个应用程序不曾发起过非预期的请求。要开始测试那些通过 HttpClient 发起的请求,就要导入 HttpClientTestingModule 模块,并把它加到 TestBed 的设置中,再继续设置被测服务。在测试中发起的这些请求会发给这些测试用的后端,而不是标准的后端。这种设置还会调用 TestBed.inject()方法来获取注入的 HttpClient 服务和对象的控制器 HttpTestingController,以便在测试期间引用它们。

8.2.5 拦截器的配置

许多拦截器都需要进行配置。例如,对于一个重试失败请求的拦截器,在默认情况下该拦截器可能会重试请求三次。但是对于特别容易出错或敏感的请求,可能要改写重试次数。HttpClient 请求包含一个上下文,该上下文可以携带有关请求的元数据。该上下文可供拦截器读取或修改,尽管发送请求时它并不会传输到后端服务器。这允许应用或其他拦截器使用配置参数来标记这些请求,例如重试请求的次数。与 HttpRequest 实例的大多数其他内容不同,请求上下文是可变的。并且,它在请求的其他不可变转换过程中仍然存在。这就使得拦截器可以通过此上下文来协调操作。

微课视频

8.3 HTTP 客户端服务的应用

8.3.1 服务

在项目 src\examples 根目录下创建 httpexamples 子目录,在 src\examples\httpexamples 目录下创建文件 hero.ts,代码如例 8-1 所示。

【例 8-1】 创建文件 hero.ts 的代码,定义接口。

```
export interface Hero {
  id: number;
  name: string;
}
```

在 src\examples\httpexamples 目录下创建文件 heroes. service. ts，代码如例 8-2 所示。

【例 8-2】 创建文件 heroes. service. ts 的代码，定义类。

```
import {Injectable} from "@angular/core";
import {HttpClient, HttpHeaders} from "@angular/common/http";
import {Observable, of, tap} from "rxjs";
import {Hero} from "./hero";
import {catchError} from "rxjs/operators";
@Injectable({providedIn: 'root'})
export class HeroService {
  httpOptions = {
    headers: new HttpHeaders({'Content-Type': 'application/json'})
  };
  private heroesUrl = 'api/heroes';
  constructor(private http: HttpClient) {
  };
  getHeroes(): Observable<Hero[]> {
    return this.http.get<Hero[]>(this.heroesUrl)
      .pipe(
        tap(
          _ => console.log('fetched heroes')),
        catchError(this.handleError('getHeroes', []))
      )
  }
  addHero(hero: Hero): Observable<Hero> {
    return this.http.post<Hero>(this.heroesUrl, hero, this.httpOptions)
      .pipe(
        tap((newHero: Hero) => console.log(`添加 hero w/ id = ${newHero.id}`)),
        catchError(this.handleError<Hero>('addHero'))
      );
  }
  deleteHero(hero: Hero | number): Observable<Hero> {
    const id = typeof hero === 'number' ? hero : hero.id;
    const url = `${this.heroesUrl}/${id}`;
    return this.http.delete<Hero>(url, this.httpOptions)
      .pipe(
        tap(_ => console.log(`删除 hero  id = ${id}`)),
        catchError(this.handleError<any>('deleteHero', hero))
      );
  }
  updateHero(hero: Hero): Observable<Hero> {
    hero.name = hero.name + (hero.id + 1)
    return this.http.put<Hero>(this.heroesUrl, hero, this.httpOptions)
      .pipe(
        tap(_ => console.log(`更新 hero  id = ${hero.id}`)),
        catchError(this.handleError<any>('updateHero', hero))
      );
  }
  private handleError<T>(operation = 'operation', result?: T) {
```

```
      return (error: any): Observable<T> => {
        console.log(`$(operation) failed : $(error.message)`);
        return of(result as T);
      }
    }
  }
```

在 src\examples\httpexamples 目录下创建文件 inmemhero. service. ts,代码如例 8-3 所示。

【例 8-3】 创建文件 inmemhero. service. ts 的代码,定义类。

```
import {InMemoryDbService} from 'angular-in-memory-web-api';     //需要安装此包
export class InmemheroService implements InMemoryDbService {
  createDb() {
    let heroes = [
      {
        id: 1, name: 'zsf'
      },
      {
        id: 2, name: 'ls'
      },
    ];
    return {
      heroes
    }
  }
}
```

8.3.2 组件

在 src\examples\httpexamples 目录下创建文件 testhttp. component. ts,代码如例 8-4 所示。

【例 8-4】 创建文件 testhttp. component. ts 的代码,定义组件。

```
import {Component, OnInit} from '@angular/core';
import {HttpClient} from "@angular/common/http";
import {Observable} from "rxjs";
import {Hero} from "./hero";
@Component({
  selector: 'testhttp',
  template: `
    <div>从 JSON 文件得到人名的数据</div>
    <p>序号 - 人名</p>
    <p *ngFor="let hero of heroes">{{hero.id}} - {{hero.name}}</p>
  `,
})
export class TesthttpComponent implements OnInit {
  heroes: Hero[] | undefined;
  private heroesUrl = 'api/heroes'
  constructor(private http: HttpClient) {
  }
  ngOnInit(): void {
    this.getHeroes().subscribe(
```

```
      data => this.heroes = data
    )
  }
  private getHeroes(): Observable<Hero[]> {
    return this.http.get<Hero[]>(this.heroesUrl)
  }
}
```

在 src\examples\httpexamples 目录下创建文件 httphome.component.ts，代码如例 8-5 所示。

【例 8-5】 创建文件 httphome.component.ts 的代码，定义组件。

```
import {Component} from '@angular/core';
import {Hero} from './hero';
import {FormBuilder, FormGroup, Validators} from "@angular/forms";
import {HeroService} from "./hero.service";
@Component({
  selector: 'heroes-home',
  template: `
    <div>
      <table>
        <tr>
          <th>ID</th>
          <th>Name</th>
          <th>操作</th>
        </tr>
        <tr *ngFor="let hero of heroes">
          <td>{{ hero.id }}</td>
          <td>{{hero.name}}</td>
          <td>
            <button (click)="deleteHero(hero.id)">删除</button>
          </td>
          <td>
            <button (click)="updateHero(hero)">更新</button>
          </td>
        </tr>
      </table>
      <br>
      <form [formGroup]="formGroup" (ngSubmit)="onSubmit()">
        <div class="block"><label>id:</label><input formControlName="id"></div>
        <div class="block"><label>name:</label><input formControlName="name"></div>
        <input type="submit" value="添加" [disabled]="!formGroup.valid">
        <br><br>
        表单是否有效：{{formGroup.valid}}<br>
        表单完整数据：{{formGroup.value | json}}
      </form>
    </div>
  `,
})
export class HttphomeComponent {
  heroes: Hero[] | undefined;
  formGroup: FormGroup = this.fb.group(
    {
      id: this.fb.control('', Validators.required),
```

```
      name: this.fb.control('', Validators.required)
    });
  constructor(private heroService: HeroService, private fb: FormBuilder) {
  }
  ngOnInit() {
    this.getHeroes();
    this.formGroup = this.fb.group(
      {
        id: this.fb.control('', Validators.required),
        name: this.fb.control('', Validators.required)
      }
    )
  }
  getHeroes() {
    this.heroService.getHeroes().subscribe(
      (data) => this.heroes = data
    );
  }
  updateHero(hero: Hero) {
    this.heroService.updateHero(hero).subscribe(
      (data: any) => {
        console.log("Hero Update:", data);
        this.getHeroes();
      }
    )
  }
  deleteHero(id: number) {
    this.heroService
      .deleteHero(id)
      .subscribe((data) => {
          console.log("Hero deleted:", data);
          this.getHeroes();
        }
      );
  }
  onSubmit() {
    const hero = this.formGroup.value;
    hero.id = Number(hero.id);
    this.heroService.addHero(hero).subscribe(hero => {
      if (hero) {
        this.getHeroes();
      } else {
        alert('发生错误')
      }
      this.formGroup.reset();
    });
  }
}
```

在 src\examples\httpexamples 目录下创建文件 httpexample.component.ts，代码如例 8-6 所示。

【例 8-6】 创建文件 httpexample.component 的代码，定义组件。

```
import {Component} from '@angular/core';
```

```
@Component({
  selector: 'root',
  template: `
    <div>HTTP 示例</div>
    <testhttp></testhttp>
    <hr>
    <heroes-home></heroes-home>
  `,
})
export class HttpexampleComponent {
}
```

8.3.3 模块和运行结果

在 src\examples\httpexamples 目录下创建文件 app-http-example.module.ts，代码如例 8-7 所示。

【例 8-7】 创建文件 app-http-example.module.ts 的代码，声明组件并导入模块。

```
import {BrowserModule} from '@angular/platform-browser';
import {NgModule} from '@angular/core';
import {HttpexampleComponent} from "./httpexample.component";
import {TesthttpComponent} from "./testhttp.component";
import {HttpClientModule} from "@angular/common/http";
import {HttpClientInMemoryWebApiModule} from "angular-in-memory-web-api";
import {InmemheroService} from "./inmemhero.service";
import {HttphomeComponent} from "./httphome.component";
import {ReactiveFormsModule} from "@angular/forms";
@NgModule({
  declarations: [
    HttpexampleComponent,
    TesthttpComponent,
    HttphomeComponent,
  ],
  imports: [
    BrowserModule,
    HttpClientModule,
    HttpClientInMemoryWebApiModule.forRoot(InmemheroService),
    ReactiveFormsModule,
  ],
})
export class AppHttpExampleModule {
}
```

修改 src\examples 目录下的文件 examplesmodules1.module.ts，代码如例 8-8 所示。

【例 8-8】 修改文件 examplesmodules1.module.ts 的代码，设置启动组件。

```
import {NgModule} from '@angular/core';
import {HttpexampleComponent} from './httpexamples/httpexample.component';
import {AppHttpExampleModule} from "./httpexamples/app-http-example.module";
@NgModule({
  imports: [
    AppHttpExampleModule
  ],
  bootstrap: [HttpexampleComponent]
```

```
})
export class ExamplesmodulesModule1 {}
```

保持其他文件不变并成功运行程序后,在浏览器地址栏中输入 localhost:4200,自动跳转到 localhost:4200/name-list,结果如图 8-1 所示。

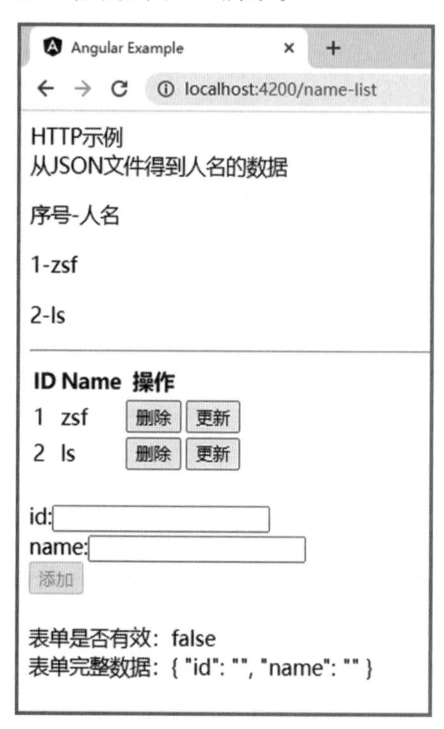

图 8-1 成功运行程序后在浏览器地址栏中输入 localhost:4200 的结果

8.4 拦截器的应用开发

8.4.1 拦截器的简单使用

在项目 src\examples 根目录下创建 interceptorexamples 子目录,在 src\examples\interceptorexamples 目录下创建文件 heroes.service.ts,代码如例 8-9 所示。

【例 8-9】 创建文件 heroes.service.ts 的代码,定义接口和类。

```
import {Injectable} from '@angular/core';
import {HttpClient, HttpHeaders, HttpParams} from '@angular/common/http';
import {Observable} from 'rxjs';
import {catchError} from 'rxjs/operators';
import {HandleError, HttpErrorHandler} from "./http-error-handler.service";
export interface Hero {
  id: number;
  name: string;
}
const httpOptions = {
  headers: new HttpHeaders({
    'Content-Type': 'application/json',
    Authorization: 'my-auth-token'
```

```
  })
};
@Injectable()
export class HeroesService {
  heroesUrl = 'api/heroes';  //Web API 的 URL
  private handleError: HandleError;
  constructor(
    private http: HttpClient,
    httpErrorHandler: HttpErrorHandler) {
    this.handleError = httpErrorHandler.createHandleError('HeroesService');
  }
  getHeroes(): Observable<Hero[]> {
    return this.http.get<Hero[]>(this.heroesUrl)
      .pipe(
        catchError(this.handleError('getHeroes', []))
      );
  }
  searchHeroes(term: string): Observable<Hero[]> {
    term = term.trim();
    const options = term ?
      {params: new HttpParams().set('name', term)} : {};
    return this.http.get<Hero[]>(this.heroesUrl, options)
      .pipe(
        catchError(this.handleError<Hero[]>('searchHeroes', []))
      );
  }
  /** POST 方法 */
  addHero(hero: Hero): Observable<Hero> {
    return this.http.post<Hero>(this.heroesUrl, hero, httpOptions)
      .pipe(
        catchError(this.handleError('addHero', hero))
      );
  }
  /** DELETE 方法 */
  deleteHero(id: number): Observable<unknown> {
    const url = `${this.heroesUrl}/${id}`;
    return this.http.delete(url, httpOptions)
      .pipe(
        catchError(this.handleError('deleteHero'))
      );
  }
  /** PUT 方法 */
  updateHero(hero: Hero): Observable<Hero> {
    httpOptions.headers =
      httpOptions.headers.set('Authorization', 'my-new-auth-token');
    return this.http.put<Hero>(this.heroesUrl, hero, httpOptions)
      .pipe(
        catchError(this.handleError('updateHero', hero))
      );
  }
}
```

在 src\examples\interceptorexamples 目录下创建文件 http-error-handler.service.ts，代码如例 8-10 所示。

【例 8-10】 创建文件 http-error-handler. service. ts 的代码,定义类。

```
import {Injectable} from '@angular/core';
import {HttpErrorResponse} from '@angular/common/http';
import {Observable, of} from 'rxjs';
import {MessageService} from "./message.service";
/** 返回类型为 HttpErrorHandler.createHandleError */
export type HandleError =
  <T> (operation?: string, result?: T) => (error: HttpErrorResponse) => Observable<T>;
@Injectable()
export class HttpErrorHandler {
  constructor(private messageService: MessageService) {
  }
  createHandleError = (serviceName = '') =>
    <T>(operation = 'operation', result = {} as T) =>
      this.handleError(serviceName, operation, result);
  handleError<T>(serviceName = '', operation = 'operation', result = {} as T) {
    return (error: HttpErrorResponse): Observable<T> => {
      console.error(error);
      const message = (error.error instanceof ErrorEvent) ?
        error.error.message :
        `server returned code ${error.status} with body "${error.error}"`;
      this.messageService.add(`${serviceName}: ${operation} failed: ${message}`);
      return of(result);
    };
  }
}
```

在 src\examples\interceptorexamples 目录下创建文件 heroes. component. ts,代码如例 8-11 所示。

【例 8-11】 创建文件 heroes. component. ts 的代码,定义组件。

```
import {Component, OnInit} from '@angular/core';
import {Hero, HeroesService} from "./heroes.service";
@Component({
  selector: 'app-heroes',
  template: `
    <div class="search">
      <label for="hero-name">人名: </label>
      <input type="text" #heroName id="hero-name">
      <!-- (click) passes input value to add() and then clears the input -->
      <button (click)="add(heroName.value); heroName.value=''">
        增加人物
      </button>
      <button (click)="search(heroName.value)">
        搜索
      </button>
    </div>
    <ul class="heroes">
      <li *ngFor="let hero of heroes">
        <a (click)="edit(hero)">
          <span class="badge">{{ hero.id || -1 }}</span>
          <span *ngIf="hero!==editHero">{{hero.name}}</span>
          <input type="text"
```

```
                *ngIf = "hero === editHero"
                [(ngModel)] = "hero.name"
                (blur) = "update()"
                (keyup.enter) = "update()">
        </a>
        <button class = "delete" title = "delete hero"
                (click) = "delete(hero)"> x
        </button>
      </li>
    </ul>
  `,
  providers: [HeroesService],
  styles: ['/* HeroesComponent\'s private CSS styles */\n' +
  '.search input {\n' +
  '  margin: 1rem 0;\n' +
  '}\n' +
  '.heroes {\n' +
  '  list-style-type: none;\n' +
  '  padding: 0;\n' +
  '}\n' +
  '.heroes li {\n' +
  '  position: relative;\n' +
  '}\n' +
  '.heroes li:hover {\n' +
  '  left: .1em;\n' +
  '}\n' +
  '.heroes a {\n' +
  '  color: black;\n' +
  '  display: block;\n' +
  '  font-size: 1.2rem;\n' +
  '  background-color: #eee;\n' +
  '  margin: .5em 0;\n' +
  '  padding: .5em 0;\n' +
  '  border-radius: 4px;\n' +
  '}\n' +
  '.heroes a:hover {\n' +
  '  color: #2c3a41;\n' +
  '  background-color: #e6e6e6;\n' +
  '}\n' +
  '.heroes .badge {\n' +
  '  padding: .5em .6em;\n' +
  '  color: white;\n' +
  '  background-color: #435B60;\n' +
  '  min-width: 16px;\n' +
  '  margin-right: .8em;\n' +
  '  border-radius: 4px 0 0 4px;\n' +
  '}\n' +
  'button.delete {\n' +
  '  position: absolute;\n' +
  '  right: -8px;\n' +
  '  top: 5px;\n' +
  '  background-color: gray;\n' +
  '  color: white;\n' +
  '  padding: 5px 8px;\n' +
```

```
    '  width: 2em;\n' +
    '}\n' +
    '.heroes input {\n' +
    '  max-width: 12rem;\n' +
    '  padding: .25rem;\n' +
    '  position: absolute;\n' +
    '  top: 8px;\n' +
    '}\n']
})
export class HeroesComponent implements OnInit {
  heroes: Hero[] = [];
  editHero: Hero | undefined;
  constructor(private heroesService: HeroesService) {
  }
  ngOnInit() {
    this.getHeroes();
  }
  getHeroes(): void {
    this.heroesService.getHeroes()
      .subscribe(heroes => (this.heroes = heroes));
  }
  add(name: string): void {
    this.editHero = undefined;
    name = name.trim();
    if (!name) {
      return;
    }
    const newHero: Hero = {name} as Hero;
    this.heroesService
      .addHero(newHero)
      .subscribe(hero => this.heroes.push(hero));
  }
  delete(hero: Hero): void {
    this.heroes = this.heroes.filter(h => h !== hero);
    this.heroesService
      .deleteHero(hero.id)
      .subscribe();
    /*
    this.heroesService.deleteHero(hero.id);
    */
  }
  edit(hero: Hero) {
    this.editHero = hero;
  }
  search(searchTerm: string) {
    this.editHero = undefined;
    if (searchTerm) {
      this.heroesService
        .searchHeroes(searchTerm)
        .subscribe(heroes => (this.heroes = heroes));
    }
  }
  update() {
    if (this.editHero) {
```

```
        this.heroesService
          .updateHero(this.editHero)
          .subscribe(hero => {
            const ix = hero ? this.heroes.findIndex(h => h.id === hero.id) : -1;
            if (ix > -1) {
              this.heroes[ix] = hero;
            }
          });
        this.editHero = undefined;
      }
    }
  }
```

8.4.2 信息处理

在 src\examples\interceptorexamples 目录下创建文件 message.service.ts，代码如例 8-12 所示。

【例 8-12】 创建文件 message.service.ts 的代码，定义类。

```
import {Injectable} from '@angular/core';
@Injectable()
export class MessageService {
  messages: string[] = ["hello" ,'hi'];
  add(message: string) {
    this.messages.push(message);
  }
  clear() {
    this.messages = [];
  }
}
```

在 src\examples\interceptorexamples 目录下创建文件 messages.component.ts，代码如例 8-13 所示。

【例 8-13】 创建文件 messages.component.ts 的代码，定义组件。

```
import {Component} from '@angular/core';
import {MessageService} from "./message.service";
@Component({
  selector: 'app-messages',
  template: `
    <div *ngIf="messageService.messages.length">
      <div>信息</div>
      <button class="clear" (click)="messageService.clear()">清除</button>
      <br>
      <ol>
        <li *ngFor='let message of messageService.messages'> {{message}} </li>
      </ol>
    </div>
  `
})
export class MessagesComponent {
  constructor(public messageService: MessageService) {}
}
```

8.4.3 配置

在 src\examples\interceptorexamples 目录下创建文件 config. service. ts,代码如例 8-14 所示。

【例 8-14】 创建文件 config. service. ts 的代码,定义接口和类。

```
import {Injectable} from '@angular/core';
import {HttpClient} from '@angular/common/http';
import {HttpErrorResponse, HttpResponse} from '@angular/common/http';
import {Observable, throwError} from 'rxjs';
import {catchError, retry} from 'rxjs/operators';
export interface Config {
  heroesUrl: string;
  textfile: string;
  date: any;
}
@Injectable()
export class ConfigService {
  configUrl = '/examples/interceptorexamples/config.json';
  constructor(private http: HttpClient) { }
  getConfig() {
    return this.http.get<Config>(this.configUrl)
      .pipe(
        retry(3), //请求失败重试 3 次
        catchError(this.handleError)
      );
  }
  getConfigResponse(): Observable<HttpResponse<Config>> {
    return this.http.get<Config>(
      this.configUrl, { observe: 'response' });
  }
  private handleError(error: HttpErrorResponse) {
    if (error.status === 0) {
      //客户端错误或网络错误时
      console.error('An error occurred:', error.error);
    } else {
      console.error(
        `Backend returned code ${error.status}, body was: `, error.error);
    }
    return throwError(
      'Something bad happened; please try again later.');
  }
  makeIntentionalError() {
    return this.http.get('not/a/real/url')
      .pipe(
        catchError(this.handleError)
      );
  }
}
```

在 src\examples\interceptorexamples 目录下创建文件 config. component. ts,代码如例 8-15 所示。

【例 8-15】 创建文件 config. component. ts 的代码,定义组件。

```
import {Component} from '@angular/core';
import {Config, ConfigService} from './config.service';
@Component({
  selector: 'app-config',
  template: `
    <div>读取 JSON 文件的配置信息</div>
    <div>
      <button (click)="clear(); showConfig()">get</button>
      <button (click)="clear(); showConfigResponse()">getResponse</button>
      <button (click)="clear()">clear</button>
      <button (click)="clear(); makeError()">error</button>
      <span *ngIf="config">
    <p>Heroes API URL is "{{config.heroesUrl}}"</p>
    <p>Textfile URL is "{{config.textfile}}"</p>
    <p>Date is "{{config.date.toDateString()}}" ({{getType(config.date)}})</p>
    <div *ngIf="headers">
      Response headers:
      <ul>
        <li *ngFor="let header of headers">{{header}}</li>
      </ul>
    </div>
  </span>
    </div>
    <p *ngIf="error" class="error">{{error | json}}</p>
  `,
  providers: [ConfigService],
  styles: ['.error { color: #b30000; }']
})
export class ConfigComponent {
  error: any;
  headers: string[] = [];
  config: Config | undefined;
  constructor(private configService: ConfigService) {}
  clear() {
    this.config = undefined;
    this.error = undefined;
    this.headers = [];
  }
  showConfig() {
    this.configService.getConfig()
      .subscribe(
        (data: Config) => this.config = { ...data },
        error => this.error = error
      );
  }
  showConfigResponse() {
    this.configService.getConfigResponse()
      .subscribe(resp => {
        const keys = resp.headers.keys();
        this.headers = keys.map(key =>
          `${key}: ${resp.headers.get(key)}`);
        this.config = { ...resp.body! };
```

```
      });
  }
  makeError() {
    this.configService.makeIntentionalError().subscribe(null, error => this.error = error);
  }
  getType(val: any): string {
    return val instanceof Date ? 'date' : Array.isArray(val) ? 'array' : typeof val;
  }
}
```

8.4.4 上传文件

在 src\examples\interceptorexamples 目录下创建文件 uploader.service.ts，代码如例 8-16 所示。

【例 8-16】 创建文件 uploader.service.ts 的代码，定义类。

```
import {Injectable} from '@angular/core';
import {
  HttpClient, HttpEvent, HttpEventType, HttpRequest, HttpErrorResponse
} from '@angular/common/http';
import {of} from 'rxjs';
import {catchError, last, map, tap} from 'rxjs/operators';
import {MessageService} from "./message.service";
@Injectable()
export class UploaderService {
  constructor(
    private http: HttpClient,
    private messenger: MessageService) {}
  upload(file: File) {
    if (!file) {return of<string>();}
    const req = new HttpRequest('POST', '/upload/file', file, {
      reportProgress: true
    });
    return this.http.request(req).pipe(
      map(event => this.getEventMessage(event, file)),
      tap(message => this.showProgress(message)),
      last(),
      catchError(this.handleError(file))
    );
  }
  /** Return distinct message for sent, upload progress, & response events */
  private getEventMessage(event: HttpEvent<any>, file: File) {
    switch (event.type) {
      case HttpEventType.Sent:
        return `Uploading file "${file.name}" of size ${file.size}.`;
      case HttpEventType.UploadProgress:
        const percentDone = Math.round(100 * event.loaded / (event.total ?? 0));
        return `File "${file.name}" is ${percentDone}% uploaded.`;
      case HttpEventType.Response:
        return `File "${file.name}" was completely uploaded!`;
      default:
        return `File "${file.name}" surprising upload event: ${event.type}.`;
```

```
      }
    }
    private handleError(file: File) {
      const userMessage = `${file.name} upload failed.`;
      return (error: HttpErrorResponse) => {
        console.error(error);
        const message = (error.error instanceof Error) ?
          error.error.message :
         `server returned code ${error.status} with body "${error.error}"`;
        this.messenger.add(`${userMessage} ${message}`);
        return of(userMessage);
      };
    }
    private showProgress(message: string) {
      this.messenger.add(message);
    }
  }
```

在 src\examples\interceptorexamples 目录下创建文件 uploader.component.ts，代码如例 8-17 所示。

【例 8-17】 创建文件 uploader.component.ts 的代码，定义组件。

```
import {Component} from '@angular/core';
import {UploaderService} from './uploader.service';
@Component({
  selector: 'app-uploader',
  template: `
    <div>上传文件</div>
    <form enctype="multipart/form-data" method="post">
      <div>
        <label for="picked">选择一个文件上传: </label>
        <input type="file"
               id="picked"
               #picked
               (click)="message=''"
               (change)="onPicked(picked)">
      </div>
      <div *ngIf="message">{{message}}</div>
    </form>
  `,
  styles: ['input[type=file] {font-size: 1.2rem; margin-top: 1rem; display: block;}'],
  providers: [ UploaderService ]
})
export class UploaderComponent {
  message = '';
  constructor(private uploaderService: UploaderService) {}
  onPicked(input: HTMLInputElement) {
    const file = input.files?.[0];
    if (file) {
      this.uploaderService.upload(file).subscribe(
        msg => {
          input.value = '';
          this.message = msg;
        }
```

```
        );
      }
    }
  }
```

在 src\examples\interceptorexamples 目录下创建文件 upload-interceptor.ts，代码如例 8-18 所示。

【例 8-18】 创建文件 upload-interceptor.ts 的代码，定义类。

```
import {Injectable} from '@angular/core';
import {
  HttpEvent, HttpInterceptor, HttpHandler,
  HttpRequest, HttpResponse,
  HttpEventType, HttpProgressEvent
} from '@angular/common/http';
import {Observable} from 'rxjs';
/** Simulate server replying to file upload request */
@Injectable()
export class UploadInterceptor implements HttpInterceptor {
  intercept(req: HttpRequest<any>, next: HttpHandler): Observable<HttpEvent<any>> {
    if (req.url.indexOf('/upload/file') === -1) {
      return next.handle(req);
    }
    const delay = 300;
    return createUploadEvents(delay);
  }
}
/** Create simulation of upload event stream */
function createUploadEvents(delay: number) {
  const chunks = 5;
  const total = 12345678;
  const chunkSize = Math.ceil(total / chunks);
  return new Observable<HttpEvent<any>>(observer => {
    observer.next({type: HttpEventType.Sent});
    uploadLoop(0);
    function uploadLoop(loaded: number) {
        setTimeout(() => {
          loaded += chunkSize;
          if (loaded >= total) {
            const doneResponse = new HttpResponse({
              status: 201,
            });
            observer.next(doneResponse);
            observer.complete();
            return;
          }
          const progressEvent: HttpProgressEvent = {
            type: HttpEventType.UploadProgress,
            loaded,
            total
          };
          observer.next(progressEvent);
          uploadLoop(loaded);
        }, delay);
```

```
    }
  });
}
```

在 src\examples\interceptorexamples 目录下创建文件 index. ts,代码如例 8-19 所示。

【例 8-19】 创建文件 index. ts 的代码,定义常量。

```
import {HTTP_INTERCEPTORS} from '@angular/common/http';
import {UploadInterceptor} from "./upload-interceptor";
export const httpInterceptorProviders = [
  {provide: HTTP_INTERCEPTORS, useClass: UploadInterceptor, multi: true},
];
```

8.4.5 组件、模块和运行结果

在 src\examples\interceptorexamples 目录下创建文件 interceptorhome. component. ts,代码如例 8-20 所示。

【例 8-20】 创建文件 interceptorhome. component. ts 的代码,定义组件。

```
import {Component} from '@angular/core';
@Component({
  selector: 'root',
  template: `
    <div>
      <input type="checkbox" id="heroes" [checked]="showHeroes" (click)="toggleHeroes()">
      <label for="heroes">人物</label>
      <input type="checkbox" id="config" [checked]="showConfig" (click)="toggleConfig()">
      <label for="config">配置</label>
      <input type="checkbox" id="uploader" [checked]="showUploader" (click)="
toggleUploader()">
      <label for="uploader">上传文件</label>
    </div>
    <hr/>
    <app-heroes *ngIf="showHeroes"></app-heroes>
    <hr/>
    <app-messages></app-messages>
    <hr />
    <app-config *ngIf="showConfig"></app-config>
    <hr />
    <app-uploader *ngIf="showUploader"></app-uploader>
    `
})
export class InterceptorhomeComponent {
  showHeroes = true;
  showConfig = true;
  showUploader = true;
  toggleUploader() {this.showUploader = !this.showUploader;}
  toggleHeroes() {this.showHeroes = !this.showHeroes;}
  toggleConfig() {this.showConfig = !this.showConfig;}
}
```

在 src\examples\interceptorexamples 目录下创建文件 app-interceptor-example. module. ts,代码如例 8-21 所示。

【例 8-21】 创建文件 app-interceptor-example. module. ts 的代码,声明组件并导入块等操作。

```
import {BrowserModule} from '@angular/platform-browser';
import {NgModule} from '@angular/core';
import {HttpClientModule} from "@angular/common/http";
import {RouterModule} from "@angular/router";
import {HeroesComponent} from "./heroes.component";
import {InterceptorhomeComponent} from "./interceptorhome.component";
import {FormsModule} from "@angular/forms";
import {MessagesComponent} from "./messages.component";
import {MessageService} from "./message.service";
import {HttpErrorHandler} from "./http-error-handler.service";
import {httpInterceptorProviders} from "./index";
import {ConfigComponent} from "./config.component";
import {UploaderComponent} from "./uploader.component";
@NgModule({
  declarations: [
    HeroesComponent,
    InterceptorhomeComponent,
    MessagesComponent,
    ConfigComponent,
    UploaderComponent
  ],
  imports: [
    BrowserModule,
    HttpClientModule,
    RouterModule.forRoot([
      {path: 'users', component: InterceptorhomeComponent},
    ]),
    FormsModule
  ],
  providers: [
    HttpErrorHandler,
    MessageService,
    httpInterceptorProviders
  ],
})
export class AppInterceptorModule {
}
```

在 src\examples\interceptorexamples 目录下创建文件 config. json,代码如例 8-22 所示。

【例 8-22】 创建文件 config. json 的代码,配置信息。

```
{
  "heroesUrl": "api/heroes",
  "textfile": "assets/textfile.txt",
  "date": "2020-01-29"
}
```

修改 src\examples 目录下的文件 examplesmodules1. module. ts,代码如例 8-23 所示。

【例 8-23】 修改文件 examplesmodules1. module. ts 的代码,放置启动组件。

```
import {NgModule} from '@angular/core';
import {InterceptorhomeComponent} from './interceptorexamples/interceptorhome.component';
```

```
import {AppInterceptorModule} from "./interceptorexamples/app-interceptor-example.module";
@NgModule({
  imports: [
    AppInterceptorModule
  ],
  bootstrap: [InterceptorhomeComponent]
})
export class ExamplesmodulesModule1 {}
```

保持其他文件不变并成功运行程序后，在浏览器地址栏中输入 localhost:4200，自动跳转到 localhost:4200/name-list；也可以在浏览器地址栏中输入 localhost:4200/users，结果如图 8-2 所示。

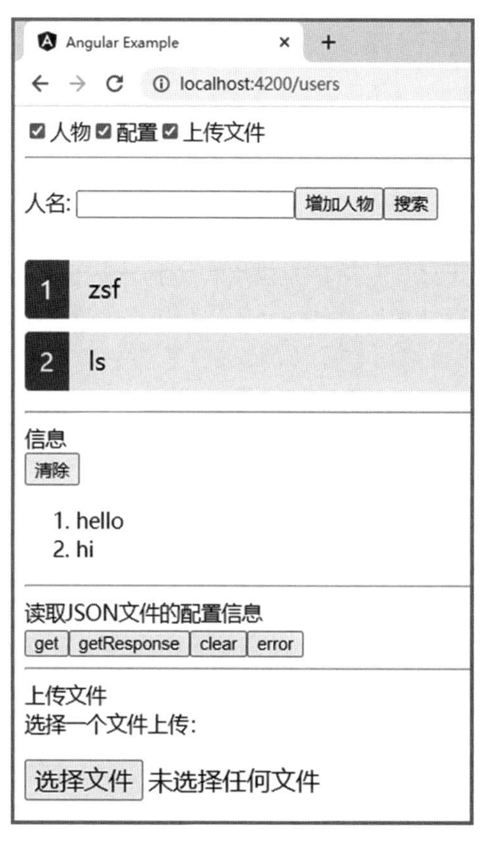

图 8-2 成功运行程序后在浏览器地址栏中输入 localhost:4200 的结果

习题 8

一、简答题

1. 简述对 HTTP 客户端服务的理解。
2. 简述对拦截器的理解。

二、实验题

1. 完成 HTTP 客户端服务的应用开发。
2. 完成拦截器的应用开发。

第 9 章

国际化及其应用

9.1 国际化概述

9.1.1 国际化的含义和实现

国际化(Internationalization,或称为 i18n)是设计和准备应用程序时以便在世界各地的不同语言环境使用的过程,而本地化是为应用程序构建适合本地语言环境的版本的过程。本地化过程包括提取文本以翻译成不同的语言、格式化特定语言环境的数据、语言环境标识(用于标识人们使用特定语言或语言变体的区域)等环节。可能的区域包括国家和地理区域。语言环境的不同决定了度量单位、日期和时间、数字、货币、翻译名称、语言和国家等详细信息的格式和解析存在差异。可以使用内置管道方法以本地格式显示日期、数字、百分比和货币等;可以在组件模板中对标签文本以进行翻译;可以标记出要翻译的表达式的复数形式;对标签替代文本进行翻译。

国际化的实现步骤如下。

(1) 为每种语言制作一份源语言文件的副本,并将所有文件进行翻译。在为一种或多种语言环境构建应用程序时,合并已完成的翻译文件。

(2) 为所有目标语言环境创建一个强适应性的用户界面,要考虑到不同语言间的差异。

(3) 要利用 Angular 的本地化功能,可以将@angular/localize 包添加到应用程序中。

9.1.2 通过 ID 引用语言环境

应用程序使用 Unicode 语言环境标识符(简称为语言环境 ID)来查找正确的语言环境数据,以实现文本字符串的国际化。语言环境 ID 遵循 Unicode 通用本地语言环境数据仓库(Common Locale Data Repository,CLDR)核心规范。CLDR 和 Angular 以 IETF BCP 47 标准(由匹配语言标签和识别语言标签组成,支持 Unicode 本地语言环境数据标记语言 LDML)标签作为语言环境 ID 的基础。语言环境 ID 可以指定语言、国家/地区和其他变体或细分的可选代码。语言环境 ID 由语言标识符、破折号(-)字符和语言环境扩展组成。为了准确地翻译 Angular 项目,必须选择国际化目标的语言和地区。许多国家(或地区)使用相同的语言,但用法上有些差异。这些差异包括语法、标点符号、货币格式、十进制数字、日期等。默认情况下,Angular 使用 en-US 作为项目的源语言环境。

Angular 提供了内置的数据转换管道。数据转换管道会使用 LOCALE_ID 标识符来根据每个语言环境的规则来格式化数据。可以利用管道 DatePipe 对日期值进行格式化；或利用管道 CurrencyPipe 将数字转换为货币字符串，也可以用管道 DecimalPipe 将数道转换成十进制数字字符串，还可以用管道 PercentPipe 将数值转换成百分比字符串。

9.2 翻译

9.2.1 翻译模板

应用程序中翻译内容的准备步骤：使用 i18n 属性标记组件模板中的文本；使用 i18n 属性在组件模板中标记属性文本字符串；使用带 $localize 标记的消息字符串标记组件代码中的文本字符串。

i18n 属性是供 Angular 工具和编译器识别的自定义属性。在组件模板中，i18n 元数据就是 i18n 属性的值。使用 i18n 属性在组件模板中标记静态文本消息以进行翻译。将它放在每个包含要翻译的固定文本的元素标签上。使用< ng-container >元素来为特定文本关联翻译行为，而不会改变文本的显示方式。每个 HTML 元素都会创建一个新的 DOM 元素。要想避免创建新的 DOM 元素，可以将文本包裹在< ng-container >元素中。在组件模板中，i18n 的元数据是 i18n-{attribute_name}属性的值，{attribute_name}为属性的名称。HTML 元素的属性包括那些要和组件模板中显示的其他文本一起翻译的文本。

在组件代码中，如果要翻译的源文本和元数据被反引号(`)字符包围，则可以使用 $localize 标记的消息字符串在代码中标记出要翻译的字符串。在 $localize 标记的消息字符串中可以包含插值文本。i18n 元数据{i18n_metadata}包裹在两个冒号(:)字符中，并放在翻译源文本之前。

将自定义 ID 与描述(description)、含义(meaning)结合使用，以进一步帮助翻译。自定义 ID 提供自定义标识符，用于描述提供额外的信息或背景。含义提供文本在特定上下文中的含义或意图。要准确翻译文本消息，就要为翻译人员提供额外信息或上下文。为 i18n 属性的值或 $localize 标记的消息字符串添加文本消息的描述。翻译时可能还需要了解应用上下文中文本消息的含义或意图，以便以与具有相同含义的其他文本相同的方式对其进行翻译。把含义放在 i18n 属性值最前面，并用“|”字符将其与描述分开，如{meaning}|{description}。

9.2.2 翻译方法

Angular 提取工具会为模板中的每个 i18n 属性生成一个翻译单元条目；会根据含义和描述为每个翻译单元分配一个唯一的 ID；具有不同含义的相同文本元素会以不同的 ID 提取。例如，如果单词 right 在两个不同的位置有两个含义(“正确”“右边”)，则该单词将被以不同的方式翻译并作为不同的翻译条目。如果相同的文本元素满足含义或定义相同、描述不同的条件，则只会提取一次文本元素并使用相同的 ID。只要出现相同的文本元素，该翻译条目就会合并回应用程序。

ICU(International Components for Unicode)表达式可以在组件模板中标记出某些条件下的替代文本。ICU 表达式包括一个组件属性、一个 ICU 子句以及由左花括号({)和右花括号(})字符包围的 case 语句。组件属性定义了变量，而 ICU 子句定义了条件文本的类型。其中,plural 子句标记复数的使用,select 子句根据定义的字符串值标记出替代文本的一些选择。为了简化翻译,可以使用带有正则表达式的 Unicode 子句(ICU 子句)的国际化组件。ICU 子句遵循 Unicode 通用本地语言环境数据仓库(CLDR)复数规则中指定的 ICU 消息格式。

不同的语言有不同的复数规则,这增加了翻译的难度。因为其他语言环境表达基数的方式不同,可能需要设置与英语不一致的复数类别。当使用 plural 子句来标记逐字翻译时可能没有意义的。在复数类别之后,输入由左花括号({)和右花括号(})包围的默认文本(英文)。zero、one、two、few、many、other 等复数类别适用于英语,可能需要根据语言环境而变化。如果不能匹配任何复数类别,Angular 就会使用 other 来匹配缺失类别的标准后备值。许多语言环境不支持某些复数类别,如默认的语言环境(en-US)使用一个非常简单的 plural 子句,不支持 few 复数类别;另一个具有简单 plural 子句的语言环境是 es。

select 子句根据定义的字符串值标记替代文本的选择。翻译所有替代项以根据变量的值显示替代文本。在选择类别后,输入由左花括号({)和右花括号(})字符包围的文本(英文)。不同的语言环境具有不同的语法结构,这增加了翻译的难度。

9.2.3 翻译文件

准备好要翻译的内容后,使用 Angular CLI 的 extract -i18n 命令将组件中的标记文本提取到源语言文件中。已标记的文本包括标记为 i18n 的文本、标记为 i18n-{attribute_name}的属性和标记为 $localize 的文本。extract-i18n 命令在项目的根目录中创建一个名为 messages. xlf 的源语言文件。

创建和更新翻译文件的步骤包括提取源语言文件,可能需要更改位置、格式和名称;复制源语言文件以便为每种语言创建一个翻译文件;翻译每个翻译文件;分别翻译复数和替代表达式。其中,翻译复数和替代表达式包括翻译复数、翻译替代表达式、翻译嵌套表达式等工作。其中,提取源语言文件的操作步骤又包括打开终端窗口;切换到应用程序根目录;运行 Angular CLI 的 extract-i18n 命令。使用 extract-i18n 命令选项可以更改源语言文件位置、格式和文件名。其中,format 选项用于设置输出文件的格式,outFile 选项用于设置输出文件的名称,output-path 选项用于设置输出目录的路径。format 选项的文件格式包括应用资源包 ARB 文件、JSON 文件、XML 本地化交换文件格式文件(XLIFF 1.2 版和 XLIFF 2 版)、XML 消息包(XMB)文件。XMB 格式生成. xmb 扩展名的源语言文件,但生成. xtb 扩展名的翻译文件。

9.3 将翻译结果合并到应用中

1. 构建

要将完成的翻译结果合并到应用中,可以使用 localize 选项将所有 i18n 消息替换为有效的翻译并构建本地化的应用变体。应用变体就是为单个语言环境翻译的应用的可分发文件的完整副本。合并翻译后,可使用服务器端语言检测或不同的子目录来提供 Angular 应用的每个可分发副本。

对于应用的编译期转换,构建过程会使用预先(Ahead-Of-Time,AOT)编译来生成小型、快速、可立即运行的应用。构建过程适用于.xlf格式或Angular能理解的另一种格式的翻译文件，例如.xtb。

要为每个语言环境构建应用的单独可分发副本，在项目配置文件angular.json中可以构建配置定义语言环境。此方法不需要为每个语言环境执行完整的应用构建，从而缩短了构建过程。

要为每个语言环境生成应用变体，使用项目配置文件angular.json中的localize选项以表明语言环境标识符到翻译文件的映射表。要从命令行开始构建应用程序，可以使用带有localize选项的build命令。若将localize设置为先前定义的语言环境标识符子集的数组，则可以单独构建语言环境版本；若将localize设置为false，则可以禁用本地化并且不生成任何特定于语言环境的版本。由于i18n部署的复杂性和最小化重建时间的需要，开发服务器一次仅支持本地化单个语言环境。如果将localize选项设置为true，定义了多个语言环境，并使用ng serve命令，就会发生错误。如果要针对特定语言环境进行开发，可以将localize选项设置为特定的语言环境。

2. 使用

Angular CLI加载并注册语言环境数据，将每个生成的版本放置在特定语言环境的目录中以使其与其他语言环境版本分开，并将其目录放在为此应用程序项目配置的outputPath中。对于每个应用变体，将HTML元素的lang属性设置为其语言环境。Angular CLI还通过将语言环境添加到所配置的baseHref中来调整每个应用版本的HTML baseHref。将localize属性设置为共享配置以有效继承所有配置。此外，会将该属性设置为覆盖其他配置。

当缺少翻译时，可以构建成功但会生成警告。为此，可以配置Angular编译器生成的警告级别，如error(抛出错误，构建失败)、ignore(什么也不做)或warning(在控制台或shell中显示默认警告)。

将源代码编译为Angular应用时，i18n属性的实例将被替换为$localize标记的消息字符串的实例。这意味着应用程序会在编译后被翻译。这也意味着可以创建Angular应用的本地化版本，而无须为每个语言环境重新编译整个Angular项目。在翻译Angular应用时，翻译转换会用翻译集合中的字符串替换和重新排序模板文字字符串的内容(静态字符串和表达式)。

3. 部署

如果myapp是包含应用程序项目可分发文件的目录，通常会在语言环境目录中为不同的语言环境提供不同的版本，例如法语版的myapp/fr和西班牙语版的myapp/es。

带有href属性的HTML base标签指定了相对链接的基本URI或URL。如果将配置文件angular.json中的localize选项设置为true或特定语言环境ID数组，Angular CLI会为应用的每个版本调整base标签的href属性。若要为应用的每个版本调整base标签的href属性，则Angular CLI会将语言环境添加到配置的baseHref中。若在配置文件angular.json中为每个语言环境指定baseHref，则要在编译时声明base Href，可以使用带有--baseHref选项的ng build命令。

多语言的典型部署方式是为来自不同子目录的每种语言提供服务。使用HTTP标头Accept-Language将用户重定向到浏览器中定义的首选语言。如果用户未定义首选语言，或者首选语言不可用,则服务器将回退到默认语言。如果要更改语言,就要转到另一个子目录。子目录的更改通常应用在应用程序的菜单中。

9.4 可选的国际化实践

1. ID

@angular/common 包中包含语言环境数据文件。语言环境数据的全局变体来自@angular/common/locales/global。Angular 的初始安装中已经包含了美国英语(en-US)的语言环境数据(即默认的语言环境)。将--localize 选项与 ng build 命令一起使用时,Angular CLI 会自动包含特定语言环境数据并设置 LOCALE_ID 值。

除了要将应用程序运行时的语言环境手动设置成默认值以外,还有另一种语言环境的设置步骤:在 Angular 代码仓库中的 language-locale 环境组合中搜索 Unicode 语言环境 ID;设置 LOCALE_ID 令牌。Angular 提取器会生成一个文件,其中包含组件模板中的每个 i18n 属性、组件代码中每个 $localize 标记的消息字符串等每个实例的翻译单元条目。更改可翻译文本时,提取器会为该翻译单元生成一个新 ID。在大多数情况下,源文本中的更改还需要更改翻译结果。因此,使用新 ID 可使文本更改与翻译保持同步。

2. 自定义 ID

某些翻译系统需要用到 ID 的特定形式或语法。要满足此要求,可以使用自定义 ID 来标记文本。大多数开发人员不需要使用自定义 ID。如果想使用独特的语法来传达额外的元数据,可以使用自定义 ID。其他元数据可能会出现在文本的库、组件或应用程序的区块中。要在 i18n 属性或以 $localize 标记的消息字符串中指定自定义 ID 时,要使用@@前缀。如果更改文本,那么提取器不会更改 ID。这导致不得不用额外的步骤来更新其翻译。使用自定义 ID 的缺点是在更改文本后进行的翻译可能与新更改的源文本不同步。

如果对两个不同的文本元素使用相同的 ID,则提取工具只会提取第一个,而且 Angular 会使用其翻译来代替两个原始文本元素。此时的这两个原始文本元素进行相同的翻译处理,因为它们都是使用同一个自定义 ID 定义的。

微课视频

9.5 国际化应用

9.5.1 服务和管道

在项目 src\examples 根目录下创建 i18nexamples 子目录,在 src\examples\i18nexamples 目录下创建文件 translate.service.ts,代码如例 9-1 所示。注意,在 assets 目录下准备 en.json、zh_cn.json 文件,文件内容和例 9-4、例 9-5 对应。

【例 9-1】 创建文件 translate.service.ts 的代码,定义类。

```
import {Injectable} from '@angular/core';
import {HttpClient} from '@angular/common/http';
import {ENGLISH_CONFIG} from './english.json';
import {JAPAN_CONFIG} from './japan.json';
import {CHINESE_CONFIG} from './chinese.json';
@Injectable({
```

```
  providedIn: 'root'
})
export class TranslateService {
  data: any = {};
  constructor(private http: HttpClient) {
  }
  //无法有效读取 JSON 文件
  use(lang: string): Promise<{}> {
    return new Promise<{}>((resolve, reject) => {
      const langPath = `assets/i18n/${lang || 'en'}.json`;
      this.http.get(langPath).subscribe(
        translation => {
          this.data = Object.assign({}, translation || {});
          resolve(this.data);
          console.log("success")
        },
        error => {
          this.data = {};
          resolve(this.data);
        }
      );
    });
  }
  useMessage(lang: string) {
    switch(lang) {
      case 'en':
        return ENGLISH_CONFIG.TITLE;
      default:
        return CHINESE_CONFIG.TITLE;
    }
  }
}
```

在 src\examples\i18nexamples 目录下创建文件 translate. pipe. ts，代码如例 9-2 所示。

【例 9-2】 创建文件 translate. pipe. ts 的代码，定义类。

```
import {Pipe, PipeTransform} from '@angular/core';
import {TranslateService} from './translate.service';
@Pipe({
  name: 'translate',
  pure: false
})
export class TranslatePipe implements PipeTransform {
  constructor(private translate: TranslateService) {
  }
  transform(key: any): any {
    return this.translate.data[key] || key;
  }
}
```

9.5.2 组件

在 src\examples\i18nexamples 目录下创建文件 i18nexample. component. ts，代码如例 9-3 所示。

【例 9-3】 创建文件 i18nexample.component.ts 的代码,定义组件。

```
import {Component} from '@angular/core';
import {TranslateService} from "./translate.service";
@Component({
  selector: 'root',
  template: `
    <!-- The content below is only a placeholder and can be replaced. -->
    <div style="text-align:center">
      <h1>{{ message |translate }}</h1>
      <img width="300" alt="Angular Logo"
src="data:image/svg+xml;base64,PHN2ZyB4bWxucz0iaHR0cDovL3d3dy53My5vcmcvMjAwMC9zdmciIHZp
ZXdCb3g9IjAgMCAyNTAgMjUwIj4KICAgIDxwYXRoIGZpbGw9IiNERDAwMzEiIGQ9Ik0xMjUgMzBMMzEuOSA2My4yb
DE0LjIgMTIzLjFMMTI1IDIzMGw3OC45LTQzLjcgMTQuMi0xMjMuMXoiIC8+CiAgICA8cGF0aCBmaWxsPSIjQzMwM
DJGIiBkPSJNMTI1IDMwdjIyLjItLjFWMjMwbDc4LjktNDMuNyAxNC4yLTEyMy4xTDEyNSAzMHoiIC8+CiAgICA8c
GF0aCAgZmlsbD0iI0ZGRkZGRiIgZD0iTTEyNSA1Mi4xTDY2LjggMTgyLjZoMjEuN2wxMS43LTI5LjJoNDkuNGwxMS
43IDI5LjJIMTgzTDEyNSA1Mi4xem0xNyA4My4zaC0zNGwxNy00MC45IDE3IDQwLjl6IiAvPgogIDwvc3ZnPg==">
    </div>
    <div>
      <button (click)="setLangMessage('en')">英语标题</button>
      <button (click)="setLangMessage('zh_cn')">汉语标题</button>
    </div>
    <div>开始学习的学习资源:</div>
    <ul>
      <li>
<div><a target="_blank" rel="noopener" href="https://angular.io/tutorial">Tour of
Heroes</a></div>
      </li>
      <li>
<div><a target="_blank" rel="noopener" href="https://angular.io/cli">CLI Documentation
</a></div>
      </li>
      <li>
        <div><a target="_blank" rel="noopener" href="https://blog.angular.io/">
Angular blog</a></div>
      </li>
    </ul>
    <!-- router-outlet></router-outlet -->
  `,
  styles: ['']
})
export class I18nexampleComponent {
  message = ""
  constructor(private translate: TranslateService) {
  }
  setLangMessage(lang: string) {
    this.message = this.translate.useMessage(lang);
  }
  setLang(lang: string) {
    this.translate.use(lang);
  }
}
```

9.5.3 国际化文本内容

在 src\examples\i18nexamples 目录下创建文件 english. json. ts，代码如例 9-4 所示。

【例 9-4】 创建文件 english. json. ts 的代码，定义英文字符串常量。

```
export const ENGLISH_CONFIG  = {
  "TITLE": "i18n Application (EN)"
}
```

在 src\examples\i18nexamples 目录下创建文件 chinese. json. ts，代码如例 9-5 所示。

【例 9-5】 创建文件 chinese. json. ts 的代码，定义中文字符串常量。

```
export const CHINESE_CONFIG  = {
    "TITLE": "国际化应用(中文)"
}
```

9.5.4 模块和运行结果

在 src\examples\i18nexamples 目录下创建文件 app-i18n-example. module. ts，代码如例 9-6 所示。

【例 9-6】 创建文件 app-i18n-example. module. ts 的代码，声明组件并导入模块等操作。

```
import {BrowserModule} from '@angular/platform-browser';
import {APP_INITIALIZER, NgModule} from '@angular/core';
import {RouterModule} from "@angular/router";
import {I18nexampleComponent} from "./i18nexample.component";
import {FormsModule} from "@angular/forms";
import {TranslateService} from "./translate.service";
import {TranslatePipe} from "./translate.pipe";
export function setupTranslateFactory(service: TranslateService): Function {
  return () => service.use('en');
}
@NgModule({
  declarations: [
    I18nexampleComponent,
    TranslatePipe,
  ],
  imports: [
    BrowserModule,
    RouterModule.forRoot([
      {path: 'i18n', component: I18nexampleComponent},
    ]),
    FormsModule,
  ],
  providers: [
    TranslateService,
    {
      provide: APP_INITIALIZER,
      useFactory: setupTranslateFactory,
      deps: [TranslateService],
      multi: true
```

```
    }
  ],
})
export class AppI18nExampleModule {
}
```

修改 src\examples 目录下的文件 examplesmodules1. module. ts，代码如例 9-7 所示。

【例 9-7】 修改文件 examplesmodules1. module. ts 的代码，设置启动组件。

```
import {NgModule} from '@angular/core';
import {AppI18nExampleModule} from "./i18nexamples/app-i18n-example.module";
import {I18nexampleComponent} from "./i18nexamples/i18nexample.component";
@NgModule({
  imports: [
    AppI18nExampleModule,
    AppInterceptorModule,
  ],
  bootstrap: [I18nexampleComponent]
})
export class ExamplesmodulesModule1 {}
```

保持其他文件不变并成功运行程序后，在浏览器地址栏中输入 localhost:4200，自动跳转到 localhost:4200/name-list；也可以在浏览器地址栏中输入 localhost:4200/i18n，结果如图 9-1 所示。

图 9-1 成功运行程序后在浏览器地址栏中输入 localhost:4200 的结果

习题 9

一、简答题

简述国际化的理解。

二、实验题

完成国际化的应用开发。

第10章

动画及其应用

10.1 动画概述

动画用于提供运动的幻觉：让用户在使用应用程序时认为 HTML 元素是随着时间改变而改变的。精心设计的动画可以让应用程序更有趣，更易用，提升用户体验，而不仅仅是装饰性的。动画使得 Web 页面的转场不会显得突兀和不协调；动画可以让用户察觉到应用程序对他们的操作作出了快速响应；良好的动画可以直观地把用户的注意力吸引到要留意的地方；典型的动画会涉及多种元素随时间变化而变化的转换。例如，HTML 元素可以移动、变换颜色、增加或缩小、隐藏或从页面中滑出。这些变化可以同时发生或按照顺序发生，还可以控制每次转换的持续时间。

Angular 动画系统是基于 CSS 功能构建的，这意味着可以变动浏览器认为可动的任何属性，包括位置、大小、变形、颜色、边框等。W3C 的 CSS Transitions(转场)页中维护了可动属性的列表。Angular 主要动画模块是@angular/animations 和@angular/platform-browser。使用 Angular CLI 创建应用程序时，这些依赖会自动添加到项目中。

为了把 Angular 动画添加到项目中，需要把与动画相关的模块和标准的 Angular 功能一起导入进来。首先，启用动画模块。导入 BrowserAnimationsModule，它能把动画能力引入 Angular 应用的根模块中。其次，把动画功能导入组件文件中。如果准备在组件文件中使用特定的动画函数，那么可从@angular/animations 中导入这些函数。最后，添加动画的元数据属性。在组件的@Component()装饰器中，添加一个名叫 animations 的元数据属性。可以把用来定义动画的引发器放进 animations 元数据属性中。

要想创建可复用的动画，使用 animation()方法可以在独立的.ts 文件中定义动画，并把该动画的定义声明为一个导出的 const 变量。然后就可以在应用的组件代码中通过 useAnimation()方法来导入并复用它了。

10.2 转场动画

10.2.1 转场动画含义和实现

在 HTML 中，颜色、透明度(opacity)都使用普通的 CSS 样式。可以设置多个样式而不必用动画。如果没有进一步细化，按钮的转换会立即完成，没有渐隐、没有收缩，也没有其他的可视化效果来指出正在发生的变化。要让这些变化不那么突兀,就要定义一个转场动画来要求这些状态之间的变化在一段时间内发生。Angular 应用程序中可以使用 style()函数来指定一组与指定的状态名相关的用作动画的 CSS 样式。名称里带中线的样式属性必须是小驼峰格式的,或者把它们包裹到引号里。使用 Angular 的 state()函数可以定义不同的状态,供每次转场结束时调用。state()函数接受一个唯一的名字和一个函数 style()两个参数。

transition()函数接受两个参数:第一个参数接受一个表达式,它定义两个转场状态之间的方向;第二个参数接受一个或多个 animate()函数。使用 animate()函数可以定义长度、延迟和缓动效果,并指定一个样式函数,以定义转场过程中的样式;为多步动画定义函数 keyframes()。

animate()函数可以接受 timings 和 styles 参数。timings 参数可以接受一个数字或由三部分组成的字符串:第一部分持续时间(duration)是必需的。这个持续时间可以表示成一个不带引号的纯数字(表示毫秒,如 100),或一个带引号的有单位的时间(表示秒数,如'0.1s')。第二部分延时(delay)的语法与持续时间一样。第三部分缓动速度(easing)用于控制动画在运行期间如何进行加速和减速。例如,ease-in 表示动画开始时很慢,然后逐渐加速。

10.2.2 触发器

动画需要触发器,以便知道该在何时开始执行。trigger()函数会把一些状态和转场组合在一起,并为这个动画命名,这样就可以在 HTML 模板中把它附加到想要触发动画的元素上了。trigger()函数描述了监听变化时要使用的触发器名称。当这个触发器名称所绑定的值发生了变化时,触发器就会启动它所定义的操作。这些操作可能是转场,也可能是其他功能。

控制 HTML 元素如何运动的动画是在组件的元数据中定义的。要在@Component()装饰器的 animations 属性下用代码定义要用的动画。为组件定义好这些动画触发器之后,可以给触发器名称加上@前缀并包在方括号里,把它附加到组件模板中的元素上。然后,使用 Angular 标准属性绑定语法可以把这个触发器绑定到模板表达式上。如< div [@triggerName] ="expression">...</div >;中 triggerName 是触发器的名称,expression 求值结果是前面定义过的动画状态之一。当该表达式的值变成新的状态时,动画就会执行或者叫触发。对于进入或离开页面的元素(插入 DOM 中或从中移除),可以让动画变成有条件的触发。例如,在 HTML 模板中可以和 * ngIf 一起使用动画触发器。

10.2.3 转场状态

转场状态可以通过 state()函数进行显式定义,或使用预定义的 * (通配符)状态和 void 状态。星号(*)可以匹配任何一个动画状态(包括 void)。它可用来定义不用在乎 HTML 元素的起始状态或结束状态的转场动画。如转场 open => * 可应用在当元素的状态从 open 变成任何其他状态的场景。当一个特定状态下的元素可能变更为多个潜在状态时,通配符状态会更好用。当动画在任意两个状态之间切换时,转场 * => * 都会生效。转场会按照其定义的顺序进行匹配。因此,可以把那些更特殊(具体)的转场放在转场 * => * 前面。

使用带样式的通配符以使动画使用当前的状态值，并用它进行动画处理。通配符是一个后备值，如果未在触发器中声明动画状态，就会使用这个值。使用 void 状态可以为进入或离开页面的元素配置转场动画。当元素离开视图时，就会触发转场 * => void（称为：leave 转场）；当元素进入视图时，就会触发转场 void => *（称为：enter 转场）。

元素进入或离开视图等价于从 DOM 中插入或删除元素。可以使用：enter 转场和：leave 转场来定位要从视图中插入或删除的 HTML 元素。当任何 *ngIf 或 *ngFor 中的视图进入页面中时，会执行：enter 转场；当移除这些视图时，就会执行：leave 转场动画。

10.2.4　触发机制

如果某个触发器以逻辑型的值作为绑定值，那么就可以使用能与 true 和 false 或 1 和 0 相比较的 transition()函数表达式来匹配这个值。可以为组件定义多个动画触发器并将这些动画触发器附着到不同的元素上，这些元素之间的父子关系会影响动画的运行方式和时机。

在每次触发动画时，父动画始终会优先触发，而子动画会被阻塞。为了运行子动画，父动画必须查询出包含子动画的每个元素，然后使用 animateChild()函数来运行它们。可以把一个名叫@.disabled 的动画控制绑定放在 HTML 元素上，以禁用该元素及其子元素上的动画。无法有选择地单独禁用单个元素上的多个动画。

选择性的子动画可以用两种方式在已禁用的父元素上运行：父动画可以用 query()函数来收集 HTML 模板中位于禁止动画区域内部的元素，而这些元素仍然可以播放动画；子动画可以被父动画查询并在稍后使用 animateChild()函数来播放它。

当动画启动和终止时，trigger()函数会发出一些回调。动画回调的潜在用途之一是用来覆盖比较慢的 API 调用（如查阅数据库）。例如，可以建立一个 InProgress 按钮并让它拥有自己的循环动画；当后端系统操作完成时，它会播放脉动效果或其他一些视觉动作；再在当前动画结束时可以调用另一个动画。

keyframe()函数类似于 CSS 的关键帧。关键帧允许在单个时间段内进行多种样式更改。关键帧包括一个用来定义动画中每个样式何时开始更改的偏移（offset）属性。偏移是个 0～1 的相对值，分别标记动画的开始和结束时间；只要使用了 offset 属性，就要同样应用于这个关键帧的每个步骤。定义关键帧的偏移量是可选的，如果省略它们，就会自动分配均匀间隔的偏移。通过在整个动画中定义特定偏移处理的样式，可以使用关键帧在动画中创建脉动效果。

query()函数允许查找正在播放动画的元素内部的元素。此函数会针对父组件中的特定 HTML 元素并把动画单独应用于其中的每个子元素上。Angular 会智能地处理初始化、收尾和清理工作，因为它负责协调页面中的这些元素。stagger()函数允许定义每个查询出的动画条目之间的时间间隔，从而让这些条目动画彼此错开一定的延迟。

若想配置一些并行的动画，如为同一个元素的两个 CSS 属性设置使用不同的 easing()函数时，可以使用 group()函数。group()函数用于对动画步骤进行分组。它不是针对动画元素。

复杂动画中可以同时发生很多事情。要创建一个需要让几个子动画逐个执行的父动画时，可以使用 group()函数来同时并行运行多个动画。sequence()函数可以一个接一个地执行这些动画，动画步骤由 style()或 animate()函数的函数调用组成。

Angular 中这些用于多元素动画的函数，都要从 query()函数开始。stagger()、group()和 sequence()函数等会以级联方式或自定义逻辑来控制要如何执行多个动画步骤。

10.3 路由转换动画

为这种路由转换添加动画，将极大地提升用户体验。Angular 路由器带有高级动画功能，它可以为路由变化时在视图之间设置转场动画。要想在路由转换时生成动画序列，需要首先定义出嵌套的动画序列。从宿主视图的顶层组件开始，在这些内嵌视图的宿主组件中嵌套添加其他动画。在转场期间，新视图将直接插入在旧视图后面，并且这两个元素会同时出现在屏幕上。要防止这种同时出现的情况发生，就要修改宿主视图，改用相对定位。然后，把已移除或已插入的子视图改用绝对定位。在这些视图中添加样式，就可以让容器就地播放动画，并防止某个视图影响页面中其他视图的位置。

启用路由转场动画需要的步骤包括：为应用程序导入路由模块，并创建一个路由配置来定义可能的路由。添加路由器出口告诉路由器要把激活的组件放在 DOM 中的什么位置。在 AppModule(根模块)中使用 RouterModule. forRoot()方法来注册一些顶层应用路由和提供者。对于特性模块，则改用 RouterModule. forChild()方法。配置好路由之后，还要告诉路由器当路由匹配时，要把视图渲染到哪里。可以通过在根组件 AppComponent 的模板中插入一个<router-outlet>容器来指定路由出口的位置。<router-outlet>指令持有当前活动路由的一组自定义数据，它可以通过该指令的 activatedRouteData 属性来访问，可以使用这些数据来播放路由转场动画。路由转场动画也可以直接在组件中定义。

微课视频

10.4 动画的应用开发

10.4.1 切换动画

在项目 src\examples 根目录下创建 animationexamples 子目录，在 src\examples\animationexamples 目录下创建文件 home. component. ts，代码如例 10-1 所示。

【例 10-1】 创建文件 home. component. ts 的代码，定义组件。

```
import {Component, OnInit} from '@angular/core';
@Component({
  selector: 'app-home',
  template: `
    <p>
      欢迎使用 Angular 动画!Welcome to Animations in Angular!
    </p>
  `,
})
export class HomeComponent implements OnInit {
  ngOnInit() {  }
}
```

在 src\examples\animationexamples 目录下创建文件 about. component. ts，代码如

例 10-2 所示。

【例 10-2】 创建文件 about. component. ts 的代码，定义组件。

```
import {Component, OnInit} from '@angular/core';
@Component({
  selector: 'app-about',
  template: `
    <p> Angular 的动画库可以轻松定义、应用页面和列表等动画效果转变
   Angular's animations library makes it easy to define and apply animation effects such as page
and list
      transitions.
    </p>
  `,
})
export class AboutComponent implements OnInit {
  ngOnInit() {  }
}
```

在 src\examples\animationexamples 目录下创建文件 open-close. component. ts，代码如例 10-3 所示。

【例 10-3】 创建文件 open-close. component. ts 的代码，定义组件。

```
import {Component, Input} from '@angular/core';
import {animate, AnimationEvent, state, style, transition, trigger} from '@angular/animations';
@Component({
  selector: 'app-open-close',
  animations: [                              //定义动画,后面的用法相同
    trigger('openClose', [                   //触发动画,后面的用法相同
      state('open', style({
        height: '200px',
        opacity: 1,
        backgroundColor: 'yellow'
      })),
      state('closed', style({
        height: '100px',
        opacity: 0.8,
        backgroundColor: 'blue'
      })),
      transition('open => closed', [         //转场动画
        animate('1s')
      ]),
      transition('closed => open', [
        animate('0.5s')
      ]),
      transition('* => closed', [
        animate('1s')
      ]),
      transition('* => open', [
        animate('0.5s')
      ]),
      transition('open <=> closed', [
        animate('0.5s')
      ]),
      transition('* => open', [
```

```
        animate('1s',
          style({opacity: '*'}),
        ),
      ]),
      transition('* => *', [
        animate('1s')
      ]),
    ]),
  ],
  template: `
    <button (click) = "toggle()">切换开/关 Toggle Open/Close </button>
    <div [@openClose] = "isOpen ? 'open' : 'closed'"
         (@openClose.start) = "onAnimationEvent( $event)"
         (@openClose.done) = "onAnimationEvent( $event)"
         class = "open-close-container">
      <p> The box is now {{isOpen ? 'Open' : 'Closed' }}!</p>
    </div>
  `,
  styles: [':host {display: block;  margin-top: 1rem;}' +
  '.open-close-container {' +
  '  border: 1px solid #dddddd;' +
  '  margin-top: 1em;' +
  '  padding: 20px 20px 0px 20px;' +
  '  color: #000000;' +
  '  font-weight: bold;' +
  '  font-size: 20px;}']
})
export class OpenCloseComponent {
  @Input() logging = false;
  isOpen = true;
  toggle() {
    this.isOpen = !this.isOpen;
  }
  onAnimationEvent(event: AnimationEvent) {
    if (!this.logging) {
      return;
    }
    console.warn(`Animation Trigger: ${event.triggerName}`);
    console.warn(`Phase: ${event.phaseName}`);
    console.warn(`Total time: ${event.totalTime}`);
    console.warn(`From: ${event.fromState}`);
    console.warn(`To: ${event.toState}`);
    console.warn(`Element: ${event.element}`);
  }
}
```

在 src\examples\animationexamples 目录下创建文件 open-close-page. component. ts，代码如例 10-4 所示。

【例 10-4】 创建文件 open-close-page. component. ts 的代码，定义组件。

```
import {Component} from '@angular/core';
@Component({
  selector: 'app-open-close-page',
  template: `
```

```
    < section >
      < h4 >开关组件 Open Close Component </h4 >
        < input type = " checkbox" id = " log - checkbox" [ checked ] = " logging" (click) =
"toggleLogging()"/>
< label for = "log - checkbox">浏览器中控制台日志输出动画事件 Console Log Animation Events </
label >
      < app - open - close [logging] = "logging"></app - open - close >
    </section >
  `
})
export class OpenClosePageComponent {
  logging = false;
  toggleLogging() {
    this.logging = !this.logging;
  }
}
```

10.4.2 状态滑动

在 src\examples\animationexamples 目录下创建文件 status-slider. component. ts，代码如例 10-5 所示。

【例 10-5】 创建文件 status-slider. component. ts 的代码，定义组件。

```
import {Component} from '@angular/core';
import {animate, keyframes, state, style, transition, trigger} from '@angular/animations';
@Component({
  selector: 'app - status - slider',
  template: `
    < nav >
      < button (click) = "toggle()">状态切换 Toggle Status </button >
    </nav >
    < div [@slideStatus] = "status" class = "box">
      {{status == 'active'? 'Active' : 'Inactive' }}
    </div >
  `,
  styles: [':host {display: block;}' +
  '.box {\n' +
  '  width: 300px;\n' +
  '  border: 5px solid black;\n' +
  '  display: block;\n' +
  '  line - height: 300px;\n' +
  '  text - align: center;\n' +
  '  font - size: 50px;\n' +
  '  color: white;\n' +
  '}\n'],
  animations: [
    trigger('slideStatus', [
      state('inactive', style({backgroundColor: 'blue'})),
      state('active', style({backgroundColor: '#754600'})),
      transition('* => active', [
        animate('2s', keyframes([
          style({backgroundColor: 'blue', offset: 0}),
          style({backgroundColor: 'red', offset: 0.8}),
```

```
          style({backgroundColor: '#754600', offset: 1.0})
        ])),
      ]),
      transition('* => inactive', [
        animate('2s', keyframes([
          style({backgroundColor: '#754600', offset: 0}),
          style({backgroundColor: 'red', offset: 0.2}),
          style({backgroundColor: 'blue', offset: 1.0})
        ]))
      ]),
      transition('* => active', [
        animate('2s', keyframes([
          style({backgroundColor: 'blue'}),
          style({backgroundColor: 'red'}),
          style({backgroundColor: 'orange'})
        ]))
      ]),
    ])
  ]
})
export class StatusSliderComponent {
  status: 'active' | 'inactive' = 'inactive';
  toggle() {
    if (this.status === 'active') {
      this.status = 'inactive';
    } else {
      this.status = 'active';
    }
  }
}
```

在 src\examples\animationexamples 目录下创建文件 status-slider-page. component. ts,代码如例 10-6 所示。

【例 10-6】 创建文件 status-slider-page. component. ts 的代码,定义组件。

```
import {Component} from '@angular/core';
@Component({
  selector: 'app-status-slider-page',
  template: `
    <section>
      <h4>状态滑动 Status Slider</h4>
      <app-status-slider></app-status-slider>
    </section>
  `
})
export class StatusSliderPageComponent {
}
```

在 src\examples\animationexamples 目录下创建文件 open-close. component. 4. ts,代码如例 10-7 所示。

【例 10-7】 创建文件 open-close. component. 4. ts 的代码,定义组件。

```
import {Component} from '@angular/core';
import {animate, state, style, transition, trigger} from '@angular/animations';
```

```
@Component({
  selector: 'app-open-close-toggle',
  template: `
  <nav>
  <button (click)="toggleAnimations()">切换动画 Toggle Animations</button>
  <button (click)="toggle()">切换开/关 Toggle Open/Closed</button>
  </nav>
  <div [@.disabled]="isDisabled">
  <div [@childAnimation]="isOpen ? 'open' : 'closed'"
    class="open-close-container">
    <p>The box is now {{ isOpen ? 'Open' : 'Closed' }}!</p>
  </div>
  </div>
  `,
  styles: [':host {display: block;margin-top: 1rem;}' +
  '.open-close-container {' +
  '  border: 1px solid #dddddd;' +
  '  margin-top: 1em;' +
  '  padding: 20px 20px 0px 20px;' +
  '  color: #000000;' +
  '  font-weight: bold;' +
  '  font-size: 20px;}'],
  animations: [
    trigger('childAnimation', [
        state('open', style({
        width: '250px',
        opacity: 1,
        backgroundColor: 'yellow'
      })),
      state('closed', style({
        width: '100px',
        opacity: 0.8,
        backgroundColor: 'blue'
      })),
      transition('* => *', [
        animate('1s')
      ]),
    ]),
  ],
})
export class OpenCloseChildComponent {
  isDisabled = false;
  isOpen = false;
  toggleAnimations() {
    this.isDisabled = !this.isDisabled;
  }
  toggle() {
    this.isOpen = !this.isOpen;
  }
}
```

在 src\examples\animationexamples 目录下创建文件 toggle-animations-page.component.ts，代码如例 10-8 所示。

【例 10-8】 创建文件 toggle-animations-page.component.ts 的代码，定义组件。

```
import {Component} from '@angular/core';
@Component({
  selector: 'app-toggle-animations-child-page',
  template: `
    <section>
      <h4>切换动画 Toggle Animations</h4>
      <app-open-close-toggle></app-open-close-toggle>
    </section>
  `
})
export class ToggleAnimationsPageComponent {
}
```

10.4.3 进入与离开

在 src\examples\animationexamples 目录下创建文件 mock-heroes. ts，代码如例 10-9 所示。

【例 10-9】 创建文件 mock-heroes. ts 的代码，定义接口和数组。

```
export interface Hero {
  id: number;
  name: string;
}
export const HEROES: Hero[] = [
  {id: 1, name: '张三丰'},
  {id: 2, name: '李斯'},
  {id: 3, name: '王阳明'},
  {id: 4, name: '赵孟頫'},
  {id: 5, name: '钱通'},
  {id: 6, name: '孙武'},
  {id: 7, name: '李白'},
  {id: 8, name: '周瑜'},
  {id: 9, name: '郑成功'},
  {id: 10, name: '陈抟'}
];
```

在 src\examples\animationexamples 目录下创建文件 hero-list-enter-leave. component. ts，代码如例 10-10 所示。

【例 10-10】 创建文件 hero-list-enter-leave. component. ts 的代码，定义组件。

```
import {Component, EventEmitter, Input, Output} from '@angular/core';
import {animate, state, style, transition, trigger} from '@angular/animations';
import {Hero} from "./mock-heroes";
@Component({
  selector: 'app-hero-list-enter-leave',
  template: `
    <ul class="heroes">
      <li *ngFor="let hero of heroes"
          [@flyInOut]="'in'" (click)="removeHero(hero.id)">
        <div class="inner">
          <span class="badge">{{ hero.id }}</span>
          <span>{{ hero.name }}</span>
        </div>
```

```
    </li>
  </ul>
`,
styles: ['.heroes {\n' +
'  list-style-type: none;\n' +
'  padding: 0;\n' +
'}\n' +
'\n' +
'.heroes li {\n' +
'  overflow:hidden;\n' +
'  margin: .5em 0;\n' +
'}\n' +
'\n' +
'.heroes li > .inner {\n' +
'  cursor: pointer;\n' +
'  background-color: #EEE;\n' +
'  padding: .3rem 0;\n' +
'  height: 1.6rem;\n' +
'  border-radius: 4px;\n' +
'}\n' +
'\n' +
'.heroes li:hover > .inner {\n' +
'  color: black;\n' +
'  background-color: #DDD;\n' +
'  transform: translateX(.1em);\n' +
'}\n' +
'\n' +
'.heroes .badge {\n' +
'  display: inline-block;\n' +
'  font-size: small;\n' +
'  color: white;\n' +
'  padding: 0.8em 0.7em 0 0.7em;\n' +
'  background-color: #3d5157;\n' +
'  position: relative;\n' +
'  left: -1px;\n' +
'  top: -4px;\n' +
'  height: 1.8em;\n' +
'  min-width: 16px;\n' +
'  text-align: right;\n' +
'  margin-right: .8em;\n' +
'  border-radius: 4px 0 0 4px;\n' +
'}\n' +
'\n' +
'label {\n' +
'  display: block;\n' +
'  padding-bottom: .5rem;\n' +
'}\n' +
'\n' +
'input {\n' +
'  font-size: 100%;\n' +
'  margin-bottom: 1rem;\n' +
'}\n'],
animations: [
  trigger('flyInOut', [
```

```
        state('in', style({transform: 'translateX(0)'})),
        transition('void => * ', [
          style({transform: 'translateX( - 100 % )'}),
          animate(100)
        ]),
        transition(' *  => void', [
          animate(100, style({transform: 'translateX(100 % )'}))
        ])
      ])
    ]
})
export class HeroListEnterLeaveComponent {
  @Input() heroes: Hero[] = [];
  @Output() remove = new EventEmitter < number >();
  removeHero(id: number) {
    this.remove.emit(id);
  }
}
```

在 src\examples\animationexamples 目录下创建文件 hero-list-enter-leave-page.component.ts,代码如例 10-11 所示。

【例 10-11】 创建文件 hero-list-enter-leave-page.component.ts 的代码,定义组件。

```
import {Component} from '@angular/core';
import {HEROES} from './mock - heroes';
@Component({
  selector: 'app - hero - list - enter - leave - page',
  template: `
    < section >
      < h4 >进入/离开 Enter/Leave </h4 >
< app - hero - list - enter - leave [heroes] = "heroes" (remove) = "onRemove( $ event)"></app - hero - list - enter - leave >
    </section >
  `
})
export class HeroListEnterLeavePageComponent {
  heroes = HEROES.slice();
  onRemove(id: number) {
    this.heroes = this.heroes.filter(hero => hero.id !== id);
  }
}
```

10.4.4 自动计算

在 src\examples\animationexamples 目录下创建文件 hero-list-auto.component.ts,代码如例 10-12 所示。

【例 10-12】 创建文件 hero-list-auto.component.ts 的代码,定义组件。

```
import {Component, EventEmitter, Input, Output} from '@angular/core';
import {animate, state, style, transition, trigger} from '@angular/animations';
import {Hero} from "./mock - heroes";
@Component({
  selector: 'app - hero - list - auto',
```

```
template: `
  <ul class = "heroes">
    <li *ngFor = "let hero of heroes"
        [@shrinkOut] = "'in'" (click) = "removeHero(hero.id)">
      <div class = "inner">
        <span class = "badge">{{ hero.id }}</span>
        <span>{{ hero.name }}</span>
      </div>
    </li>
  </ul>
`,
styles: ['.heroes {\n' +
'  list-style-type: none;\n' +
'  padding: 0;\n' +
'}\n' +
'\n' +
'.heroes li {\n' +
'  overflow:hidden;\n' +
'  margin: .5em 0;\n' +
'}\n' +
'\n' +
'.heroes li > .inner {\n' +
'  cursor: pointer;\n' +
'  background-color: #EEE;\n' +
'  padding: .3rem 0;\n' +
'  height: 1.6rem;\n' +
'  border-radius: 4px;\n' +
'}\n' +
'\n' +
'.heroes li:hover > .inner {\n' +
'  color: black;\n' +
'  background-color: #DDD;\n' +
'  transform: translateX(.1em);\n' +
'}\n' +
'\n' +
'.heroes .badge {\n' +
'  display: inline-block;\n' +
'  font-size: small;\n' +
'  color: white;\n' +
'  padding: 0.8em 0.7em 0 0.7em;\n' +
'  background-color: #3d5157;\n' +
'  position: relative;\n' +
'  left: -1px;\n' +
'  top: -4px;\n' +
'  height: 1.8em;\n' +
'  min-width: 16px;\n' +
'  text-align: right;\n' +
'  margin-right: .8em;\n' +
'  border-radius: 4px 0 0 4px;\n' +
'}\n' +
'\n' +
'label {\n' +
'  display: block;\n' +
'  padding-bottom: .5rem;\n' +
```

```
    '}\n' +
    '\n' +
    'input {\n' +
    '  font - size: 100 % ;\n' +
    '  margin - bottom: 1rem;\n' +
    '}\n'],
    animations: [
      trigger('shrinkOut', [
        state('in', style({height: ' * '})),
        transition(' *  => void', [
          style({height: ' * '}),
          animate(250, style({height: 0}))
        ])
      ])
    ]
})
export class HeroListAutoComponent {
  @Input() heroes: Hero[ ] = [ ];
  @Output() remove = new EventEmitter < number >();
  removeHero(id: number) {
    this.remove.emit(id);
  }
}
```

在 src\examples\animationexamples 目录下创建文件 hero-list-auto-page. component. ts，代码如例 10-13 所示。

【例 10-13】 创建文件 hero-list-auto-page. component. ts 的代码，定义组件。

```
import {Component} from '@angular/core';
import {HEROES} from './mock - heroes';
@Component({
  selector: 'app - hero - list - auto - page',
  template: `
    < section >
      < h4 >自动计算 Automatic Calculation </h4 >
      < app - hero - list - auto [heroes] = "heroes" (remove) = "onRemove( $ event)"></app - hero -
list - auto >
    </section >
  `
})
export class HeroListAutoCalcPageComponent {
  heroes = HEROES.slice();
  onRemove(id: number) {
    this.heroes = this.heroes.filter(hero => hero.id !== id);
  }
}
```

10.4.5 过滤与交错

在 src\examples\animationexamples 目录下创建文件 hero-list-page. component. ts，代码如例 10-14 所示。

【例 10-14】 创建文件 hero-list-page. component. ts 的代码，定义组件。

```
import {Component, HostBinding, OnInit} from '@angular/core';
import {animate, query, stagger, style, transition, trigger} from '@angular/animations';
import {Hero, HEROES} from "./mock-heroes";
@Component({
  selector: 'app-hero-list-page',
  template: `
    <h4>过滤/交错 Filter/Stagger</h4>
    <label for="search">查找人物(Search heroes): </label>
    <input type="text" id="search" #criteria
           (input)="updateCriteria(criteria.value)"
           placeholder="查找人物 Search heroes">
    <ul class="heroes" [@filterAnimation]="heroesTotal">
      <li *ngFor="let hero of heroes" class="hero">
        <div class="inner">
          <span class="badge">{{ hero.id }}</span>
          <span>{{ hero.name }}</span>
        </div>
      </li>
    </ul>
  `,
  styles: ['.heroes {\n' +
  '  list-style-type: none;\n' +
  '  padding: 0;\n' +
  '}\n' +
  '\n' +
  '.heroes li {\n' +
  '  overflow:hidden;\n' +
  '  margin: .5em 0;\n' +
  '}\n' +
  '\n' +
  '.heroes li > .inner {\n' +
  '  cursor: pointer;\n' +
  '  background-color: #EEE;\n' +
  '  padding: .3rem 0;\n' +
  '  height: 1.6rem;\n' +
  '  border-radius: 4px;\n' +
  '}\n' +
  '\n' +
  '.heroes li:hover > .inner {\n' +
  '  color: black;\n' +
  '  background-color: #DDD;\n' +
  '  transform: translateX(.1em);\n' +
  '}\n' +
  '\n' +
  '.heroes .badge {\n' +
  '  display: inline-block;\n' +
  '  font-size: small;\n' +
  '  color: white;\n' +
  '  padding: 0.8em 0.7em 0 0.7em;\n' +
  '  background-color: #3d5157;\n' +
  '  position: relative;\n' +
  '  left: -1px;\n' +
  '  top: -4px;\n' +
  '  height: 1.8em;\n' +
```

```
    '  min-width: 16px;\n' +
    '  text-align: right;\n' +
    '  margin-right: .8em;\n' +
    '  border-radius: 4px 0 0 4px;\n' +
    '}\n' +
    '\n' +
    'label {\n' +
    '  display: block;\n' +
    '  padding-bottom: .5rem;\n' +
    '}\n' +
    '\n' +
    'input {\n' +
    '  font-size: 100%;\n' +
    '  margin-bottom: 1rem;\n' +
    '}\n'],
    animations: [
      trigger('pageAnimations', [
        transition(':enter', [
          query('.hero', [
            style({opacity: 0, transform: 'translateY(-100px)'}),
            stagger(30, [
              animate('500ms cubic-bezier(0.35, 0, 0.25, 1)',
                style({opacity: 1, transform: 'none'}))
            ])
          ])
        ])
      ]),
      trigger('filterAnimation', [
        transition(':enter, * => 0, * => -1', []),
        transition(':increment', [
          query(':enter', [
            style({opacity: 0, width: 0}),
            stagger(50, [
              animate('300ms ease-out', style({opacity: 1, width: '*'})),
            ]),
          ], {optional: true})
        ]),
        transition(':decrement', [
          query(':leave', [
            stagger(50, [
              animate('300ms ease-out', style({opacity: 0, width: 0})),
            ]),
          ])
        ]),
      ]),
    ]
})
export class HeroListPageComponent implements OnInit {
  @HostBinding('@pageAnimations')
  public animatePage = true;
  heroesTotal = -1;
  private _heroes: Hero[] = [];
  get heroes() {
    return this._heroes;
```

```
  }
  ngOnInit() {
    this._heroes = HEROES;
  }
  updateCriteria(criteria: string) {
    criteria = criteria ? criteria.trim() : '';
     this._heroes = HEROES.filter(hero => hero.name.toLowerCase().includes(criteria.
toLowerCase()));
    const newTotal = this.heroes.length;
    if (this.heroesTotal !== newTotal) {
      this.heroesTotal = newTotal;
    } else if (!criteria) {
      this.heroesTotal = -1;
    }
  }
}
```

10.4.6　列表与集合

在 src\examples\animationexamples 目录下创建文件 hero-list-groups.component.ts，代码如例 10-15 所示。

【例 10-15】 创建文件 hero-list-groups.component.ts 的代码，定义组件。

```
import {Component, EventEmitter, Input, Output} from '@angular/core';
import {animate, group, state, style, transition, trigger} from '@angular/animations';
import {Hero} from "./mock-heroes";
@Component({
  selector: 'app-hero-list-groups',
  template: `
    <ul class="heroes">
      <li *ngFor="let hero of heroes"
          [@flyInOut]="'in'" (click)="removeHero(hero.id)">
        <div class="inner">
          <span class="badge">{{ hero.id }}</span>
          <span>{{ hero.name }}</span>
        </div>
      </li>
    </ul>
  `,
  styles: ['.heroes {\n' +
  '  list-style-type: none;\n' +
  '  padding: 0;\n' +
  '}\n' +
  '\n' +
  '.heroes li {\n' +
  '  overflow:hidden;\n' +
  '  margin: .5em 0;\n' +
  '}\n' +
  '\n' +
  '.heroes li > .inner {\n' +
  '  cursor: pointer;\n' +
  '  background-color: #EEE;\n' +
  '  padding: .3rem 0;\n' +
```

```
'  height: 1.6rem;\n' +
'  border-radius: 4px;\n' +
'}\n' +
'\n' +
'.heroes li:hover > .inner {\n' +
'  color: black;\n' +
'  background-color: #DDD;\n' +
'  transform: translateX(.1em);\n' +
'}\n' +
'\n' +
'.heroes .badge {\n' +
'  display: inline-block;\n' +
'  font-size: small;\n' +
'  color: white;\n' +
'  padding: 0.8em 0.7em 0 0.7em;\n' +
'  background-color: #3d5157;\n' +
'  position: relative;\n' +
'  left: -1px;\n' +
'  top: -4px;\n' +
'  height: 1.8em;\n' +
'  min-width: 16px;\n' +
'  text-align: right;\n' +
'  margin-right: .8em;\n' +
'  border-radius: 4px 0 0 4px;\n' +
'}\n' +
'\n' +
'label {\n' +
'  display: block;\n' +
'  padding-bottom: .5rem;\n' +
'}\n' +
'\n' +
'input {\n' +
'  font-size: 100%;\n' +
'  margin-bottom: 1rem;\n' +
'}\n'],
animations: [
  trigger('flyInOut', [
    state('in', style({
      width: 120,
      transform: 'translateX(0)', opacity: 1
    })),
    transition(':enter', [
      style({width: 10, transform: 'translateX(50px)', opacity: 0}),
      group([
        animate('0.3s 0.1s ease', style({
          transform: 'translateX(0)',
          width: 120
        })),
        animate('0.3s ease', style({
          opacity: 1
        }))
      ])
    ]),
    transition(':leave', [
```

```
          group([
            animate('0.3s ease', style({
              transform: 'translateX(50px)',
              width: 10
            })),
            animate('0.3s 0.2s ease', style({
              opacity: 0
            }))
          ])
        ])
      ])
    ]
})
export class HeroListGroupsComponent {
  @Input() heroes: Hero[] = [];
  @Output() remove = new EventEmitter<number>();
  removeHero(id: number) {
    this.remove.emit(id);
  }
}
```

在 src\examples\animationexamples 目录下创建文件 hero-list-group-page.component.ts，代码如例 10-16 所示。

【例 10-16】 创建文件 hero-list-group-page.component.ts 的代码，定义组件。

```
import {Component} from '@angular/core';
import {HEROES} from './mock-heroes';
@Component({
  selector: 'app-hero-list-groups-page',
  template: `
    <section>
      <h4>人物名单组 Hero List Group</h4>
      <app-hero-list-groups [heroes]="heroes" (remove)="onRemove($event)"></app-
hero-list-groups>
    </section>
  `
})
export class HeroListGroupPageComponent {
  heroes = HEROES.slice();
  onRemove(id: number) {
    this.heroes = this.heroes.filter(hero => hero.id !== id);
  }
}
```

10.4.7 插入与删除

在 src\examples\animationexamples 目录下创建文件 insert-remove.component.ts，代码如例 10-17 所示。

【例 10-17】 创建文件 insert-remove.component.ts 的代码，定义组件。

```
import {Component} from '@angular/core';
import {animate, style, transition, trigger} from '@angular/animations';
@Component({
```

```
  selector: 'app-insert-remove',
  animations: [
    trigger('myInsertRemoveTrigger', [
      transition(':enter', [
        style({opacity: 0}),
        animate('100ms', style({opacity: 1})),
      ]),
      transition(':leave', [
        animate('100ms', style({opacity: 0}))
      ])
    ]),
  ],
  template: `
    <h4>插入/删除 Insert/Remove</h4>
    <nav>
      <button (click)="toggle()">切换插入/删除 Toggle Insert/Remove</button>
    </nav>
    <div @myInsertRemoveTrigger *ngIf="isShown" class="insert-remove-container">
      <p>盒子被插入 The box is inserted</p>
    </div>
  `,
  styles: [':host {display: block;}' +
  '.insert-remove-container {\n' +
  '  border: 1px solid #dddddd;\n' +
  '  margin-top: 1em;\n' +
  '  padding: 20px 20px 0px 20px;\n' +
  '  color: #000000;\n' +
  '  font-weight: bold;\n' +
  '  font-size: 20px;\n' +
  '}']
})
export class InsertRemoveComponent {
  isShown = false;
  toggle() {
    this.isShown = !this.isShown;
  }
}
```

10.4.8 服务组件

在 src\examples\animationexamples 目录下创建文件 animations.ts,代码如例 10-18 所示。

【例 10-18】 创建文件 animations.ts 的代码,定义动画和类。

```
import {animate, animateChild, animation, group, query, style, transition, trigger} from '@angular/animations';
export const transitionAnimation = animation([
  style({
    height: '{{ height }}',
    opacity: '{{ opacity }}',
    backgroundColor: '{{ backgroundColor }}'
  }),
  animate('{{ time }}')
```

```
]);
export const slideInAnimation =
  trigger('routeAnimations', [
    transition('HomePage <=> AboutPage', [
      style({position: 'relative'}),
      query(':enter, :leave', [
        style({
          position: 'absolute',
          top: 0,
          left: 0,
          width: '100%'
        })
      ]),
      query(':enter', [
        style({left: '-100%'})
      ]),
      query(':leave', animateChild()),
      group([
        query(':leave', [
          animate('300ms ease-out', style({left: '100%'}))
        ]),
        query(':enter', [
          animate('300ms ease-out', style({left: '0%'}))
        ])
      ]),
      query(':enter', animateChild()),
    ]),
    transition('* <=> FilterPage', [
      style({position: 'relative'}),
      query(':enter, :leave', [
        style({
          position: 'absolute',
          top: 0,
          left: 0,
          width: '100%'
        })
      ]),
      query(':enter', [
        style({left: '-100%'})
      ]),
      query(':leave', animateChild()),
      group([
        query(':leave', [
          animate('200ms ease-out', style({left: '100%'}))
        ]),
        query(':enter', [
          animate('300ms ease-out', style({left: '0%'}))
        ])
      ]),
      query(':enter', animateChild()),
    ])
  ]);
```

在 src\examples\animationexamples 目录下创建文件 animationhome.component.ts，

代码如例 10-19 所示。

【例 10-19】 创建文件 animationhome. component. ts 的代码,定义组件。

```
import {Component, HostBinding} from '@angular/core';
import {RouterOutlet} from '@angular/router';
import {slideInAnimation} from './animations';
@Component({
  selector: 'root',
  template: `
    <h3>动画 Animations</h3>
    <input type="checkbox"
           id="animation-toggle"
           [checked]="!animationsDisabled"
           (click)="toggleAnimations()">
    <label for="animation-toggle">切换所有动画 Toggle All Animations</label>
    <nav>
      <a id="home" routerLink="/home" routerLinkActive="active">主页 Home</a>
      <a id="about" routerLink="/about" routerLinkActive="active">说明 About</a>
      <a id="open-close" routerLink="/open-close" routerLinkActive="active">开/关
Open/Close</a>
      <a id="status" routerLink="/status" routerLinkActive="active">状态滑动 Status
Slider</a>
      <a id="toggle" routerLink="/toggle" routerLinkActive="active">切换动画 Toggle
Animations</a>
      <a id="enter-leave" routerLink="/enter-leave" routerLinkActive="active">进入/离
开 Enter/Leave</a>
      <a id="auto" routerLink="/auto" routerLinkActive="active">自动计算 Auto
Calculation</a>
      <a id="heroes" routerLink="/heroes" routerLinkActive="active">过滤/交错 Filter/
Stagger</a>
<a id="hero-groups" routerLink="/hero-groups" routerLinkActive="active">人物团体 Hero
Groups</a>
<a id="insert-remove" routerLink="/insert-remove" routerLinkActive="active">插入/删除
Insert/Remove</a>
    </nav>
    <hr>
    <div [@routeAnimations]="prepareRoute(outlet)">
      <router-outlet #outlet="outlet"></router-outlet>
    </div>
  `,
  styles: ['nav a {padding: .7rem;}'],
  animations: [
    slideInAnimation
  ]
})
export class AnimationhomeComponent {
  @HostBinding('@.disabled')
  public animationsDisabled = false;
  prepareRoute(outlet: RouterOutlet) {
    return outlet?.activatedRouteData?.['animation'];
  }
  toggleAnimations() {
    this.animationsDisabled = !this.animationsDisabled;
  }
}
```

10.4.9　模块和运行结果

在 src\examples\animationexamples 目录下创建文件 app-animation. module. ts，代码如例 10-20 所示。

【例 10-20】 创建文件 app-animation. module. ts 的代码，定义路由并声明组件。

```
import {NgModule} from '@angular/core';
import {BrowserModule} from '@angular/platform-browser';
import {BrowserAnimationsModule} from '@angular/platform-browser/animations';
import {RouterModule} from '@angular/router';
import {AnimationhomeComponent} from "./animationhome.component";
import {OpenCloseComponent} from './open-close.component';
import {OpenClosePageComponent} from './open-close-page.component';
import {OpenCloseChildComponent} from './open-close.component.4';
import {ToggleAnimationsPageComponent} from './toggle-animations-page.component';
import {StatusSliderComponent} from './status-slider.component';
import {StatusSliderPageComponent} from './status-slider-page.component';
import {HeroListPageComponent} from './hero-list-page.component';
import {HeroListGroupPageComponent} from './hero-list-group-page.component';
import {HeroListGroupsComponent} from './hero-list-groups.component';
import {HeroListEnterLeavePageComponent} from './hero-list-enter-leave-page.component';
import {HeroListEnterLeaveComponent} from './hero-list-enter-leave.component';
import {HeroListAutoCalcPageComponent} from './hero-list-auto-page.component';
import {HeroListAutoComponent} from './hero-list-auto.component';
import {HomeComponent} from './home.component';
import {AboutComponent} from './about.component';
import {InsertRemoveComponent} from './insert-remove.component';
@NgModule({
  imports: [
    BrowserModule,
    BrowserAnimationsModule,
    RouterModule.forRoot([
      {path: '', pathMatch: 'full', redirectTo: '/home'},
      {path: 'open-close', component: OpenClosePageComponent},
      {path: 'status', component: StatusSliderPageComponent},
      {path: 'toggle', component: ToggleAnimationsPageComponent},
      {
        path: 'heroes', component: HeroListPageComponent,
        data: {animation: 'FilterPage'}
      },
      {path: 'hero-groups', component: HeroListGroupPageComponent},
      {path: 'enter-leave', component: HeroListEnterLeavePageComponent},
      {path: 'auto', component: HeroListAutoCalcPageComponent},
      {path: 'insert-remove', component: InsertRemoveComponent},
      {path: 'home', component: HomeComponent, data: {animation: 'HomePage'}},
      {path: 'about', component: AboutComponent, data: {animation: 'AboutPage'}},
    ])
  ],
  declarations: [
    AnimationhomeComponent,
    StatusSliderComponent,
    OpenCloseComponent,
    OpenCloseChildComponent,
```

```
        OpenClosePageComponent,
        StatusSliderPageComponent,
        ToggleAnimationsPageComponent,
        HeroListPageComponent,
        HeroListGroupsComponent,
        HeroListGroupPageComponent,
        HeroListEnterLeavePageComponent,
        HeroListEnterLeaveComponent,
        HeroListAutoCalcPageComponent,
        HeroListAutoComponent,
        HomeComponent,
        InsertRemoveComponent,
        AboutComponent
    ],
})
export class AppAnimationModule {
}
```

修改 src\examples 目录下的文件 examplesmodules1. module. ts,代码如例 10-21 所示。

【例 10-21】 修改文件 examplesmodules1. module. ts 的代码,设置启动组件。

```
import {NgModule} from '@angular/core';
import {AnimationhomeComponent} from './animationexamples/animationhome.component';
import {AppAnimationModule} from "./animationexamples/app-animation.module";
@NgModule({
  imports: [
    AppAnimationModule//插入
  ],
  bootstrap: [AnimationhomeComponent]
})
export class ExamplesmodulesModule1 {}
```

保持其他文件不变并成功运行程序后,在浏览器地址栏中输入 localhost:4200,自动跳转到 localhost:4200/home,结果如图 10-1 所示。

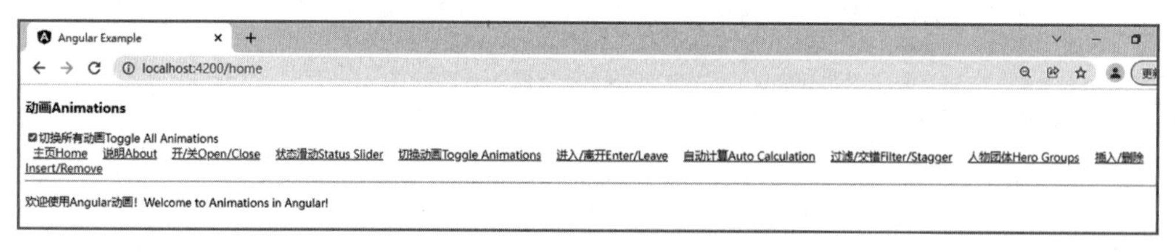

图 10-1 成功运行程序后在浏览器地址栏中输入 localhost:4200/home 的结果

习题 10

一、简答题

简述对动画的理解。

二、实验题

完成动画的应用开发。

第 11 章

PWA、Service Worker、Web Worker

11.1 PWA 概述

1. 含义

渐进式 Web 应用(Progressive Web Apps,PWA)使用现代 API 进行构建和增强 Web 应用,提供增强的功能,具有可靠性和可安装性。它是经过设计的 Web 应用,它们的功能强大、可靠且可安装。这 3 个特点给它们带来类似于与平台无关或为平台量身定做的使用体验。如今的 Web 本身就非常强大。例如,可以使用 WebRTC、地理定位和推送通知构建超本地化的视频聊天应用;可以使该应用程序可安装,并通过 WebGL 和 WebVR 进行虚拟对话。随着 Web Assembly 的引入,开发人员可以利用其他生态系统(如 C、C++和 Rust),将工作经验和应用功能也带到 Web 中。

在现代 API、Web Assembly 以及未推出的新 API 之间,Web 应用程序的功能比以往任何时候都强大,而且这些功能与日俱增。无论网络如何,可靠的渐进式 Web 应用程序都会让人感觉快速且可以依赖。用户喜欢在眨眼间就能及时获得响应交互的应用以及可以依赖的使用体验。随着页面加载时间变短,用户跳离的概率会减少。Web 应用程序的性能会影响用户体验。用户期望应用程序在网络连接缓慢或不稳定甚至离线时也能启动。即使在很难向服务器发出请求的情况下,也仍然可以访问和使用最近交互过的离线内容(如媒体曲目或者门票和旅游日程)。当无法发出请求时,就会告知用户有问题,而不是让用户遭遇应用程序的启动失败或崩溃。

2. PWA 运行

安装后的 PWA 是在独立窗口中运行的,而不是在浏览器标签页中运行。PWA 可以从用户的主屏幕(即最初始的页面)、程序坞(放置程序快捷方式的快速启动栏)、任务栏或工具栏上启动,也可以在设备上搜索它们,还可以使用应用切换器实现跳转。安装 Web 应用后,新功能会开启。在浏览器中运行时通常保留的键盘快捷键将变得可用。PWA 可以注册或接收来自其他应用的内容,或者成为处理不同类型文件的默认应用程序。

当 PWA 移出标签页并进入独立应用窗口时,它转变了用户对它的看法以及与之交互的方式。通过渐进式增强,现代浏览器中启用了新功能。通过文件系统访问、媒体控制、应用标记和完整的剪贴板支持等功能来扩展 Web 的能力。所有这些功能都使用 Web 以用户为中心且安全的权限模型进行构建,确保访问网站对用户来说永远不会是一件可怕的事情。

利用服务工作进程和 Web 应用清单，Web 应用变得可靠且可安装。如果新功能不可用，用户仍然可以获得好的体验。

11.2 Service Worker 概述

1. 含义

Service Worker 可以增强传统的 Web 发布模式，并使应用程序能够提供可与原生代码（运行在操作系统和硬件层面）媲美的高可靠性、高性能的用户体验。为 Angular 应用添加 Service Worker 是把应用程序转换成渐进式应用的步骤之一。简单来说，Service Worker 就是一段运行在 Web 浏览器中，并为应用程序管理缓存的脚本。

Service Worker 的功能就像一个网络代理。它们会拦截所有由应用程序发出的 HTTP 请求，并选择如何给出响应。例如，它们可以查询局部缓存，如果有缓存的响应数据就用缓存的数据作出响应。这种代理行为不会局限于通过程序调用 API（如 fetch）发起的请求，还包括 HTML 中对资源的引用，甚至是对文件 index. html 的首次请求。基于 Service Worker 的缓存是完全可编程的，并且不依赖于服务端指定的那些控制缓存策略的头。

不像应用程序中的其他脚本（如 Angular 的应用包），Service Worker 在用户关闭浏览器页标签时仍然会被保留。在浏览器下次加载本应用程序时，Service Worker 会首先被加载，然后拦截加载本应用程序时对每一项资源的请求。如果这个 Service Worker 就是为此而设计的，它就能完全满足应用程序加载时的需求，而不需要依赖网络。使用 Service Worker 来减少对网络的依赖可以显著改善用户体验。

2. 实现

作为单页面应用，Angular 应用可以受益于 Service Worker 的优势。从 Angular 5 开始，Angular 提供了一份 Service Worker 的实现。开发人员可以利用 Service Worker，并受益于其增强的可靠性和性能，而无须再针对底层 API 写代码。

Angular 的 Service Worker 的设计目标是优化那些使用慢速、不可靠网络的终端用户的体验，同时还要尽可能减小提供过期内容的风险。为了达成此目标，Angular 的 Service Worker 的行为需要遵循一些设计原则，这些原则包括：像安装原生应用程序一样缓存应用程序，并作为整体被缓存，它的所有文件作为整体进行更新；正在运行的应用程序使用所有文件的同一版本继续运行，不要突然开始接收来自新版本的、可能不兼容的缓存文件；当用户刷新本应用程序时会看到最新的被完全缓存的版本，新的页标签中会加载最新的缓存代码；更改发布后相对较快的代码在后台进行更新，在一次完整的更新完成之前仍然使用应用的上一个版本；Service Worker 会尽量节省带宽，它只会下载那些发生了变化的资源。要支持这些行为，Service Worker 会从服务器上下载一个 manifest 文件。这个 manifest 文件（不同于 Web 应用程序的 manifest）描述要缓存的资源，并包含每个文件内容的哈希值。每当发布应用程序的新版本时，manifest 的内容就会改变，通知 Service Worker 应该下载并缓存应用程序的新版本。

安装 Service Worker 就像引入一个 NgModule 一样简单。除了使用浏览器注册的 Service Worker 之外，还要制作一些可供注入的服务，它们可以与 Service Worker 交互并控制它。例如，应用程序可以要求当新的更新已经就绪时通知自己，或要求 Service Worker

检查服务器看是否有可用的更新。

3. 应用外壳

应用外壳是一种在构建期间借助路由渲染部分应用的方法。它可以通过快速启动一个静态渲染页面（所有页面的公共骨架）来改善用户体验。与此同时，浏览器会下载完整的客户端版本，并在代码加载后自动切换到完整版。这能让用户快速看到应用程序中第一个有意义的画面，因为浏览器可以渲染出 HTML 和 CSS，而无须初始化任何 JavaScript。

应用外壳的开发步骤包括：先准备应用程序，对于既有应用程序手动添加 RouterModule 并在应用程序中定义< router-outlet >。再创建一个应用外壳。接着，验证该应用程序是否使用应用外壳的内容构建的。

4. 与 Service Worker 通信

把 ServiceWorkerModule 导入根模块（如 AppModule）中不仅会注册 Service Worker，还会提供一些服务，使得应用程序能和 Service Worker 通信，并控制应用缓存。

SwUpdate 服务使得应用程序能访问一些事件，这些事件会指出 Service Worker 何时发现了应用程序可用的更新或者一个更新何时可以被激活，这意味着它可以通过更新后的应用程序提供服务了。

SwUpdate 服务支持四个独立的操作：获取出现可用应用程序更新的通知，如果要刷新页面，这些就是可加载的新版本；获取更新被激活的通知，这时 Service Worker 可以立即使用这些新版本提供服务；要求 Service Worker 向服务器查询是否有新版本；要求 Service Worker 为当前标签页激活该应用程序的最新版本。

可以要求 Service Worker 检查是否有任何更新的应用程序已经发布到了服务器上。Service Worker 会在初始化和每次导航请求（也就是用户导航到应用中的另一个地址）时检查更新。不过，如果站点更新非常频繁，或者需要按计划进行更新，则可能会选择手动检查更新。

为了避免影响页面的首次渲染，在注册 Service Worker 脚本之前，ServiceWorkerModule 默认会在应用程序达到稳定态之前等待最多 30s。如果不断轮询更新（如调用 setInterval() 方法或 RxJS 的 interval() 方法）就会阻止应用程序达到稳定态，则直到 30s 结束之前都不会往浏览器中注册 Service Worker 脚本。注意，应用程序中所执行的各种轮询都会阻止它达到稳定态。可以通过在开始轮询更新之前先等应用程序达到稳定态来消除这种延迟。还可以为 Service Worker 定义不一样的注册策略。在某些情况下，Service Worker 用来为客户端提供服务的应用版本可能处于损坏状态，如果不重新加载整个页面，则无法恢复应用程序的正常状态。

5. 通知

推送通知是吸引用户的一种引人注目的方法。通过 Service Worker 的强大功能，即使应用程序不在焦点上，也可以将通知发送到设备。Angular 的 Service Worker 支持显示推送通知和处理通知单击事件。使用 Service Worker 时推送通知交互是使用 SwPush 服务来处理的。

可以通过推送具有有效负载的消息来调用推送通知。在 Chrome 中，可以在没有后端的情况下测试推送通知。打开 Devtools -> Application -> Service Workers 并使用 Push 输入发送 JSON 通知负载。

6. 操作

操作提供了一种自定义用户如何与应用通知交互的方法。Angular 的 Service Worker 支持 openWindow、focusLastFocusedOrOpen、navigateLastFocusedOrOpen 等操作。openWindow 操作在指定的 URL 处打开一个新选项卡,该选项卡相对于 Service Worker 范围进行解析。focusLastFocusedOrOpen 操作聚焦最后一个聚焦的客户端。如果没有客户端打开,则它会在指定的 URL 处打开一个新选项卡,该选项卡是在 Service Worker 范围内解析的。navigateLastFocusedOrOpen 操作聚焦最后一个聚焦的客户端并将其导航到指定的 URL,该 URL 在 Service Worker 范围内进行解析。如果没有打开的客户端,则它会在指定的 URL 处打开一个新选项卡。

11.3 生产环境下的 Service Worker

1. 整体性

从概念上说,可以把 Service Worker 想象成一个转发式缓存或装在最终用户浏览器中的 CDN 边缘。Service Worker 的工作是从本地缓存中满足 Angular 应用对资源或数据的请求,而不用等待网络响应。和所有缓存一样,它有一些规则来决定内容该如何过期或更新。

在 Service Worker 的语境下,版本是指用来表示 Angular 应用程序的某一次构建成果的一组资源。发布一个新的构建时,Service Worker 就把它看作应用的一个新版本。就算只修改了一个文件,也同样如此。在任一给定的时间,Service Worker 可能会在它的缓存中拥有此应用的多个版本,这几个版本也都能用于提供服务。

要保持应用程序的整体性,Service Worker 会用所有的文件共同组成一个版本。组成版本的这些文件通常包括 HTML、JavaScript 和 CSS 文件。把这些文件分成一组至关重要,因为它们会互相引用,并且依赖于一些特定内容。

当使用惰性加载模块时,文件的整体性就显得格外重要。某个 JavaScript 包可能引用很多惰性块,而这些惰性块的文件名在应用程序的每次构建时都是唯一的。如果运行应用的 X 版本视图加载一个惰性块,但该块的服务器已经升级到了 X+1 版本,这次惰性加载操作就会失败。借助于 Service Worker 的这种版本控制行为,应用服务器就可以确保 Angular 应用程序中的这组文件始终保持一致。

每当用户打开或刷新应用时,Service Worker 都会通过查看清单文件 ngsw.json 是否更新来检查该应用的更新情况。如果它找到了更新,就会自动下载并缓存这个更新的版本并在下次加载应用程序时提供。文件 ngsw.json 清单中唯一带哈希值的资源就是构建清单时 dist 目录中的资源。而其他资源,特别是从 CDN 加载的资源,其内容在构建时是未知的,或者会比应用程序的部署变更得更加频繁。

长周期缓存的潜在副作用之一就是可能无意中缓存了无效的资源。在普通的 HTTP 缓存中,硬刷新或缓存过期限制了缓存这种无效文件导致的负面影响。而 Service Worker 会忽略这样的约束,事实上会对整个应用程序进行长期缓存。因此,让 Service Worker 获得正确的内容就显得至关重要。

2. 验证

为了确保资源的整体性,Service Worker 会验证所有带哈希的资源的哈希值。如果某个特定的文件未能通过验证，Service Worker 就会尝试用 cache-busting（缓存清除）URL 为参数重新获取内容以消除浏览器或中间缓存的影响。如果新内容也未能通过验证，则 Service Worker 会认为该应用程序的整个版本都无效，并停止用它提供服务。如有必要，Service Worker 会进入安全模式，这些请求将退化为直接访问网络。如果服务无效、损坏或内容过期的风险很高，则会选择不使用缓存。

导致哈希值不匹配的原因包括：在源服务器和最终用户之间缓存图层可能会提供陈旧的内容；非原子化的部署可能会导致 Service Worker 看到部分更新后的内容；构建过程中的错误可能会导致更新了资源却没有更新文件 ngsw.json，没有更新资源却更新了文件 ngsw.json 等。

如果 Service Worker 没有哈希值可以验证给定的资源，则它仍然会缓存其内容，但会使用重新验证时失效的策略来承认 HTTP 缓存头。也就是说，当被缓存资源的 HTTP 缓存头指出该资源已过期时，Service Worker 将继续提供内容，并尝试在后台刷新资源。这样，那些被破坏的非哈希资源留在缓存中的时间就不会超出为它配置的生命周期。

3. 版本

如果应用程序的资源版本突然发生了变化或没有给出警告，就可能会有问题。Service Worker 会保证正在运行的应用程序会继续运行和当前应用程序相同的版本。如果在新的 Web 浏览器选项卡中打开了该应用程序的另一个实例，则会提供该应用程序的最新版本。因此，这个新标签可以和原始标签同时运行不同版本的应用。如果没有 Service Worker，则不能保证稍后在这个正在运行的应用程序中惰性加载的代码和其初始代码的版本是一样的。

Service Worker 可能会更改运行中应用程序的版本的原因包括：由于哈希验证失败，以致当前版本变成了无效的；某个无关的错误导致 Service Worker 进入了安全模式（或者说它被暂时禁用了）。Service Worker 能知道在任何指定的时刻正在使用应用程序哪些版本，并清除那些没有被任何选项卡使用的版本。另一些可能导致 Service Worker 在运行期间改变应用程序版本的因素的一些正常事件包括：页面被重新加载/刷新；页面通过 SwUpdate 服务请求立即激活更新。

4. 异常

Service Worker 是一个运行在 Web 浏览器中的小脚本。有时，Service Worker 也可能会需要更新，以修复错误和增强特性。首次打开应用程序时或在一段非活动时间之后再访问应用时，就会下载 Service Worker。如果 Service Worker 发生了变化，Service Worker 就会在后台进行更新。Service Worker 的大部分更新对应用来说都是透明的（旧缓存仍然有效，其内容仍然能正常使用）。但是，在 Service Worker 中可能偶尔会有错误修复或更新功能，需要让旧的缓存失效。这时，应用程序就会从网络上透明地进行刷新。

某些情况下，可能想要完全绕过 Service Worker，转而让浏览器处理请求。例如，用到某些 Service Worker 尚不支持的特性时（如报告文件上传的进度），要想绕过 Service Worker，可以设置一个名叫 ngsw-bypass 的请求头或查询参数（这个请求头或查询参数的值会被忽略，可以把它设为空字符串或略去）。

偶尔，可能会需要检查运行中的 Service Worker，以调查问题或确保它在按预期的方式设计运行。浏览器提供了用于调试 Service Worker 的内置工具，而且 Service Worker 本身也包含了一些有用的调试功能。

驱动程序的 NORMAL 状态表示 Service Worker 正在正常运行，并且没有处于降级运行的状态。EXISTING_CLIENTS_ONLY 状态表示 Service Worker 没有应用程序的最新版本的干净副本。较旧的缓存版本可以被安全地使用，所以现有的选项卡将继续使用较旧的版本运行本应用程序，但新的应用将从网络上加载。SAFE_MODE 状态表示 Service Worker 不能保证使用缓存数据的安全性，发生了意外错误或所有缓存版本都无效。这时所有的流量都将从网络提供，尽量少运行 Service Worker 中的代码。EXISTING_CLIENTS_ONLY、SAFE_MODE 这两种状态都是暂时的；它们仅在 Service Worker 实例的生命周期内保存。浏览器有时会终止空闲的 Service Worker，以节省内存和处理能力，并创建一个新的 Service Worker 实例来响应网络事件。无论先前实例的状态如何，新实例均以 NORMAL 状态启动。

空闲任务队列是 Service Worker 中所有在后台发生的未决任务的队列。如果这个队列中存在任何任务，则列出它们的描述。在 Service Worker 中出现的任何错误都会记录在调试日志里。

5. 更新

Chrome 等浏览器提供了能与 Service Worker 交互的开发者工具。使用开发人员工具时，Service Worker 将继续在后台运行，并且不会重新启动。这可能会导致开着 DevTools 时的行为与用户实际遇到的行为不一样。如果查看缓存存储器的查看器，缓存就会经常过期。右键单击缓存存储器的标题并刷新缓存，在 Service Worker 页停止并重新启动，Service Worker 将会触发一次更新检查。

像任何复杂的系统一样，错误或损坏的配置可能会导致 Service Worker 以不可预知的方式工作。虽然它在设计时就尝试将此类问题的影响降至最低，但是，如果管理员需要快速停用 Service Worker，则需要多种故障保护机制，具体操作是删除或重命名文件 ngsw.json。当 Service Worker 对 ngsw.json 的请求返回 404 时，Service Worker 就会删除它的所有缓存并注销自己，这本质上就是自毁。

@angular/service-worker 包中还包含一个小脚本 safety-worker.js，当该脚本文件被加载时，就会把它自己从浏览器中注销，并移除此 Service Worker 的缓存。这个脚本可以作为终极武器来摆脱那些已经安装在客户端页面上的不想要的 Service Worker。不能直接注册安全(Safety)的 Service Worker，必须在想要注销 Service Worker 脚本 URL 中提供 safety-worker.js 的内容，直到确定所有用户都已成功注销了原有的 Service Worker。对大多数网站而言，这意味着应该永远为旧的 Service Worker URL 提供安全 Service Worker。这个脚本可以用来停用@angular/service-worker(并移除相应的缓存)以及任何其他曾在站点上提供过的 Service Worker。

Service Worker 无法在重定向后工作。如果设置了从旧位置到新位置的重定向，则 Service Worker 将停止工作。对于完全从 Service Worker 加载该网站的用户，不会触发重定向。如果不得不更改应用的位置，就可能会出现问题。老的 Service Worker 会尝试更新并将请求发送到原来的位置，该位置重定向到新位置就会导致错误。为了解决这个问题，可能需要用故障安全或安全 Service Worker 的方法移除老的 Service Worker。

11.4　Service Worker 配置

1. 配置文件

配置文件 ngsw-config. json 指定了 Service Worker 应该缓存哪些文件和数据的 URL，以及如何更新缓存的文件和数据。Angular CLI 会在 ng build 期间处理配置文件。配置文件 ngsw-config. json 使用 JSON 格式。所有文件路径都必须以/开头，也就是应用部署目录。如无特别说明，这些模式都使用受限的 glob 格式：** 匹配 0 到多段路径；* 匹配 0 个或更多个除/之外的字符；? 匹配除/之外的一个字符；! 前缀表示该模式是反的，也就是说只包含与该模式不匹配的文件。

2. 资产性资源

资产(Asset)是与应用一起更新的应用版本的一部分。它们可以包含从页面的同源地址加载的资源以及从 CDN 和其他外部 URL 加载的第三方资源。由于在构建时可能没法提前知道所有这些外部 URL，因此也可以指定 URL 的模式。当 Service Worker 处理请求时，它将按照资源组在文件 ngsw-config. json 中出现的顺序对其进行检查。与所请求的资源匹配的第一个资源组处理该请求。建议将更具体的资源组放在列表中较高的位置。例如，与/foo. js 匹配的资源组应出现在与 *. js 匹配的资源组之前。每个资产组都会指定一组资源和一个管理它们的策略。此策略用来决定何时获取资源以及当检测到更改时该怎么做。

资产接口 AssetGroup 包含 name、installMode、updateMode、resources、cacheQueryOptions 等成员。name 是强制性的，用来标识资产组。installMode 决定了资源最初的缓存方式(prefetch 或者 lazy)，默认为 prefetch。其中，prefetch 表明 Service Worker 在缓存当前版本的应用程序时要获取每一个列出的资源。这是一个带宽密集型的模式，但可以确保这些资源在请求时可用，即使浏览器正处于离线状态。lazy 表明 Service Worker 不会预先缓存任何资源，只缓存它收到请求的资源。这是一种按需缓存模式，不会请求的资源也不会被缓存。这对于类似为不同分辨率提供图片的资源很有用，Service Worker 为特定的屏幕和设备方向缓存正确的资源。对于已经存在于缓存中的资源，updateMode 会决定在发现了新版本应用后的缓存行为(prefetch 或者 lazy)的默认值为 installMode。自上一版本以来更改过的所有组中资源都会根据 updateMode 进行更新。prefetch 表明 Service Worker 立即下载并缓存更新过的资源。lazy 表明 Service Worker 不要缓存这些资源，而是先把它们看作未被请求的，等到它们再次被请求时才进行更新。lazy 这个 updateMode 只有在 installMode 也同样是 lazy 时才有效。resources 描述要缓存的资源，分为 files 和 urls。其中，files 列出了与 dist 目录中的文件相匹配的模式，它们可以是单个文件，也可以是能匹配多个文件的类似 glob 的模式；urls 包括要在运行时进行匹配的 URL 和 URL 模式。这些资源不是直接获取的，也没有内容散列，但它们会根据 HTTP 标头进行缓存。这对于类似 Google Fonts 服务的 CDN 非常有用。不支持 glob 的逆模式，将按字面匹配，也就是说它不会匹配除了。之外的任何字符。cacheQueryOptions 这些选项用来修改对请求进行匹配的行为。它们会传给浏览器的函数 Cache # match()。目前，该函数只支持 ignoreSearch，表明忽略查询参数，默认为 false。

3. 数据请求

与资产性资源不同,数据请求不会随应用一起版本化。它们会根据手动配置的策略进行缓存,这些策略对 API 请求和所依赖的其他数据等情况会更有用。接口 DataGroup 包含 name、urls、cacheConfig、cacheQueryOptions 等成员。每个数据组都有一个 name,用作它的唯一标识。urls 是一个 URL 模式的列表,匹配这些模式的 URL 将会根据数据组的策略进行缓存。只有非修改型的请求(如 GET)才会进行缓存。不支持 glob 中的否定模式;? 只做字面匹配,也就是说,它只能匹配? 字符。version 是个整型字段,默认为 0。API 有时可能会以不向后兼容的方式更改格式。新版本的应用可能与旧的 API 格式不兼容,因此也就与该 API 中目前已缓存的资源不兼容。version 提供了一种机制,用于指出这些被缓存的资源已经通过不向后兼容的方式进行了更新,并且旧的缓存条目(即来自以前版本的缓存条目)应该被丢弃。

4. 配置项

cacheConfig 包括 maxSize、maxAge、timeout、strategy 等成员。其中,必需的 maxSize 表示缓存的最大条目数或响应数。开放式缓存可以无限增长,并最终超过存储配额,建议适时清理。必需的 maxAge 表示在响应因失效而要清除之前允许在缓存中留存的时间。maxAge 是一个表示持续时间的字符串,可使用天、小时、分钟、秒、微秒等单位作为后缀。timeout 表示持续时间的字符串,用于指定网络超时时间。如果配置了网络超时时间,Service Worker 就会先等待这么长时间再使用缓存,可使用天、小时、分钟、秒、微秒等单位作为后缀。Service Worker 可以使用 performance、freshness 两种缓存策略之一来获取数据资源。performance 是默认值,表明为尽快给出响应而优化。如果缓存中存在某个资源,则使用这个缓存版本,而不再发起网络请求。它允许资源有一定的陈旧性(取决于 maxAge)以换取更好的性能。适用于那些不经常改变的资源,如用户头像。freshness 为数据的即时性而优化,优先从网络获取请求的数据。只有当网络超时时,请求才会根据 timeout 的设置回退到缓存中。这对于那些频繁变化的资源很有用,如账户余额。还可以模拟第三种策略 staleWhileRevalidate,它会返回缓存的数据(如果可用),也会在后台从网络上获取新数据以供下次使用。要使用 staleWhileRevalidate 策略,需要在 cacheConfig 中把 strategy 设置为 freshness,并且把 timeout 设置为 0ms(毫秒)。本质上说,它会做如下工作:首先尝试从网络上获取;如果网络请求没有在 0ms 内(也就是立刻)完成,就用缓存作为后备(忽略缓存有效期);一旦网络请求完成,就更新缓存以供将来的请求使用;如果指定的资源在缓存中不存在,就等待网络请求。

对于没有匹配上任何资产组或数据组的导航请求,Service Worker 会把它们重定向到指定的索引文件。下列请求将会视为导航请求:它的模式是 navigation;它接受 text/html 响应(根据 Accept 头的值决定);它的 URL 符合特定(不能包含__且最后一段路径中不能包含文件扩展名)的条件。

有时可能希望忽略一些特定的路由(它们可能不是 Angular 应用的一部分),而是把它们传给服务器。urls 包含一个将要在运行期间匹配的 URL 和类似 glob 的 URL 模式。它既可以包含正向模式,也可以包含反向模式(如用! 开头的模式)。只有那些能匹配任意正向 URL 或 URL 模式并且不匹配任何一个反向模式的 URL 才会视为导航请求。当匹配时,URL 查询将会被忽略。

5. 用 Web Worker 处理后台进程

Web Worker 允许在后台线程中运行 CPU 密集型计算，解放主线程以更新用户界面。如果应用程序需要进行很多计算，如生成 CAD 图纸或进行繁重的几何计算，使用 Web Worker 可以提高应用的性能。

在应用的任何位置添加 Web Worker。某些环境或平台（如服务端渲染中使用的@angular/platform-server）不支持 Web Worker，为了确保应用能够在这些环境中工作，必须提供一个回退机制来执行本来要由 Web Worker 执行的计算。

11.5 PWA 的应用开发

微课视频

11.5.1 创建文件 sw.js

在项目 src\examples 根目录下创建 pwaexamples 子目录，在 src\examples\pwaexamples 目录下创建文件 sw.js，代码如例 11-1 所示。

【例 11-1】 创建文件 sw.js 的代码，Service Worker 示例。

```
self.addEventListener("install", event => {
  self.skipWaiting()      //跳过等待
});
self.addEventListener("activate", event => {
  clients.claim()         //立即受控
});
```

11.5.2 创建文件 index.html

在 src\examples\pwaexamples 目录下创建文件 index.html，代码如例 11-2 所示。

【例 11-2】 创建文件 index.html 的代码，应用 PWA 示例。

```
<!DOCTYPE html>
<html>
<head>
  <meta charset="UTF-8" />
  <meta name="viewport" content="width=device-width, initial-scale=1.0" />
  <title>第一个 PWA</title>
</head>
<body>
  <h1>第一个 PWA</h1>
  <img src="images/network.jpg" />
  <script type="module">
    navigator.serviceWorker
      .register("sw.js")
      .then(() => {
        console.log("sw.js 注册成功");
      })
      .catch(e => {
        console.log("sw.js 注册失败", e);
      });
  </script>
</body>
</html>
```

11.5.3 运行文件 index. html

用如例 11-3 所示的命令安装 HTTP Server 工具 http-server,命令如例 11-3 所示。

【例 11-3】 安装工具 http-server 的命令,使用 npm 来安装工具 http-server。

```
npm install http - server  - g
```

在 src\examples\pwaexamples 目录下启动 Windows 命令工具 CMD,用如例 11-4 所示的命令启动工具 http-server,命令如例 11-4 所示。

【例 11-4】 启动工具 http-server 的命令,启动 http-server 作为服务器。

```
http - server
```

运行结果如图 11-1 所示。

```
D:\angularworkspace\firstangular\src\examples\pwaexamples>http-server
Starting up http-server, serving ./

http-server version: 14.1.0

http-server settings:
CORS: disabled
Cache: 3600 seconds
Connection Timeout: 120 seconds
Directory Listings: visible
AutoIndex: visible
Serve GZIP Files: false
Serve Brotli Files: false
Default File Extension: none

Available on:
  http://192.168.218.193:8080
  http://169.254.159.253:8080
  http://192.168.56.1:8080
  http://192.168.99.1:8080
  http://192.168.10.2:8080
  http://192.168.137.1:8080
  http://127.0.0.1:8080
Hit CTRL-C to stop the server
```

图 11-1 启动工具 http-server 的结果

在 src\examples\pwaexamples 目录下创建 images 子目录,在 src\examples\pwaexamples\images 目录下准备一张图片 network.jpg,在浏览器地址栏中输入 localhost:8080,运行文件 index.html,结果如图 11-2 所示。打开 Chrome 的内置工具 DevTools,结果如图 11-3 所示。

图 11-2 在浏览器地址栏中输入 localhost:8080 的结果

图 11-3 在浏览器地址栏中输入 localhost：8080 后 DevTools 工具控制台的结果

11.5.4 组件

在 src\examples\pwaexamples 目录下创建文件 img-card. component. ts，代码如例 11-5 所示。注意，确保 src\assets 目录下有文件 logo. png、offline. jpg 和 network. jpg。

【例 11-5】 创建文件 img-card. component. ts 的代码，定义组件。

```
import {Component} from '@angular/core';
@Component({
  selector: 'app-img-card',
  template: `
    <div>图片组件</div>
    <img [src] = "imgsrc"/>
    <img src = 'assets/offline.jpg'/>
    <img src = '{{imgsrc}}'/>
  `,
})
export class ImgCardComponent  {
   imgsrc  = 'assets/logo.png';
}
```

在 src\examples\pwaexamples 目录下创建文件 pwaexample. component. ts，代码如例 11-6 所示。

【例 11-6】 创建文件 pwaexample. component. ts 的代码，定义组件。

```
import {Component} from '@angular/core';
@Component({
  selector: 'root',
  template: `
    {{title}}
    <h1>第一个 PWA</h1>
    <img src = "assets/network.jpg" />
    <script type = "module">
      navigator.serviceWorker
        .register("sw.js")
        .then(() => {
          console.log("sw.js 注册成功");
```

```
        })
        .catch(e => {
          console.log("sw.js 注册失败", e);
        });
    </script>
    <app-img-card></app-img-card>
  `,
})
export class PwaexampleComponent {
  title = 'Progressive Web Application';
}
```

11.5.5 模块和运行结果

在 src\examples\pwaexamples 目录下创建文件 app-pwa-example. module. ts,代码如例 11-7 所示。

【例 11-7】 创建文件 app-pwa-example. module. ts 的代码,定义路径并声明组件。

```
import {BrowserModule} from '@angular/platform-browser';
import {NgModule} from '@angular/core';
import {RouterModule} from "@angular/router";
import {PwaexampleComponent} from "./pwaexample.component";
import {ImgCardComponent} from "./img-card.component";
@NgModule({
  declarations: [
    ImgCardComponent,
    PwaexampleComponent
  ],
  imports: [
    BrowserModule,
    RouterModule.forRoot([
      {path: 'pwa', component: PwaexampleComponent},
    ]),
  ],
})
export class AppPwaExampleModule {
}
```

修改 src\examples 目录下的文件 examplesmodules1. module. ts,代码如例 11-8 所示。

【例 11-8】 修改文件 examplesmodules1. module. ts 的代码,设置启动组件。

```
import {NgModule} from '@angular/core';
import {AppPwaExampleModule} from "./pwaexamples/app-pwa-example.module";
import {PwaexampleComponent} from "./pwaexamples/pwaexample.component";
@NgModule({
  imports: [
    AppPwaExampleModule,
  ],
    bootstrap: [PwaexampleComponent]
})
export class ExamplesmodulesModule1 {}
```

保持其他文件不变并成功运行程序后,在浏览器地址栏中输入 localhost:4200,自动跳转到 localhost:4200/home,也可以在浏览器地址栏中输入 localhost:4200/pwa,结果如

图 11-4 所示。

图 11-4　成功运行程序后在浏览器地址栏中输入 localhost:4200 的结果

习题 11

一、简答题

1. 简述对 PWA 的理解。
2. 简述对 Service Worker 的理解。
3. 简述对 Web Worker 的理解。

二、实验题

完成 PWA 的应用开发。

第12章

测试及其应用

12.1 测试概述

12.1.1 含义

Angular CLI 创建应用程序项目后可生成 Jasmine 和 Karma 的配置文件。可以通过编辑 src 目录下文件 karma.conf.js 和文件 test.ts 微调测试选项。文件 karma.conf.js 是 Karma 配置文件的一部分。Angular CLI 会基于文件 angular.json 中指定的项目结构和文件 karma.conf.js 在内存中构建出完整的运行时配置。此外，还可以使用其他的测试库和测试运行器来对 Angular 应用程序进行单元测试。每个库和运行器都有自己特有的安装过程、配置项和语法。

测试文件的扩展名必须是.spec.ts，也叫规约(spec)文件。原文件(*.ts)和规约文件(*.spec.ts)位于同一个文件夹中，其根文件名部分(如 app.component)都是一样的。这种约定的好处包括：这些测试很容易找到；一眼就能看到应用中是否缺少一些测试；邻近的测试可以揭示一个部件会如何在上下文中工作；移动源代码时(在所难免)不会忘了同时移动测试源代码；重命名源文件时(在所难免)不会忘了重命名测试文件。

单击某一行测试可以单独重跑这个测试，或者重跑所选测试组(测试套件)中的那些测试。同时，ng test 命令还会监听这些变化。应用程序的集成测试规范可以测试跨文件夹和模块的多个部分之间的交互。它们并不属于任何一个特定的部分，可以在 tests 目录下为它们创建一个合适的文件夹。避免应用程序项目出 Bug 的最佳方式之一就是使用测试套件。持续集成服务器可以配置项目的代码仓库，以便每次提交和收到 Pull Request 时就会运行测试。可以使用 Jenkins 或其他软件来搭建免费持续集成服务器。当要用 Angular CLI 命令 ng test 运行持续集成测试时，可能需要再调整一下配置，以运行 Chrome 浏览器测试。这个配置文件是给 Karma 测试运行器使用的，必须改为不用沙箱的 Chrome 启动方式。

12.1.2 服务测试

服务往往是最容易进行单元测试的文件。服务通常依赖于 Angular 在构造函数中注入的其他服务。几乎总是使用 Angular 依赖注入机制来将服务注入到应用类中，应该有一些测试来体现这种使用模式。在很多情况下，调用服务的构造函数时，可手动创建和注入这些依赖。

当服务有依赖时，DI会查找或创建这些被依赖的服务。如果该被依赖的服务还有自己的依赖，DI也会查找或创建它们。作为服务的测试人员，至少要考虑第一层的服务依赖，用TestBed测试实用工具来提供和创建服务时，可以让Angular DI来创建服务并处理构造函数的参数顺序。

TestBed是Angular测试实用工具中最重要的测试工具之一，它的API很庞大。TestBed创建了一个动态构造的Angular测试模块，用来模拟一个Angular的模块。TestBed.configureTestingModule()方法接受一个元数据对象，它可以拥有模块的大部分属性。

大多数测试套件都会调用beforeEach()方法来为每一个it()方法测试设置前置条件，并依赖TestBed来创建类和注入服务。还有另一种测试，不是调用beforeEach()方法，也不是使用TestBed，而是显式地创建类。

对远程服务器进行HTTP调用的数据服务通常会注入并委托给Angular的HttpClient服务进行XMLHttpRequest调用。数据服务和HttpClient之间的扩展交互较复杂，而HttpClientTestingModule可以让这些测试场景更易于管理。

12.1.3 组件测试

组件是结合了HTML模板和TypeScript的类，由模板和类一起工作。在很多情况下，单独测试组件(不需要DOM参与)能以更简单、更明显的方式验证组件的大部分行为。可以像测试服务类那样来测试一个组件本身。对组件的测试时只测试一个单元。

要测试没有依赖的组件的步骤包括使用关键字new创建一个组件；调用它的API；对其公开状态的期望值进行断言。当组件有依赖时，可能要先使用TestBed来同时创建该组件及其依赖，再测试。

组件会与DOM以及其他组件进行交互。很多组件都与模板中描述的DOM元素进行了复杂的交互，导致一些HTML内容会在组件状态发生变化时出现和消失。对类的测试可以验证类的行为，但无法表明这个组件是否能正确渲染、响应用户输入，或是集成到它的父组件和子组件中，所以必须创建与组件关联的DOM元素，必须检查DOM以确认组件状态是否在适当的时候被正确地显示，并且必须模拟用户与屏幕的交互以确定这些交互是否正确，必须判断该组件的行为是否符合预期。这些测试需要在浏览器DOM中创建宿主元素，就像Angular所做的那样，然后检查组件与DOM的交互是否如模板中描述的那样工作。TestBed可以做这种测试，还可以使用TestBed的其他特性以及其他的测试辅助函数。

配置好TestBed之后就可以调用createComponent()方法；调用createComponent()方法后不能再重新配置TestBed，不能再调用任何TestBed配置方法。否则，TestBed会抛出一个错误。TestBed.createComponent<T>()会创建一个组件T的实例，并为该组件返回一个强类型的ComponentFixture。ComponentFixture是一个测试工具，用于与所创建的组件及其对应的元素进行交互。

ComponentFixture的属性和方法提供了对组件、组件的DOM和Angular环境方面的访问。ComponentFixture.nativeElement(简称nativeElement)的属性依赖于其运行时环境。Angular在编译时不知道nativeElement是什么样的HTML元素，甚至可能不是

HTML 元素。在浏览器中运行的应用中，nativeElement 的值始终是 HTMLElement 或其派生类之一，可以在测试中探索熟悉的方法和属性。当应用程序运行在非浏览器平台(如服务器或 Web Worker)上时，nativeElement 可能具有一个缩小版的 API，甚至根本不存在。

Angular 依靠 DebugElement 抽象在其支持的所有平台上安全地工作。Angular 不会创建 HTML 元素树，而会创建一个 DebugElement 树来封装运行时平台上的原生元素。nativeElement 属性会解包 DebugElement 并返回特定于平台的元素对象。从 Angular 的 core 库中可以导入 DebugElement 符号。

有些应用程序可能要在某些时候运行在不同的平台上。例如，组件可能会首先在服务器上渲染以便在连接不良的设备上更快地启动本应用。服务器端渲染器可能不支持完整的 HTML 元素 API。DebugElement 提供了适用于其支持的所有平台的查询方法。这些查询方法接受一个谓词函数，当 DebugElement 树中的一个节点与选择条件匹配时，该函数返回 true。借助从库中为运行时平台导入 By 类可以创建一个谓词，这里的 By 类是从浏览器平台导入的。静态 By. css()方法会用标准的 CSS 选择器来选择 DebugElement 中的各个节点。但使用 By. css()方法可能会有点过于复杂；而用 HTMLElement()方法(如 querySelector()方法)进行过滤通常更简单、更清晰。

12.1.4 测试指令和管道

属性型指令会修改元素、组件或其他指令的行为。它的名字反映了该指令的应用方式是作为宿主元素的一个属性而存在。此时，测试单个用例不太可能涉及指令的全部能力。要找到并测试那些使用了该指令的所有组件会很乏味、很脆弱，而且几乎不可能做到完全覆盖。纯类测试可能会有一点帮助，但像这种属性型指令往往会操纵 DOM。孤立的单元测试不会触及 DOM，因此也无法给人带来对指令功效(正确有效)的信心。更好的解决方案是创建一个人工测试组件来演示应用该指令的所有方法。

一些好的测试指令包括：By. directive 谓词是一种获取不知道类型但都附有指令的元素的好办法，伪类可以帮助找到那些没有该指令的任意元素。DebugElement. styles 提供了对元素样式的访问，即使没有真正的浏览器也是如此。如果 nativeElement 显得比使用其抽象版本更容易或更清晰，那就把它暴露出来。Angular 会在指令宿主元素的注入器中添加上该指令。DebugElement. properties 允许访问指令设置的自定义属性。

除了@Pipe 元数据和一个接口之外，大多数管道都不依赖于 Angular。编写 DOM 测试来支持管道测试，都是为了对管道进行隔离测试的。

12.1.5 Mock 测试

Angular 执行变更检测时就会发生绑定。在生产环境中，Angular 创建一个组件或者用户输入按键、异步活动(如 AJAX)完成时，都会自动进行变更检测。要让 Angular 测试环境自动运行变更检测，可以通过配置带有 ComponentFixtureAutoDetect 提供者的 TestBed 来实现这一点。先从测试工具函数库中导入 ComponentFixtureAutoDetect，再把它添加到测试模块配置的 providers 中。ComponentFixtureAutoDetect 服务会响应异步活动，如 promise、定时器和 DOM 事件；但却看不见对组件属性的直接同步更新。

很多组件都会分别用 templateUrl 和 styleUrls 属性来指定外部模板和外部样式。运行 ng test 命令测试组件不是问题，因为它会在运行测试之前编译应用。如果在非 Angular CLI 环境中运行这些测试，那么这个组件的测试可能会失败。

组件通常都有服务依赖；而待测组件不必注入真正的服务。事实上，如果它们测试的是替身（如 stub、fake、spies 或 mock）则通常测试效果会更好，而使用真正的服务可能会遇到麻烦。Angular 有一个分层注入系统，它具有多个层级的注入器，从 TestBed 创建的根注入器开始，直到组件树中的各个层级。获得注入服务的最安全的方式（始终有效）就是从被测组件的注入器中获取它。组件注入器是测试夹具所提供的 DebugElement 中的一个属性，还可以通过 TestBed. inject()方法来从根注入器获得服务。这只有当 Angular 要把根注入器中的服务实例注入测试组件时才是可行的。

在测试组件时，只有该服务的公开 API 才有意义。通常，测试本身不应该调用远程服务器，它们应该模拟这样的调用。同步 Observable 的一个关键优势是可以把异步过程转换成同步测试。

12.1.6　异步测试

使用 fakeAsync()方法可以进行异步测试。想要在应用程序的测试时使用 fakeAsync()方法，就必须在测试的环境设置文件中导入 zone. js/testing。如果测试体要进行 XMLHttpRequest 调用，则 fakeAsync()方法无效。函数 tick()是用 TestBed 导入的 Angular 测试工具函数之一。它是 fakeAsync()方法的伴生工具，只能在 fakeAsync()方法测试体内调用它。用函数 tick()来推进（虚拟）时钟，会在所有挂起的异步活动完成之前模拟时间的流逝。在这种情况下，它会等待错误处理程序中的 setTimeout()方法。fakeAsync()方法可以模拟时间的流逝，以便计算出 fakeAsync()方法里面的日期差。

Jasmine 还为模拟日期提供了 clock 特性。Angular 会在 jasmine. clock(). install()方法于 fakeAsync()方法内调用时自动运行这些测试。直到调用了 jasmine. clock(). uninstall()方法为止。fakeAsync()方法不是必需的，如果嵌套它就抛出错误。在默认情况下，此功能处于禁用状态。fakeAsync()方法使用 RxJS 调度器，就像使用 setTimeout()方法或 setInterval()方法一样，但需要导入 zone. js/plugins/zone-patch-rxjs-fake-async 来给 RxJS 调度器打补丁。fakeAsync()方法默认支持 setTimeout()方法、setInterval()方法、requestAnimationFrame()方法、webkitRequestAnimationFrame()方法、mozRequestAnimationFrame()方法等宏任务。如果运行其他宏任务，就会抛出错误。要在依赖 Zone. js 应用中使用< canvas >元素，需要导入 zone-patch-canvas 补丁（或者在 polyfills. ts 中，或者在用到< canvas >的那个文件中）。

RxJS 的 defer 操作符返回一个可观察对象。它的参数是一个返回 promise 或可观察对象的工厂函数。当某个订阅者订阅 defer 生成的可观察对象时，defer 就会调用此工厂函数生成新的可观察对象并让该订阅者订阅这个新对象。defer 操作符会把 Promise. resolve()方法转换成一个新的可观察对象，它和 HttpClient 一样只会发送一次然后立即结束（complete）。这样，当订阅者收到数据后就会自动取消订阅。还有一个类似的用来生成异步错误的测试助手。

要使用 waitForAsync()方法进行异步测试，必须在 test 的设置文件中导入 zone. js/testing。如果用 Angular CLI 创建项目，那就已经在 src/test. ts 中导入了 zone-testing。

waitForAsync()方法通过把测试代码安排在特殊的异步测试区下运行来隐藏某些用来处理异步的样板代码。不需要把 Jasmine 的 done()方法传给测试并让测试调用 done()方法，因为它在 promise 或者可观察对象的回调函数中是 undefined。在 waitForAsync()方法中使用 intervalTimer()方法时，别忘了在测试后通过 clearInterval()方法取消这个定时器，否则 waitForAsync()方法永远不会结束。虽然 waitForAsync()方法和 fakeAsync()方法可以大大简化 Angular 的异步测试，但仍然可以回退到传统技术，并给 it()传一个以 done()方法为参数的函数。但不能在 waitForAsync()方法或 fakeAsync()方法中调用 done()方法。编写带有 done()方法的测试函数要比用 waitForAsync()方法和 fakeAsync()方法的形式笨重，但有时还要编写带有 done()方法的测试函数。

在经过一段显著的延迟之后，可观察对象经常会发送很多次。组件可以用重叠的值序列和错误序列来协调多个可观察对象。RxJS 弹珠测试是一种测试可观察场景的好方法，它既简单又复杂。弹珠测试使用类似的弹珠语言来指定测试中的可观察流和期望值，它的美妙之处在于对可观察对象流的视觉定义。弹珠帧是测试时间线上的虚拟单位，每个符号(—、x、|、#)都表示经过了一帧。冷可观察对象在订阅它之前不会产生值，大多数应用中可观察对象都是冷的，所有 HttpClient 方法返回的都是冷可观察对象；而热可观察对象在订阅它之前就已经生成了这些值。用来报告路由器活动的 Router. events 可观察对象就是一种热可观察对象。

具有输入和输出属性的组件通常会出现在宿主组件的视图模板中。宿主使用属性绑定来设置输入属性，并使用事件绑定来监听输出属性引发的事件。

12.1.7 路由组件测试

所谓路由组件就是指会要求 Router 导航到其他组件的组件。一般来说，应该测试组件而不是路由器，应该只关心组件有没有根据给定的条件导航到正确的地址。路由目标组件是指 Router 导航到的目标。它测试起来可能很复杂，特别是当路由到这个组件包含参数的时候。组件的模板中通常还会有嵌套组件，嵌套组件的模板还可能包含更多组件。这棵组件树可能非常深，并且大多数时候在测试这棵树顶部的组件时，这些嵌套的组件都无关紧要。希望写个单元测试来确认这些链接是否正确使用了 RouterLink 指令时，不必用 Router 进行导航，也不必使用< router-outlet >来指出 Router 应该把路由目标组件插入什么地方。

如果声明的都是真实的组件，那么也同样要声明它们的嵌套组件，并要为这棵组件树中的任何组件提供要注入的所有服务。为了减少测试的工作量，第一种办法是可以对不需要的组件提供桩。这项技术中，要为那些在测试中无关紧要的组件或指令创建和声明一些测试桩。这些测试桩的选择器要和其对应的真实组件一致，但其模板和类是空的。然后在 TestBed 的配置中那些真正有用的组件、指令、管道之后声明它们。第二种办法就是把 NO_ERRORS_SCHEMA 添加到 TestBed. schemas 的元数据中。NO_ERRORS_SCHEMA 要求 Angular 编译器忽略不认识的元素和属性。此种方法称为浅层测试，因为此测试只包含本测试所关心的组件模板中的元素。NO_ERRORS_SCHEMA 方法在这两者中比较简单，但也不要过度使用它，因为 NO_ERRORS_SCHEMA 会阻止编译器提示因为疏忽或拼写错误而缺失的组件和属性。如果人工找出这些 bug 可能要浪费几个小时，但编译器可以立即捕获它们。桩组件方式还有其他优点。虽然这个例子中的桩是空的,但如果想要和它们用某种形式互动,也可以给它们一些裁剪过的模板和类。在实践中,可以在测试代码中组合使用这两种技术。

用 RouterLink 的桩指令进行测试可以确认带有链接和 outlet 的组件的设置的正确性,确认组件有应该有的链接,确认它们都指向了正确的方向。这些测试程序不关心用户在单击链接时的结果,也不关心应用是否会成功地导航到目标组件。对于这些有限的测试目标,使用 RouterLink 桩指令和 RouterOutlet 桩组件是最佳选择。依靠真正的路由器进行测试会让路由组件测试变得很脆弱,即失败可能与组件无关。

12.1.8 调试

如果测试没能如预期般工作,可以在浏览器中查看和调试它们。在浏览器中调试这些测试规约的方式与调试应用时相同。例如,打开 Karma 的浏览器窗口,单击 DEBUG 按钮,弹出一个新的浏览器选项卡并重新运行测试。也可以打开浏览器的 DevTools,在测试中设置一个断点,刷新浏览器,它会在这个断点处停下来。

12.1.9 代码覆盖率

Angular CLI 可以运行单元测试并创建代码覆盖率报告。代码覆盖率报告会展示代码库中可能无法通过单元测试进行正确测试的任意部位。

要生成覆盖率报告,可以在项目的根目录下运行 ng test --no-watch --code-coverage 命令。通过代码覆盖率可以估算出代码测试了多少。如果团队确定要设置单元测试的最小覆盖率,可以使用 Angular CLI 来强制实施这个最低要求。要启用此功能,修改 Karma 配置文件 karma.conf.js,并在 coverageReporter 键下添加 check 属性。check 属性可以在应用程序项目中运行单元测试时强制要求至少 80%的代码覆盖率。

12.2 TestBed 的应用开发

微课视频

12.2.1 创建组件

在项目 src\examples 根目录下创建 testexamples 子目录,在 src\examples\testexamples 目录下创建 banner 子目录,在 src\examples\testexamples\banner 目录下创建文件 banner.component.ts,代码如例 12-1 所示。

【例 12-1】 创建文件 banner.component.ts 的代码,定义组件。

```
import {Component} from '@angular/core';
@Component({
  selector: 'app-banner',
  template: '<h1>{{title}}</h1>',
  styles: ['h1 {color: green; font-size: 350%}']
})
export class BannerComponent {
  title = '主页面信息';
}
```

12.2.2 创建测试文件

在 src\examples\testexamples\banner 目录下创建文件 banner. component. spec. ts，代码如例 12-2 所示。

【例 12-2】 创建文件 banner. component. spec. ts 的代码，定义测试用例。

```
import {ComponentFixture, TestBed} from '@angular/core/testing';
import {BannerComponent} from './banner.component';
//DOM 测试
describe('BannerComponent (inline template)', () => {
  let component: BannerComponent;
  let fixture: ComponentFixture<BannerComponent>;
  let h1: HTMLElement;
  beforeEach(() => {
    TestBed.configureTestingModule({
      declarations: [ BannerComponent ],
    });
    fixture = TestBed.createComponent(BannerComponent);
    component = fixture.componentInstance;
    h1 = fixture.nativeElement.querySelector('h1');
  });
  it('no title in the DOM after createComponent()', () => {
    expect(h1.textContent).toEqual('');
  });
  it('should display original title', () => {
    fixture.detectChanges();
    expect(h1.textContent).toContain(component.title);
  });
  it('should display original title after detectChanges()', () => {
    fixture.detectChanges();
    expect(h1.textContent).toContain(component.title);
  });
  it('should display a different test title', () => {
    component.title = 'Test Title';
    fixture.detectChanges();
    expect(h1.textContent).toContain('Test Title');
  });
});
```

12.2.3 运行结果

运行文件 banner. component. spec. ts，在浏览器地址栏中输入 localhost：9876，此过程可以简称为测试 banner. component. ts，单击图 12-1 中的 DEBUG 按钮，结果如图 12-2 所示。

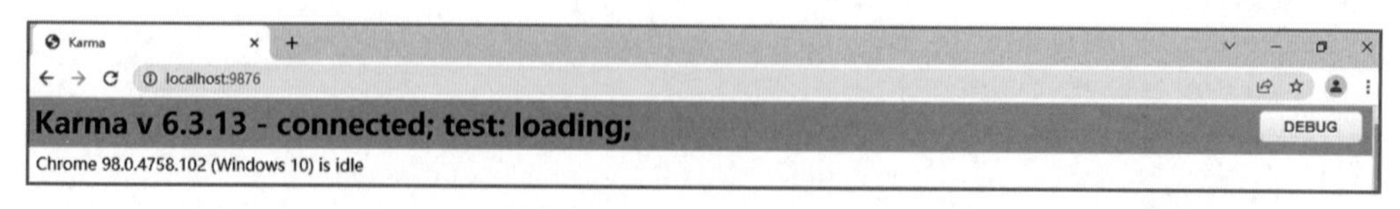

图 12-1 成功运行程序后在浏览器地址栏中输入 localhost：9876 的结果

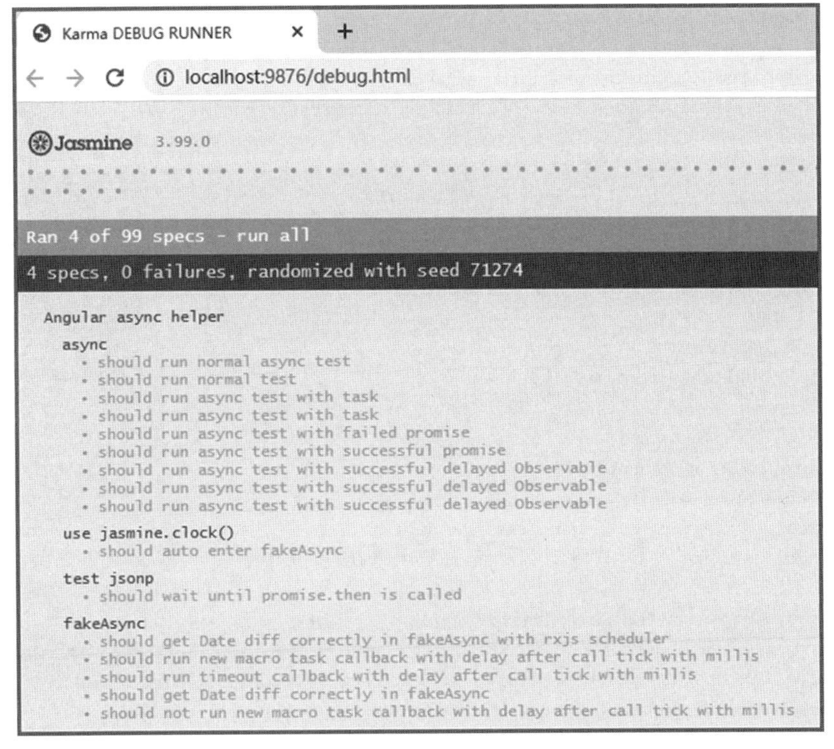

图 12-2　单击图 12-1 中的 DEBUG 按钮的结果

12.3　服务测试应用

微课视频

1. 创建服务

在 src\examples\testexamples 目录下创建 shared 子目录，在 src\examples\testexamples\shared 目录下创建文件 highlight.directive.ts，代码如例 12-3 所示。

【例 12-3】 创建文件 highlight.directive.ts 的代码，定义指令。

```
import {Directive, ElementRef, Input, OnChanges} from '@angular/core';
@Directive({selector: '[highlight]'})
export class HighlightDirective implements OnChanges {
  defaultColor =  'rgb(211, 211, 211)';
  @Input('highlight') bgColor = '';
  constructor(private el: ElementRef) {
    el.nativeElement.style.customProperty = true;
  }
  ngOnChanges() {
    this.el.nativeElement.style.backgroundColor = this.bgColor || this.defaultColor;
  }
}
```

2. 创建测试

在 src\examples\testexamples\shared 目录下创建文件 highlight.directive.spec.ts，代码如例 12-4 所示。

【例 12-4】 创建文件 highlight.directive.spec.ts 的代码，定义测试条件。

```
import {Component, DebugElement} from '@angular/core';
import {ComponentFixture, TestBed} from '@angular/core/testing';
import {By} from '@angular/platform-browser';
import {HighlightDirective} from './highlight.directive';
@Component({
  template: `
  <h2 highlight="yellow">Something Yellow</h2>
  <h2 highlight>The Default (Gray)</h2>
  <h2>No Highlight</h2>
  <input #box [highlight]="box.value" value="cyan"/>`
})
class TestComponent { }
describe('HighlightDirective', () => {
  let fixture: ComponentFixture<TestComponent>;
  let des: DebugElement[];
  let bareH2: DebugElement;
  beforeEach(() => {
    fixture = TestBed.configureTestingModule({
      declarations: [ HighlightDirective, TestComponent ]
    })
    .createComponent(TestComponent);
    fixture.detectChanges();
    des = fixture.debugElement.queryAll(By.directive(HighlightDirective));
    bareH2 = fixture.debugElement.query(By.css('h2:not([highlight])'));
  });
  it('should have three highlighted elements', () => {
    expect(des.length).toBe(3);
  });
  it('should color 1st <h2> background "yellow"', () => {
    const bgColor = des[0].nativeElement.style.backgroundColor;
    expect(bgColor).toBe('yellow');
  });
  it('should color 2nd <h2> background w/ default color', () => {
    const dir = des[1].injector.get(HighlightDirective) as HighlightDirective;
    const bgColor = des[1].nativeElement.style.backgroundColor;
    expect(bgColor).toBe(dir.defaultColor);
  });
  it('should bind <input> background to value color', () => {
    const input = des[2].nativeElement as HTMLInputElement;
    expect(input.style.backgroundColor).toBe('cyan', 'initial backgroundColor');
    input.value = 'green';
    input.dispatchEvent(new Event('input'));
    fixture.detectChanges();
    expect(input.style.backgroundColor).toBe('green', 'changed backgroundColor');
  });
  it('bare <h2> should not have a customProperty', () => {
    expect(bareH2.properties['customProperty']).toBeUndefined();
  });
  it('can inject `HighlightDirective` in 1st <h2>', () => {
    const dir = des[0].injector.get(HighlightDirective);
    expect(dir).toBeTruthy();
  });
  it('cannot inject `HighlightDirective` in 3rd <h2>', () => {
    const dir = bareH2.injector.get(HighlightDirective, null);
```

```
    expect(dir).toBe(null);
  });
  it('should have `HighlightDirective` in 1st <h2> providerTokens', () => {
    expect(des[0].providerTokens).toContain(HighlightDirective);
  });
  it('should not have `HighlightDirective` in 3rd <h2> providerTokens', () => {
    expect(bareH2.providerTokens).not.toContain(HighlightDirective);
  });
});
```

12.4 组件测试应用

1. 创建组件

在 src\examples\testexamples\shared 目录下创建文件 about.component.ts，代码如例 12-5 所示。

【例 12-5】 创建文件 about.component.ts 的代码，定义组件。

```
import {Component} from '@angular/core';
@Component({
  template: `
  <h2 highlight="skyblue">About</h2>
  <h3>Quote of the day:</h3>
  <twain-quote></twain-quote>
  `
})
export class AboutComponent { }
```

2. 创建测试文件

在 src\examples\testexamples\shared 目录下创建文件 about.component.spec.ts，代码如例 12-6 所示。

【例 12-6】 创建文件 about.component.spec.ts 的代码，定义测试条件。

```
import {CUSTOM_ELEMENTS_SCHEMA} from '@angular/core';
import {ComponentFixture, TestBed} from '@angular/core/testing';
import {AboutComponent} from './about.component';
import {HighlightDirective} from '../shared/highlight.directive';
let fixture: ComponentFixture<AboutComponent>;
describe('AboutComponent (highlightDirective)', () => {
  beforeEach(() => {
    fixture = TestBed.configureTestingModule({
      declarations: [ AboutComponent, HighlightDirective ],
      schemas:      [ CUSTOM_ELEMENTS_SCHEMA ]
    })
    .createComponent(AboutComponent);
    fixture.detectChanges();
  });
  it('should have skyblue <h2>', () => {
    const h2: HTMLElement = fixture.nativeElement.querySelector('h2');
    const bgColor = h2.style.backgroundColor;
    expect(bgColor).toBe('skyblue');
  });
});
```

12.5 Jasmine 应用

在 src\examples\testexamples 目录下创建 testing 子目录，在 src\examples\testexamples\testing 目录下创建文件 index.ts，代码如例 12-7 所示。

【例 12-7】 创建文件 index.ts 的代码，定义函数。

```
import {DebugElement} from '@angular/core';
import {tick, ComponentFixture} from '@angular/core/testing';
export * from './async-observable-helpers';
export * from './activated-route-stub';
export * from './jasmine-matchers';
export * from './router-link-directive-stub';
export function advance(f: ComponentFixture<any>): void {
  tick();
  f.detectChanges();
}
export const ButtonClickEvents = {
   left:  {button: 0},
   right: {button: 2}
};
export function click(el: DebugElement | HTMLElement, eventObj: any = ButtonClickEvents.
left): void {
  if (el instanceof HTMLElement) {
    el.click();
  } else {
    el.triggerEventHandler('click', eventObj);
  }
}
```

在 src\examples\testexamples\testing 目录下创建文件 jasmine-matchers.d.ts，代码如例 12-8 所示。

【例 12-8】 创建文件 jasmine-matchers.d.ts 的代码，定义接口。

```
declare namespace jasmine {
  interface Matchers<T> {
    // @ts-ignore
    toHaveText(actual: any, expectationFailOutput?: any): jasmine.CustomMatcher;
  }
}
```

在 src\examples\testexamples\testing 目录下创建文件 jasmine-matchers.ts，代码如例 12-9 所示。

【例 12-9】 创建文件 jasmine-matchers.ts 的代码，定义函数。

```
export function addMatchers(): void {
  jasmine.addMatchers({
    toHaveText
  });
}
function toHaveText(): jasmine.CustomMatcher {
  return {
```

```
    compare: (actual: any, expectedText: string, expectationFailOutput?: any): jasmine.
CustomMatcherResult => {
      const actualText = elementText(actual);
      const pass = actualText.indexOf(expectedText) > -1;
      const message = pass ? '' : composeMessage();
      return { pass, message };
      function composeMessage() {
        const a = (actualText.length < 100 ? actualText : actualText.substr(0, 100) + '...');
        const efo = expectationFailOutput ? ` '${expectationFailOutput}'` : '';
        return `Expected element to have text content '${expectedText}' instead of '${a}'
${efo}`;
      }
    }
  };
}
function elementText(n: any): string {
  if (n instanceof Array) {
    return n.map(elementText).join('');
  }
  if (n.nodeType === Node.COMMENT_NODE) {
    return '';
  }
  if (n.nodeType === Node.ELEMENT_NODE && n.hasChildNodes()) {
    return elementText(Array.prototype.slice.call(n.childNodes));
  }
  if (n.nativeElement) { n = n.nativeElement; }
  return n.textContent;
}
```

12.6　路由测试应用

微课视频

在 src\examples\testexamples\testing 目录下创建文件 activated-route-stub.ts，代码如例 12-10 所示。

【例 12-10】 创建文件 activated-route-stub.ts 的代码，定义类。

```
export {ActivatedRoute} from '@angular/router';
import {convertToParamMap, ParamMap, Params} from '@angular/router';
import {ReplaySubject} from 'rxjs';
export class ActivatedRouteStub {
  private subject = new ReplaySubject<ParamMap>();
  constructor(initialParams?: Params) {
    this.setParamMap(initialParams);
  }
  readonly paramMap = this.subject.asObservable();
  setParamMap(params: Params = {}) {
    this.subject.next(convertToParamMap(params));
  }
}
```

在 src\examples\testexamples\testing 目录下创建文件 async-observable-helpers.ts，代码如例 12-11 所示。

【例 12-11】 创建文件 async-observable-helpers.ts 的代码，定义函数。

```
import {defer} from 'rxjs';
export function asyncData<T>(data: T) {
  return defer(() => Promise.resolve(data));
}
export function asyncError<T>(errorObject: any) {
  return defer(() => Promise.reject(errorObject));
}
```

在 src\examples\testexamples\testing 目录下创建文件 router-link-directive-stub. ts，代码如例 12-12 所示。

【例 12-12】 创建文件 router-link-directive-stub. ts 的代码，定义指令。

```
import {Directive, Input, HostListener} from '@angular/core';
export {RouterLink} from '@angular/router';
@Directive({
  selector: '[routerLink]'
})
export class RouterLinkDirectiveStub {
  @Input('routerLink') linkParams: any;
  navigatedTo: any = null;
  @HostListener('click')
  onClick() {
    this.navigatedTo = this.linkParams;
  }
}
import {NgModule} from '@angular/core';
@NgModule({
  declarations: [
    RouterLinkDirectiveStub
  ]
})
export class RouterStubsModule {}
```

微课视频

12.7 异步测试应用

在 src\examples\testexamples 目录下创建 twain 子目录，在 src\examples\testexamples\twain 目录下创建文件 quote. ts，代码如例 12-13 所示。

【例 12-13】 创建文件 quote. ts 的代码，定义接口。

```
export interface Quote {
  id: number;
  quote: string;
}
```

在 src\examples\testexamples\twain 目录下创建文件 twain. service. ts，代码如例 12-14 所示。

【例 12-14】 创建文件 twain. service. ts 的代码，定义类。

```
import {Injectable} from '@angular/core';
import {HttpClient, HttpErrorResponse} from '@angular/common/http';
import {Observable, of, throwError, Observer} from 'rxjs';
import {concat, map, retryWhen, switchMap, take} from 'rxjs/operators';
```

```
import {Quote} from './quote';
@Injectable()
export class TwainService {
  constructor(private http: HttpClient) { }
  private nextId = 1;
  getQuote(): Observable<string> {
    return Observable.create((observer: Observer<number>) => observer.next(this.nextId++
)).pipe(
      switchMap((id: number) => this.http.get<Quote>(`api/quotes/${id}`)),
      map((q: Quote) => q.quote),
      retryWhen(errors => errors.pipe(
        switchMap((error: HttpErrorResponse)  => {
          if (error.status === 404) {
            this.nextId = 1;
            return of(null);
          }
          console.error(error);
          return throwError('Cannot get Twain quotes from the server');
        }),
        take(2),
        concat(throwError('There are no Twain quotes'))
      ))
    );
  }
}
```

在 src\examples\testexamples\twain 目录下创建文件 twain.data.ts，代码如例 12-15 所示。

【例 12-15】 创建文件 twain.data.ts 的代码，定义数组。

```
import {Quote} from './quote';
export const QUOTES: Quote[] = [
  'Always do right. This will gratify some people and astonish the rest.',
  'I have never let my schooling interfere with my education.',
  'Don\'t go around saying the world owes you a living. The world owes you nothing. It was here first.',
  'Whenever you find yourself on the side of the majority, it is time to pause and reflect.',
  'If you tell the truth, you don\'t have to remember anything.',
  'Clothes make the man. Naked people have little or no influence on society.',
  'It\'s not the size of the dog in the fight, it\'s the size of the fight in the dog.',
  'Truth is stranger than fiction, but it is because Fiction is obliged to stick to
possibilities; Truth isn\'t.',
  'The man who does not read good books has no advantage over the man who cannot read them.',
  'Get your facts first, and then you can distort them as much as you please.',
]
.map((q, i) => ({ id: i + 1, quote: q }));
```

在 src\examples\testexamples\twain 目录下创建文件 twain.component.ts，代码如例 12-16 所示。

【例 12-16】 创建文件 twain.component.ts 的代码，定义组件。

```
import {Component, OnInit} from '@angular/core';
import {Observable, of} from 'rxjs';
import {catchError, startWith} from 'rxjs/operators';
import {TwainService} from './twain.service';
```

```
@Component({
  selector: 'twain-quote',
  template: `
    <p class="twain"><i>{{quote | async}}</i></p>
    <button (click) = "getQuote()"> Next quote </button>
    <p class="error" *ngIf = "errorMessage">{{errorMessage}}</p>`,
  styles: [
    '.twain {font-style: italic;} .error { color: red; }'
  ]
})
export class TwainComponent implements OnInit {
  errorMessage!: string;
  quote!: Observable<string>;
  constructor(private twainService: TwainService) {}
  ngOnInit(): void {
    this.getQuote();
  }
  getQuote() {
    this.errorMessage = '';
    this.quote = this.twainService.getQuote().pipe(
      startWith('...'),
      catchError( (err: any) => {
        setTimeout(() => this.errorMessage = err.message || err.toString());
        return of('...');
      })
    );
  }
}
```

在 src\examples\testexamples\twain 目录下创建文件 twain.component.spec.ts,代码如例 12-17 所示。

【例 12-17】 创建文件 twain.component.spec.ts 的代码,定义组件。

```
import {fakeAsync, ComponentFixture, TestBed, tick, waitForAsync} from '@angular/core/testing';
import {asyncData, asyncError} from '../testing';
import {of, throwError} from 'rxjs';
import {last} from 'rxjs/operators';
import {TwainComponent} from './twain.component';
import {TwainService} from './twain.service';
describe('TwainComponent', () => {
  let component: TwainComponent;
  let fixture: ComponentFixture<TwainComponent>;
  let getQuoteSpy: jasmine.Spy;
  let quoteEl: HTMLElement;
  let testQuote: string;
  const errorMessage = () => {
    const el = fixture.nativeElement.querySelector('.error');
    return el ? el.textContent : null;
  };
  beforeEach(() => {
    testQuote = 'Test Quote';
    const twainService = jasmine.createSpyObj('TwainService', ['getQuote']);
    getQuoteSpy = twainService.getQuote.and.returnValue(of(testQuote));
    TestBed.configureTestingModule({
```

```
    declarations: [TwainComponent],
    providers: [{provide: TwainService, useValue: twainService}]
  });
  fixture = TestBed.createComponent(TwainComponent);
  component = fixture.componentInstance;
  quoteEl = fixture.nativeElement.querySelector('.twain');
});
describe('when test with synchronous observable', () => {
  it('should not show quote before OnInit', () => {
    expect(quoteEl.textContent).toBe('', 'nothing displayed');
    expect(errorMessage()).toBeNull('should not show error element');
    expect(getQuoteSpy.calls.any()).toBe(false, 'getQuote not yet called');
  });
  it('should show quote after component initialized', () => {
    fixture.detectChanges();
    expect(quoteEl.textContent).toBe(testQuote);
    expect(getQuoteSpy.calls.any()).toBe(true, 'getQuote called');
  });
  it('should display error when TwainService fails', fakeAsync(() => {
        getQuoteSpy.and.returnValue(throwError('TwainService test failure'));
        fixture.detectChanges();
        tick();
        fixture.detectChanges();
        expect(errorMessage()).toMatch(/test failure/, 'should display error');
        expect(quoteEl.textContent).toBe('...', 'should show placeholder');
      }));
});
describe('when test with asynchronous observable', () => {
  beforeEach(() => {
    getQuoteSpy.and.returnValue(asyncData(testQuote));
  });
  it('should not show quote before OnInit', () => {
    expect(quoteEl.textContent).toBe('', 'nothing displayed');
    expect(errorMessage()).toBeNull('should not show error element');
    expect(getQuoteSpy.calls.any()).toBe(false, 'getQuote not yet called');
  });
  it('should still not show quote after component initialized', () => {
    fixture.detectChanges();
    expect(quoteEl.textContent).toBe('...', 'should show placeholder');
    expect(errorMessage()).toBeNull('should not show error');
    expect(getQuoteSpy.calls.any()).toBe(true, 'getQuote called');
  });
  it('should show quote after getQuote (fakeAsync)', fakeAsync(() => {
        fixture.detectChanges();
        expect(quoteEl.textContent).toBe('...', 'should show placeholder');
        tick();
        fixture.detectChanges();
        expect(quoteEl.textContent).toBe(testQuote, 'should show quote');
        expect(errorMessage()).toBeNull('should not show error');
      }));
  it('should show quote after getQuote (waitForAsync)', waitForAsync(() => {
        fixture.detectChanges();
        expect(quoteEl.textContent).toBe('...', 'should show placeholder');
        fixture.whenStable().then(() => {
```

```
          fixture.detectChanges();
          expect(quoteEl.textContent).toBe(testQuote);
          expect(errorMessage()).toBeNull('should not show error');
        });
      }));
    it('should show last quote (quote done)', (done: DoneFn) => {
      fixture.detectChanges();
      component.quote.pipe(last()).subscribe(() => {
        fixture.detectChanges();
        expect(quoteEl.textContent).toBe(testQuote);
        expect(errorMessage()).toBeNull('should not show error');
        done();
      });
    });
    it('should show quote after getQuote (spy done)', (done: DoneFn) => {
      fixture.detectChanges();
      getQuoteSpy.calls.mostRecent().returnValue.subscribe(() => {
        fixture.detectChanges();
        expect(quoteEl.textContent).toBe(testQuote);
        expect(errorMessage()).toBeNull('should not show error');
        done();
      });
    });
    it('should display error when TwainService fails', fakeAsync(() => {
          getQuoteSpy.and.returnValue(asyncError<string>('TwainService test failure'));
          fixture.detectChanges();
          tick();
          fixture.detectChanges();
          expect(errorMessage()).toMatch(/test failure/, 'should display error');
          expect(quoteEl.textContent).toBe('...', 'should show placeholder');
      }));
  });
});
```

微课视频

12.8 Mock 测试应用

在 src\examples\testexamples 目录下创建 welcome 子目录，在 src\examples\testexamples\welcome 目录下创建文件 user.service.ts，代码如例 12-18 所示。

【例 12-18】 创建文件 user.service.ts 的代码，定义类。

```
import {Injectable} from '@angular/core';
@Injectable()
export class UserService {
  isLoggedIn = true;
  user = {name: 'Sam Spade'};
}
```

在 src\examples\testexamples\welcome 目录下创建文件 welcome.component.ts，代码如例 12-19 所示。

【例 12-19】 创建文件 welcome.component.ts 的代码，定义组件。

```
import {Component, OnInit} from '@angular/core';
```

```
import {UserService} from "./user.service";
@Component({
  selector: 'app-welcome',
  template: '<h3 class="welcome"><i>{{welcome}}</i></h3>'
})
export class WelcomeComponent implements OnInit {
  welcome = '';
  constructor(private userService: UserService) { }
  ngOnInit(): void {
    this.welcome = this.userService.isLoggedIn ?
      'Welcome, ' + this.userService.user.name : 'Please log in.';
  }
}
```

在 src\examples\testexamples\welcome 目录下创建文件 welcome.component.spec.ts，代码如例 12-20 所示。

【例 12-20】 创建文件 welcome.component.spec.ts 的代码，定义组件测试用例。

```
import {ComponentFixture, inject, TestBed} from '@angular/core/testing';
import {WelcomeComponent} from './welcome.component';
import {UserService} from "./user.service";
class MockUserService {
  isLoggedIn = true;
  user = { name: 'Test User'};
}
//组件测试
describe('WelcomeComponent (class only)', () => {
  let comp: WelcomeComponent;
  let userService: UserService;
  beforeEach(() => {
    TestBed.configureTestingModule({
      providers: [
        WelcomeComponent,
        { provide: UserService, useClass: MockUserService }
      ]
    });
    comp = TestBed.inject(WelcomeComponent);
    userService = TestBed.inject(UserService);
  });
  it('should not have welcome message after construction', () => {
    expect(comp.welcome).toBe('');
  });
  it('should welcome logged in user after Angular calls ngOnInit', () => {
    comp.ngOnInit();
    expect(comp.welcome).toContain(userService.user.name);
  });
  it('should ask user to log in if not logged in after ngOnInit', () => {
    userService.isLoggedIn = false;
    comp.ngOnInit();
    expect(comp.welcome).not.toContain(userService.user.name);
    expect(comp.welcome).toContain('log in');
  });
});
describe('WelcomeComponent', () => {
```

```
    let comp: WelcomeComponent;
    let fixture: ComponentFixture<WelcomeComponent>;
    let componentUserService: UserService;
    let userService: UserService;
    let el: HTMLElement;
    let userServiceStub: Partial<UserService>;
    beforeEach(() => {
      userServiceStub = {
        isLoggedIn: true,
        user: {name: 'Test User'},
      };
      TestBed.configureTestingModule({
         declarations: [ WelcomeComponent ],
         providers: [ {provide: UserService, useValue: userServiceStub} ],
      });
      fixture = TestBed.createComponent(WelcomeComponent);
      comp     = fixture.componentInstance;
      userService = fixture.debugElement.injector.get(UserService);
      componentUserService = userService;
      userService = TestBed.inject(UserService);
      el = fixture.nativeElement.querySelector('.welcome');
    });
    it('should welcome the user', () => {
      fixture.detectChanges();
      const content = el.textContent;
      expect(content).toContain('Welcome', '"Welcome ..."');
      expect(content).toContain('Test User', 'expected name');
    });
    it('should welcome "Bubba"', () => {
      userService.user.name = 'Bubba';
      fixture.detectChanges();
      expect(el.textContent).toContain('Bubba');
    });
    it('should request login if not logged in', () => {
      userService.isLoggedIn = false;
      fixture.detectChanges();
      const content = el.textContent;
      expect(content).not.toContain('Welcome', 'not welcomed');
      expect(content).toMatch(/log in/i, '"log in"');
    });
    it("should inject the component's UserService instance",
      inject([UserService], (service: UserService) => {
      expect(service).toBe(componentUserService);
    }));
    it('TestBed and Component UserService should be the same', () => {
      expect(userService).toBe(componentUserService);
    });
});
```

微课视频

12.9 测试综合应用

12.9.1 创建文件

在 src\examples\testexamples 目录下创建文件 demo.ts,代码如例 12-21 所示。

【例 12-21】 创建文件 demo. ts 的代码,创建接口、类等。

```
import {Component, ContentChildren, Directive, EventEmitter,
         Injectable, Input, Output, Optional,
         HostBinding, HostListener,
         OnInit, OnChanges, OnDestroy,
         Pipe, PipeTransform,
         SimpleChanges} from '@angular/core';
import {of} from 'rxjs';
import {delay} from 'rxjs/operators';
export interface Hero {
  name: string;
}
@Injectable()
export class ValueService {
  value = 'real value';
  getValue() {return this.value;}
  setValue(value: string) {this.value = value;}
  getObservableValue() {return of('observable value');}
  getPromiseValue() {return Promise.resolve('promise value');}
  getObservableDelayValue() {
    return of('observable delay value').pipe(delay(10));
  }
}
//服务通常依赖于 Angular 在构造函数中注入的其他服务
@Injectable()
export class MasterService {
  constructor(private valueService: ValueService) { }
  getValue() {return this.valueService.getValue();}
 * Reverse the input string.
 */
@Pipe({ name: 'reverse' })
export class ReversePipe implements PipeTransform {
  transform(s: string) {
    let r = '';
    for (let i = s.length; i; ) { r += s[--i]; }
    return r;
  }
}
@Component({
  selector: 'bank-account',
  template: `
   Bank Name: {{bank}}
   Account Id: {{id}}
  `
})
export class BankAccountComponent {
  @Input() bank = '';
  @Input('account') id = '';
}
/** A component with attributes, styles, classes, and property setting */
@Component({
  selector: 'bank-account-parent',
  template: `
```

```
    < bank - account
        bank = "RBC"
        account = "4747"
        [style.width.px] = "width"
        [style.color] = "color"
        [class.closed] = "isClosed"
        [class.open] = "!isClosed">
    </bank - account >
  `
})
export class BankAccountParentComponent {
  width = 200;
  color = 'red';
  isClosed = true;
}
@Component({
  selector: 'lightswitch - comp',
  template: `
    < button (click) = "clicked()"> Click me!</button >
    < span >{{message}}</span >`
})
export class LightswitchComponent {
  isOn = false;
  clicked() {this.isOn = !this.isOn;}
  get message() {return `The light is ${this.isOn ? 'On' : 'Off'}`;}
}
@Component({
  selector: 'child - 1',
  template: '< span > Child - 1({{text}})</span >'
})
export class Child1Component {
  @Input() text = 'Original';
}
@Component({
  selector: 'child - 2',
  template: '< div > Child - 2({{text}})</div >'
})
export class Child2Component {
  @Input() text = '';
}
@Component({
  selector: 'child - 3',
  template: '< div > Child - 3({{text}})</div >'
})
export class Child3Component {
  @Input() text = '';
}
@Component({
  selector: 'input - comp',
  template: '< input [(ngModel)] = "name">'
})
export class InputComponent {
  name = 'John';
}
```

```
//属性
@Directive({selector: 'input[value]'})
export class InputValueBinderDirective {
  @HostBinding()
  @Input()
  value: any;
  @Output()
  valueChange: EventEmitter<any> = new EventEmitter();
  @HostListener('input', ['$event.target.value'])
  onInput(value: any) {this.valueChange.emit(value);}
}
@Component({
  selector: 'input-value-comp',
  template: `
    Name: <input [(value)]="name"> {{name}}
  `
})
export class InputValueBinderComponent {
  name = 'Sally';
}
@Component({
  selector: 'parent-comp',
  template: 'Parent(<child-1></child-1>)'
})
export class ParentComponent { }
@Component({
  selector: 'io-comp',
  template: '<div class="hero" (click)="click()">Original {{hero.name}}</div>'
})
export class IoComponent {
  @Input() hero!: Hero;
  @Output() selected = new EventEmitter<Hero>();
  click() {this.selected.emit(this.hero);}
}
@Component({
  selector: 'io-parent-comp',
  template: `
  <p *ngIf="!selectedHero"><i>Click to select a hero</i></p>
  <p *ngIf="selectedHero">The selected hero is {{selectedHero.name}}</p>
  <io-comp
    *ngFor="let hero of heroes"
    [hero]=hero
    (selected)="onSelect($event)">
  </io-comp>
  `
})
export class IoParentComponent {
  heroes: Hero[] = [ {name: 'Bob'}, {name: 'Carol'}, {name: 'Ted'}, {name: 'Alice'} ];
  selectedHero!: Hero;
  onSelect(hero: Hero) {this.selectedHero = hero;}
}
@Component({
  selector: 'my-if-comp',
  template: 'MyIf(<span *ngIf="showMore">More</span>)'
```

```
})
export class MyIfComponent {
  showMore = false;
}
@Component({
  selector: 'my-service-comp',
  template: 'injected value: {{valueService.value}}',
  providers: [ValueService]
})
export class TestProvidersComponent {
  constructor(public valueService: ValueService) {}
}
@Component({
  selector: 'my-service-comp',
  template: 'injected value: {{valueService.value}}',
  viewProviders: [ValueService]
})
export class TestViewProvidersComponent {
  constructor(public valueService: ValueService) {}
}
@Component({
  selector: 'external-template-comp',
  template: `
    <span>from external template</span>
  `
})
export class ExternalTemplateComponent implements OnInit {
  serviceValue = '';
  constructor(@Optional() private service?: ValueService) {  }
  ngOnInit() {
    if (this.service) {this.serviceValue = this.service.getValue();}
  }
}
@Component({
  selector: 'comp-w-ext-comp',
  template: `
  <h3>comp-w-ext-comp</h3>
  <external-template-comp></external-template-comp>
  `
})
export class InnerCompWithExternalTemplateComponent { }
@Component({selector: 'needs-content', template: '<ng-content></ng-content>'})
export class NeedsContentComponent {
  @ContentChildren('content') children: any;
}
@Component({
  selector: 'my-if-child-1',
  template: `
    <h4>MyIfChildComp</h4>
    <div>
      <label>Child value: <input [(ngModel)]="childValue"> </label>
    </div>
    <p><i>Change log:</i></p>
    <div *ngFor="let log of changeLog; let i=index">{{i + 1}} - {{log}}</div>`
```

```
})
export class MyIfChildComponent implements OnInit, OnChanges, OnDestroy {
  @Input() value = '';
  @Output() valueChange = new EventEmitter<string>();
  get childValue() {return this.value;}
  set childValue(v: string) {
    if (this.value === v) {return;}
    this.value = v;
    this.valueChange.emit(v);
  }
  changeLog: string[] = [];
  ngOnInitCalled = false;
  ngOnChangesCounter = 0;
  ngOnDestroyCalled = false;
  ngOnInit()    {
    this.ngOnInitCalled = true;
    this.changeLog.push('ngOnInit called');
  }
  ngOnDestroy() {
    this.ngOnDestroyCalled = true;
    this.changeLog.push('ngOnDestroy called');
  }
  ngOnChanges(changes: SimpleChanges) {
    for (const propName in changes) {
      this.ngOnChangesCounter += 1;
      const prop = changes[propName];
      const cur  = JSON.stringify(prop.currentValue);
      const prev = JSON.stringify(prop.previousValue);
      this.changeLog.push(`${propName}: currentValue = ${cur}, previousValue = ${prev}`
);
    }
  }
}
@Component({
  selector: 'my-if-parent-comp',
  template: `
    <h3>MyIfParentComp</h3>
    <label>Parent value:
      <input [(ngModel)]="parentValue">
    </label>
    <button (click)="clicked()">{{toggleLabel}} Child</button><br>
    <div *ngIf="showChild"
         style="margin: 4px; padding: 4px; background-color: aliceblue;">
      <my-if-child-1  [(value)]="parentValue"></my-if-child-1>
    </div>
  `
})
export class MyIfParentComponent implements OnInit {
  ngOnInitCalled = false;
  parentValue = 'Hello, World';
  showChild = false;
  toggleLabel = 'Unknown';
  ngOnInit() {
    this.ngOnInitCalled = true;
```

```
      this.clicked();
    }
    clicked() {
      this.showChild = !this.showChild;
      this.toggleLabel = this.showChild ? 'Close' : 'Show';
    }
  }
  @Component({
    selector: 'reverse-pipe-comp',
    template: `
      <input [(ngModel)] = "text">
      <span>{{text | reverse}}</span>
    `
  })
  export class ReversePipeComponent {
    text = 'my dog has fleas.';
  }
  @Component({template: '<div>Replace Me</div>'})
  export class ShellComponent { }
  @Component({
    selector: 'root',
    template: `
      <h1>Specs Demo</h1>
      <my-if-parent-comp></my-if-parent-comp>
      <hr>
      <h3>Input/Output Component</h3>
      <io-parent-comp></io-parent-comp>
      <hr>
      <h3>External Template Component</h3>
      <external-template-comp></external-template-comp>
      <hr>
      <h3>Component With External Template Component</h3>
      <comp-w-ext-comp></comp-w-ext-comp>
      <hr>
      <h3>Reverse Pipe</h3>
      <reverse-pipe-comp></reverse-pipe-comp>
      <hr>
      <h3>InputValueBinder Directive</h3>
      <input-value-comp></input-value-comp>
      <hr>
      <h3>Button Component</h3>
      <lightswitch-comp></lightswitch-comp>
      <hr>
      <h3>Needs Content</h3>
      <needs-content #nc>
        <child-1 #content text = "My"></child-1>
        <child-2 #content text = "dog"></child-2>
        <child-2 text = "has"></child-2>
        <child-3 #content text = "fleas"></child-3>
        <div #content>!</div>
      </needs-content>
    `
  })
  export class DemoComponent { }
```

```
export const demoDeclarations = [
  DemoComponent,
  BankAccountComponent, BankAccountParentComponent,
  LightswitchComponent,
  Child1Component, Child2Component, Child3Component,
  ExternalTemplateComponent, InnerCompWithExternalTemplateComponent,
  InputComponent,
  InputValueBinderDirective, InputValueBinderComponent,
  IoComponent, IoParentComponent,
  MyIfComponent, MyIfChildComponent, MyIfParentComponent,
  NeedsContentComponent, ParentComponent,
  TestProvidersComponent, TestViewProvidersComponent,
  ReversePipe, ReversePipeComponent, ShellComponent
];
export const demoProviders = [MasterService, ValueService];
import {NgModule} from '@angular/core';
import {BrowserModule} from '@angular/platform-browser';
import {FormsModule} from '@angular/forms';
@NgModule({
  imports: [BrowserModule, FormsModule],
  declarations: demoDeclarations,
  providers:     demoProviders,
  bootstrap:       [DemoComponent]
})
export class DemoModule { }       //模块 DemoModule 声明
```

在 src\examples\testexamples 目录下创建文件 demo. spec. ts，代码如例 12-22 所示。

【例 12-22】 创建文件 demo. spec. ts 的代码，定义测试用例。

```
import {
  LightswitchComponent,
  MasterService,
  ValueService,
  ReversePipe
} from './demo';
export class FakeValueService extends ValueService {
  override value = 'faked service value';
}
describe('demo (no TestBed):', () => {
   //测试服务
  describe('ValueService', () => {
    let service: ValueService;
    beforeEach(() => { service = new ValueService(); });
    it('#getValue should return real value', () => {
      expect(service.getValue()).toBe('real value');
    });
    it('#getObservableValue should return value from observable',
      (done: DoneFn) => {
      service.getObservableValue().subscribe(value => {
        expect(value).toBe('observable value');
        done();
      });
    });
    it('#getPromiseValue should return value from a promise',
```

```
    (done: DoneFn) => {
    service.getPromiseValue().then(value => {
      expect(value).toBe('promise value');
      done();
    });
  });
});
 //测试有依赖的服务
describe('MasterService without Angular testing support', () => {
  let masterService: MasterService;
  it('#getValue should return real value from the real service', () => {
    masterService = new MasterService(new ValueService());
    expect(masterService.getValue()).toBe('real value');
  });
  it('#getValue should return faked value from a fakeService', () => {
    masterService = new MasterService(new FakeValueService());
    expect(masterService.getValue()).toBe('faked service value');
  });
  it('#getValue should return faked value from a fake object', () => {
    const fake =  { getValue: () => 'fake value' };
    masterService = new MasterService(fake as ValueService);
    expect(masterService.getValue()).toBe('fake value');
  });
  it('#getValue should return stubbed value from a spy', () => {
    const valueServiceSpy =
      jasmine.createSpyObj('ValueService', ['getValue']);
    const stubValue = 'stub value';
    valueServiceSpy.getValue.and.returnValue(stubValue);
    masterService = new MasterService(valueServiceSpy);
    expect(masterService.getValue())
      .toBe(stubValue, 'service returned stub value');
    expect(valueServiceSpy.getValue.calls.count())
      .toBe(1, 'spy method was called once');
    expect(valueServiceSpy.getValue.calls.mostRecent().returnValue)
      .toBe(stubValue);
  });
});
describe('MasterService (no beforeEach)', () => {
  it('#getValue should return stubbed value from a spy', () => {
    const {masterService, stubValue, valueServiceSpy} = setup();
    expect(masterService.getValue())
      .toBe(stubValue, 'service returned stub value');
    expect(valueServiceSpy.getValue.calls.count())
      .toBe(1, 'spy method was called once');
    expect(valueServiceSpy.getValue.calls.mostRecent().returnValue)
      .toBe(stubValue);
  });
  function setup() {
    const valueServiceSpy =
      jasmine.createSpyObj('ValueService', ['getValue']);
    const stubValue = 'stub value';
    const masterService = new MasterService(valueServiceSpy);
    valueServiceSpy.getValue.and.returnValue(stubValue);
    return { masterService, stubValue, valueServiceSpy };
```

```
    }
  });
  describe('ReversePipe', () => {
    let pipe: ReversePipe;
    beforeEach(() => { pipe = new ReversePipe(); });
    it('transforms "abc" to "cba"', () => {
      expect(pipe.transform('abc')).toBe('cba');
    });
    it('no change to palindrome: "able was I ere I saw elba"', () => {
      const palindrome = 'able was I ere I saw elba';
      expect(pipe.transform(palindrome)).toBe(palindrome);
    });
  });
  describe('LightswitchComp', () => {
    it('#clicked() should toggle #isOn', () => {
      const comp = new LightswitchComponent();
      expect(comp.isOn).toBe(false, 'off at first');
      comp.clicked();
      expect(comp.isOn).toBe(true, 'on after click');
      comp.clicked();
      expect(comp.isOn).toBe(false, 'off after second click');
    });
    it('#clicked() should set #message to "is on"', () => {
      const comp = new LightswitchComponent();
      expect(comp.message).toMatch(/is off/i, 'off at first');
      comp.clicked();
      expect(comp.message).toMatch(/is on/i, 'on after clicked');
    });
  });
});
```

在 src\examples\testexamples 目录下创建文件 demo.testbed.spec.ts，代码如例 12-23 所示。

【例 12-23】 创建文件 demo.testbed.spec.ts 的代码，定义测试用例。

```
import {
  DemoModule,
  BankAccountComponent, BankAccountParentComponent,
  LightswitchComponent,
  Child1Component, Child2Component, Child3Component,
  MasterService,
  ValueService,
  ExternalTemplateComponent,
  InputComponent,
  IoComponent, IoParentComponent,
  MyIfComponent, MyIfChildComponent, MyIfParentComponent,
  NeedsContentComponent, ParentComponent,
  TestProvidersComponent, TestViewProvidersComponent,
  ReversePipeComponent, ShellComponent
} from './demo';
import {By} from '@angular/platform-browser';
import {Component,
        DebugElement,
        Injectable} from '@angular/core';
```

```
import {FormsModule} from '@angular/forms';
import {NgModel, NgControl} from '@angular/forms';
import {
  ComponentFixture, fakeAsync, inject, TestBed, tick, waitForAsync
} from '@angular/core/testing';
import {addMatchers} from "./testing/jasmine-matchers";
import {click} from "./testing";
export class NotProvided extends ValueService { /* example below */ }
beforeEach(addMatchers);
describe('demo (with TestBed):', () => {
  describe('ValueService', () => {
    let service: ValueService;
    //调用 beforeEach()为 it()测试设置前置条件,并依赖 TestBed 创建类和注入服务
    beforeEach(() => {
      //使用 TestBed 测试服务
      TestBed.configureTestingModule({providers: [ValueService]});
      service = TestBed.inject(ValueService);
    });
    it('should use ValueService', () => {
      service = TestBed.inject(ValueService);
      expect(service.getValue()).toBe('real value');
    });
    it('can inject a default value when service is not provided', () => {
      expect(TestBed.inject(NotProvided, null)).toBeNull();
    });
    it('test should wait for ValueService.getPromiseValue', waitForAsync(() => {
      service.getPromiseValue().then(
        value => expect(value).toBe('promise value')
      );
    }));
    it('test should wait for ValueService.getObservableValue', waitForAsync(() => {
      service.getObservableValue().subscribe(
        value => expect(value).toBe('observable value')
      );
    }));
    it('test should wait for ValueService.getObservableDelayValue', (done: DoneFn) => {
      service.getObservableDelayValue().subscribe(value => {
        expect(value).toBe('observable delay value');
        done();
      });
    });
    it('should allow the use of fakeAsync', fakeAsync(() => {
      let value: any;
      service.getPromiseValue().then((val: any) => value = val);
      tick();
      expect(value).toBe('promise value');
    }));
  });
  describe('MasterService', () => {
    let masterService: MasterService;
    let valueServiceSpy: jasmine.SpyObj<ValueService>;
    beforeEach(() => {
      const spy = jasmine.createSpyObj('ValueService', ['getValue']);
      TestBed.configureTestingModule({
```

```
        providers: [
          MasterService,
          {provide: ValueService, useValue: spy}
        ]
      });
      masterService = TestBed.inject(MasterService);
      valueServiceSpy = TestBed.inject(ValueService) as jasmine.SpyObj<ValueService>;
    });
    it('#getValue should return stubbed value from a spy', () => {
      const stubValue = 'stub value';
      valueServiceSpy.getValue.and.returnValue(stubValue);
      expect(masterService.getValue())
        .toBe(stubValue, 'service returned stub value');
      expect(valueServiceSpy.getValue.calls.count())
        .toBe(1, 'spy method was called once');
      expect(valueServiceSpy.getValue.calls.mostRecent().returnValue)
        .toBe(stubValue);
    });
  });
  describe('use inject within `it`', () => {
    beforeEach(() => {
      TestBed.configureTestingModule({ providers: [ValueService] });
    });
    it('should use modified providers',
      inject([ValueService], (service: ValueService) => {
        service.setValue('value modified in beforeEach');
        expect(service.getValue())
          .toBe('value modified in beforeEach');
      })
    );
  });
  describe('using waitForAsync(inject) within beforeEach', () => {
    let serviceValue: string;
    beforeEach(() => {
      TestBed.configureTestingModule({providers: [ValueService]});
    });
    beforeEach(waitForAsync(inject([ValueService], (service: ValueService) => {
      service.getPromiseValue().then(value => serviceValue = value);
    })));
    it('should use asynchronously modified value ... in synchronous test', () => {
      expect(serviceValue).toBe('promise value');
    });
  });
  describe('TestBed component tests', () => {
    beforeEach(waitForAsync(() => {
      TestBed
        .configureTestingModule({
          imports: [DemoModule],
        })
        .compileComponents();
    }));
    it('should create a component with inline template', () => {
      const fixture = TestBed.createComponent(Child1Component);
      fixture.detectChanges();
```

```
    // @ts-ignore
    expect(fixture).toHaveText('Child');
  });
  it('should create a component with external template', () => {
    const fixture = TestBed.createComponent(ExternalTemplateComponent);
    fixture.detectChanges();
    // @ts-ignore
    expect(fixture).toHaveText('from external template');
  });
  it('should allow changing members of the component', () => {
    const fixture = TestBed.createComponent(MyIfComponent);
    fixture.detectChanges();
    // @ts-ignore
    expect(fixture).toHaveText('MyIf()');
    fixture.componentInstance.showMore = true;
    fixture.detectChanges();
    // @ts-ignore
    expect(fixture).toHaveText('MyIf(More)');
  });
  it('should create a nested component bound to inputs/outputs', () => {
    const fixture = TestBed.createComponent(IoParentComponent);
    fixture.detectChanges();
    const heroes = fixture.debugElement.queryAll(By.css('.hero'));
    expect(heroes.length).toBeGreaterThan(0, 'has heroes');
    const comp = fixture.componentInstance;
    const hero = comp.heroes[0];
    click(heroes[0]);
    fixture.detectChanges();
    const selected = fixture.debugElement.query(By.css('p'));
    // @ts-ignore
    expect(selected).toHaveText(hero.name);
  });
  it('can access the instance variable of an `*ngFor` row component', () => {
    const fixture = TestBed.createComponent(IoParentComponent);
    const comp = fixture.componentInstance;
    const heroName = comp.heroes[0].name;
    fixture.detectChanges();
    const ngForRow = fixture.debugElement.query(By.directive(IoComponent));
    const hero = ngForRow.context.hero;
    expect(hero.name).toBe(heroName, 'ngRow.context.hero');
    const rowComp = ngForRow.componentInstance;
    expect(rowComp).toEqual(jasmine.any(IoComponent), 'component is IoComp');
    expect(rowComp.hero.name).toBe(heroName, 'component.hero');
  });
  it('should support clicking a button', () => {
    const fixture = TestBed.createComponent(LightswitchComponent);
    const btn = fixture.debugElement.query(By.css('button'));
    const span = fixture.debugElement.query(By.css('span')).nativeElement;
    fixture.detectChanges();
    expect(span.textContent).toMatch(/is off/i, 'before click');
    click(btn);
    fixture.detectChanges();
    expect(span.textContent).toMatch(/is on/i, 'after click');
  });
```

```
    it('should support entering text in input box (ngModel)', waitForAsync(() => {
      const expectedOrigName = 'John';
      const expectedNewName = 'Sally';
      const fixture = TestBed.createComponent(InputComponent);
      fixture.detectChanges();
      const comp = fixture.componentInstance;
      const input = fixture. debugElement. query ( By. css ( ' input ' )). nativeElement as
HTMLInputElement;
      expect(comp.name).toBe(expectedOrigName,
        `At start name should be ${expectedOrigName} `);
      fixture.whenStable().then(() => {
        expect(input.value).toBe(expectedOrigName,
          `After ngModel updates input box, input.value should be ${expectedOrigName} `);
        input.value = expectedNewName;
        expect(comp.name).toBe(expectedOrigName,
          `comp.name should still be ${expectedOrigName} after value change, before binding
happens`);
        input.dispatchEvent(new Event('input'));
        return fixture.whenStable();
      })
        .then(() => {
          expect(comp.name).toBe(expectedNewName,
            `After ngModel updates the model, comp.name should be ${expectedNewName} `);
        });
    }));
    it('should support entering text in input box (ngModel) - fakeAsync', fakeAsync(() => {
      const expectedOrigName = 'John';
      const expectedNewName = 'Sally';
      const fixture = TestBed.createComponent(InputComponent);
      fixture.detectChanges();
      const comp = fixture.componentInstance;
      const input = fixture. debugElement. query ( By. css ( ' input ' )). nativeElement as
HTMLInputElement;
      expect(comp.name).toBe(expectedOrigName,
        `At start name should be ${expectedOrigName} `);
      tick();
      expect(input.value).toBe(expectedOrigName,
        `After ngModel updates input box, input.value should be ${expectedOrigName} `);
      input.value = expectedNewName;
      expect(comp.name).toBe(expectedOrigName,
        `comp.name should still be ${expectedOrigName} after value change, before binding
happens`);
      input.dispatchEvent(new Event('input'));
      tick();
      expect(comp.name).toBe(expectedNewName,
        `After ngModel updates the model, comp.name should be ${expectedNewName} `);
    }));
    it('ReversePipeComp should reverse the input text', fakeAsync(() => {
      const inputText = 'the quick brown fox.';
      const expectedText = '.xof nworb kciuq eht';
      const fixture = TestBed.createComponent(ReversePipeComponent);
      fixture.detectChanges();
      const comp = fixture.componentInstance;
      const input = fixture. debugElement. query ( By. css ( ' input ' )). nativeElement as
```

```
HTMLInputElement;
      const span = fixture.debugElement.query(By.css('span')).nativeElement as HTMLElement;
      input.value = inputText;
      input.dispatchEvent(new Event('input'));
      tick();
      fixture.detectChanges();
      expect(span.textContent).toBe(expectedText, 'output span');
      expect(comp.text).toBe(inputText, 'component.text');
    }));
    it('can examine attached directives and listeners', () => {
      const fixture = TestBed.createComponent(InputComponent);
      fixture.detectChanges();
      const inputEl = fixture.debugElement.query(By.css('input'));
      expect(inputEl.providerTokens).toContain(NgModel, 'NgModel directive');
      const ngControl = inputEl.injector.get(NgControl);
      expect(ngControl).toEqual(jasmine.any(NgControl), 'NgControl directive');
      expect(inputEl.listeners.length).toBeGreaterThan(2, 'several listeners attached');
    });
    it('BankAccountComponent should set attributes, styles, classes, and properties', () => {
      const fixture = TestBed.createComponent(BankAccountParentComponent);
      fixture.detectChanges();
      const comp = fixture.componentInstance;
      const el = fixture.debugElement.children[0];
      const childComp = el.componentInstance as BankAccountComponent;
      expect(childComp).toEqual(jasmine.any(BankAccountComponent));
      expect(el.context).toBe(childComp, 'context is the child component');
      expect(el.attributes['account']).toBe(childComp.id, 'account attribute');
      expect(el.attributes['bank']).toBe(childComp.bank, 'bank attribute');
      expect(el.classes['closed']).toBe(true, 'closed class');
      expect(el.classes['open']).toBeFalsy('open class');
      expect(el.styles['color']).toBe(comp.color, 'color style');
      expect(el.styles['width']).toBe(comp.width + 'px', 'width style');
    });
  });
  describe('TestBed component overrides:', () => {
    //没有 beforeEach() 的测试
    it("should override ChildComp's template", () => {
      const fixture = TestBed.configureTestingModule({
        declarations: [Child1Component],
      })
        .overrideComponent(Child1Component, {
          set: { template: '<span>Fake</span>' }
        })
        .createComponent(Child1Component);

      fixture.detectChanges();
      // @ts-ignore
      expect(fixture).toHaveText('Fake');
    });
    it("should override TestProvidersComp's ValueService provider", () => {
      const fixture = TestBed.configureTestingModule({
        declarations: [TestProvidersComponent],
      })
        .overrideComponent(TestProvidersComponent, {
```

```
        remove: { providers: [ValueService] },
        add: { providers: [{ provide: ValueService, useClass: FakeValueService }] },
      })
      .createComponent(TestProvidersComponent);
    fixture.detectChanges();
    // @ts-ignore
    expect(fixture).toHaveText('injected value: faked value', 'text');
    const tokens = fixture.debugElement.providerTokens;
    expect(tokens).toContain(fixture.componentInstance.constructor, 'component ctor');
    expect(tokens).toContain(TestProvidersComponent, 'TestProvidersComp');
    expect(tokens).toContain(ValueService, 'ValueService');
  });
  it("should override TestViewProvidersComp's ValueService viewProvider", () => {
    const fixture = TestBed.configureTestingModule({
      declarations: [TestViewProvidersComponent],
    })
      .overrideComponent(TestViewProvidersComponent, {
        set: { viewProviders: [{ provide: ValueService, useClass: FakeValueService }] },
      })
      .createComponent(TestViewProvidersComponent);
    fixture.detectChanges();
    // @ts-ignore
    expect(fixture).toHaveText('injected value: faked value');
  });
  it("injected provider should not be same as component's provider", () => {
    @Component({ template: '<my-service-comp></my-service-comp>' })
    class TestComponent { }
    const fixture = TestBed.configureTestingModule({
      declarations: [TestComponent, TestProvidersComponent],
      providers: [ValueService]
    })
      .overrideComponent(TestComponent, {
        set: { providers: [{ provide: ValueService, useValue: {} }] }
      })
      .overrideComponent(TestProvidersComponent, {
        set: { providers: [{ provide: ValueService, useClass: FakeValueService }] }
      })
      .createComponent(TestComponent);
    let testBedProvider!: ValueService;
    inject([ValueService], (s: ValueService) => testBedProvider = s)();
    const tcProvider = fixture.debugElement.injector.get(ValueService) as ValueService;
    const tpcProvider = fixture.debugElement.children[0].injector.get(ValueService) as
FakeValueService;
    expect(testBedProvider).not.toBe(tcProvider, 'testBed/tc not same providers');
    expect(testBedProvider).not.toBe(tpcProvider, 'testBed/tpc not same providers');
    expect(testBedProvider instanceof ValueService).toBe(true, 'testBedProvider is
ValueService');
    expect(tcProvider).toEqual({} as ValueService, 'tcProvider is {}');
    expect(tpcProvider instanceof FakeValueService).toBe(true, 'tpcProvider is
FakeValueService');
  });
  it('can access template local variables as references', () => {
    const fixture = TestBed.configureTestingModule({
    declarations: [ShellComponent, NeedsContentComponent, Child1Component, Child2Component,
```

```
Child3Component],
      })
        .overrideComponent(ShellComponent, {
          set: {
            selector: 'test-shell',
            template: `
          <needs-content #nc>
            <child-1 #content text="My"></child-1>
            <child-2 #content text="dog"></child-2>
            <child-2 text="has"></child-2>
            <child-3 #content text="fleas"></child-3>
            <div #content>!</div>
          </needs-content>
          `
          }
        })
        .createComponent(ShellComponent);
      fixture.detectChanges();
      const el = fixture.debugElement.children[0];
      const comp = el.componentInstance;
      expect(comp.children.toArray().length).toBe(4,
        'three different child components and an ElementRef with #content');
      expect(el.references['nc']).toBe(comp, '#nc reference to component');
      const contentRefs = el.queryAll( de => de.references['content']);
      expect(contentRefs.length).toBe(4, 'elements w/ a #content reference');
    });
  });
  describe('nested (one-deep) component override', () => {
    beforeEach(() => {
      TestBed.configureTestingModule({
        declarations: [ParentComponent, FakeChildComponent]
      });
    });
    it('ParentComp should use Fake Child component', () => {
      const fixture = TestBed.createComponent(ParentComponent);
      fixture.detectChanges();
      // @ts-ignore
      expect(fixture).toHaveText('Parent(Fake Child)');
    });
  });
  describe('nested (two-deep) component override', () => {
    beforeEach(() => {
      TestBed.configureTestingModule({
        declarations: [ParentComponent, FakeChildWithGrandchildComponent, FakeGrandchildComponent]
      });
    });
    it('should use Fake Grandchild component', () => {
      const fixture = TestBed.createComponent(ParentComponent);
      fixture.detectChanges();
      // @ts-ignore
      expect(fixture).toHaveText('Parent(Fake Child(Fake Grandchild))');
    });
  });
  describe('lifecycle hooks w/ MyIfParentComp', () => {
```

```
let fixture: ComponentFixture<MyIfParentComponent>;
let parent: MyIfParentComponent;
let child: MyIfChildComponent;
beforeEach(() => {
  TestBed.configureTestingModule({
    imports: [FormsModule],
    declarations: [MyIfChildComponent, MyIfParentComponent]
  });
  fixture = TestBed.createComponent(MyIfParentComponent);
  parent = fixture.componentInstance;
});
it('should instantiate parent component', () => {
  expect(parent).not.toBeNull('parent component should exist');
});
it('parent component OnInit should NOT be called before first detectChanges()', () => {
  expect(parent.ngOnInitCalled).toBe(false);
});
it('parent component OnInit should be called after first detectChanges()', () => {
  fixture.detectChanges();
  expect(parent.ngOnInitCalled).toBe(true);
});
it('child component should exist after OnInit', () => {
  fixture.detectChanges();
  getChild();
  expect(child instanceof MyIfChildComponent).toBe(true, 'should create child');
});
it("should have called child component's OnInit ", () => {
  fixture.detectChanges();
  getChild();
  expect(child.ngOnInitCalled).toBe(true);
});
it('child component called OnChanges once', () => {
  fixture.detectChanges();
  getChild();
  expect(child.ngOnChangesCounter).toBe(1);
});
it('changed parent value flows to child', () => {
  fixture.detectChanges();
  getChild();
  parent.parentValue = 'foo';
  fixture.detectChanges();
  expect(child.ngOnChangesCounter).toBe(2,
    'expected 2 changes: initial value and changed value');
  expect(child.childValue).toBe('foo',
    'childValue should eq changed parent value');
});
it('changed child value flows to parent', waitForAsync(() => {
  fixture.detectChanges();
  getChild();
  child.childValue = 'bar';
  return new Promise<void>(resolve => {
    setTimeout(() => resolve(), 0);
  })
    .then(() => {
```

```
          fixture.detectChanges();
          expect(child.ngOnChangesCounter).toBe(2,
            'expected 2 changes: initial value and changed value');
          expect(parent.parentValue).toBe('bar',
            'parentValue should eq changed parent value');
        });
    }));
    it('clicking "Close Child" triggers child OnDestroy', () => {
      fixture.detectChanges();
      getChild();
      const btn = fixture.debugElement.query(By.css('button'));
      click(btn);
      fixture.detectChanges();
      expect(child.ngOnDestroyCalled).toBe(true);
    });
    function getChild() {
      let childDe: DebugElement;
      try {
        childDe = fixture.debugElement.children[4].children[0];
      } catch (err) { /* we'll report the error */ }
      childDe = fixture.debugElement
        .queryAll(de => de.componentInstance instanceof MyIfChildComponent)[0];
      childDe = fixture.debugElement
        .query(de => de.componentInstance instanceof MyIfChildComponent);
      if (childDe && childDe.componentInstance) {
        child = childDe.componentInstance;
      } else {
        fail('Unable to find MyIfChildComp within MyIfParentComp');
      }
      return child;
    }
  });
});
@Component({
  selector: 'child-1',
  template: 'Fake Child'
})
class FakeChildComponent { }
@Component({
  selector: 'child-1',
  template: 'Fake Child(<grandchild-1></grandchild-1>)'
})
class FakeChildWithGrandchildComponent { }
@Component({
  selector: 'grandchild-1',
  template: 'Fake Grandchild'
})
class FakeGrandchildComponent { }
@Injectable()
class FakeValueService extends ValueService {
  override value = 'faked value';
}
```

在 src\examples\testexamples 目录下创建文件 async-helper.spec.ts，代码如例 12-24 所示。

【例 12-24】 创建文件 async-helper. spec. ts 的代码，定义测试用例。

```
import {fakeAsync, tick, waitForAsync} from '@angular/core/testing';
import {interval, of} from 'rxjs';
import {delay, take} from 'rxjs/operators';
describe('Angular async helper', () => {
  describe('async', () => {
    let actuallyDone = false;
    beforeEach(() => {
      actuallyDone = false;
    });
    afterEach(() => {
      expect(actuallyDone).toBe(true, 'actuallyDone should be true');
    });
    it('should run normal test', () => {
      actuallyDone = true;
    });
    it('should run normal async test', (done: DoneFn) => {
      setTimeout(() => {
        actuallyDone = true;
        done();
      }, 0);
    });
    it('should run async test with task', waitForAsync(() => {
         setTimeout(() => {
           actuallyDone = true;
         }, 0);
       }));
    it('should run async test with task', waitForAsync(() => {
         const id = setInterval(() => {
           actuallyDone = true;
           clearInterval(id);
         }, 100);
       }));
    it('should run async test with successful promise', waitForAsync(() => {
         const p = new Promise(resolve => {
           setTimeout(resolve, 10);
         });
         p.then(() => {
           actuallyDone = true;
         });
       }));
    it('should run async test with failed promise', waitForAsync(() => {
         const p = new Promise((resolve, reject) => {
           setTimeout(reject, 10);
         });
         p.catch(() => {
           actuallyDone = true;
         });
       }));
    it('should run async test with successful delayed Observable', (done: DoneFn) => {
      const source = of(true).pipe(delay(10));
      source.subscribe(val => actuallyDone = true, err => fail(err), done);
    });
```

```
    it('should run async test with successful delayed Observable', waitForAsync(() => {
        const source = of(true).pipe(delay(10));
        source.subscribe(val => actuallyDone = true, err => fail(err));
      }));
    it('should run async test with successful delayed Observable', fakeAsync(() => {
        const source = of(true).pipe(delay(10));
        source.subscribe(val => actuallyDone = true, err => fail(err));
        tick(10);
      }));
  });
  describe('fakeAsync', () => {
    it('should run timeout callback with delay after call tick with millis', fakeAsync(() => {
        let called = false;
        setTimeout(() => {
          called = true;
        }, 100);
        tick(100);
        expect(called).toBe(true);
      }));
    it('should run new macro task callback with delay after call tick with millis',
      fakeAsync(() => {
        function nestedTimer(cb: () => any): void {
          setTimeout(() => setTimeout(() => cb()));
        }
        const callback = jasmine.createSpy('callback');
        nestedTimer(callback);
        expect(callback).not.toHaveBeenCalled();
        tick(0);
        expect(callback).toHaveBeenCalled();
      }));
    it('should not run new macro task callback with delay after call tick with millis',
      fakeAsync(() => {
        function nestedTimer(cb: () => any): void {
          setTimeout(() => setTimeout(() => cb()));
        }
        const callback = jasmine.createSpy('callback');
        nestedTimer(callback);
        expect(callback).not.toHaveBeenCalled();
        tick(0, {processNewMacroTasksSynchronously: false});
        expect(callback).not.toHaveBeenCalled();
        tick(0);
        expect(callback).toHaveBeenCalled();
      }));
    it('should get Date diff correctly in fakeAsync', fakeAsync(() => {
        const start = Date.now();
        tick(100);
        const end = Date.now();
        expect(end - start).toBe(100);
      }));
    it('should get Date diff correctly in fakeAsync with rxjs scheduler', fakeAsync(() => {
        let result = '';
        of('hello').pipe(delay(1000)).subscribe(v => {
          result = v;
        });
```

```
        expect(result).toBe('');
        tick(1000);
        expect(result).toBe('hello');
        const start = new Date().getTime();
        let dateDiff = 0;
        interval(1000).pipe(take(2)).subscribe(() => dateDiff = (new Date().getTime() -
start));
        tick(1000);
        expect(dateDiff).toBe(1000);
        tick(1000);
        expect(dateDiff).toBe(2000);
      }));
  });
  describe('use jasmine.clock()', () => {
    beforeEach(() => {
      jasmine.clock().install();
    });
    afterEach(() => {
      jasmine.clock().uninstall();
    });
    it('should auto enter fakeAsync', () => {
      let called = false;
      setTimeout(() => {
        called = true;
      }, 100);
      jasmine.clock().tick(100);
      expect(called).toBe(true);
    });
  });
  describe('test jsonp', () => {
    function jsonp(url: string, callback: () => void) {
    }
    it('should wait until promise.then is called', waitForAsync(() => {
        let finished = false;
        new Promise<void>(res => {
          jsonp('localhost:8080/jsonp', () => {
            finished = true;
            res();
          });
        }).then(() => {
          expect(finished).toBe(true);
        });
      }));
  });
});
```

12.9.2 模块和运行结果

在 src\examples\testexamples 目录下创建文件 app-test-example.module.ts，代码如例 12-25 所示。

【例 12-25】 创建文件 app-test-example.module.ts 的代码，声明组件和指令。

```
import {NgModule} from '@angular/core';
```

```
import {BannerComponent} from "./banner/banner.component";
import {WelcomeComponent} from "./welcome/welcome.component";
import {TwainComponent} from "./twain/twain.component";
import {AboutComponent} from "./shared/about.component";
import {HighlightDirective} from "./shared/highlight.directive";
import {TesthomeComponent} from "./testhome.component";
import {BrowserModule} from "@angular/platform-browser";
import {UserService} from "./welcome/user.service";
import {TwainService} from "./twain/twain.service";
@NgModule({
  imports: [
    BrowserModule,
  ],
  declarations: [
    BannerComponent,
    WelcomeComponent,
    TwainComponent,
    AboutComponent,
    HighlightDirective,
    TesthomeComponent
  ],
  providers: [
    UserService,
    TwainService
  ],
})
export class AppTestExampleModule {
}
```

在 src\examples\testexamples 目录下创建文件 demo-main.ts,代码如例 12-26 所示。

【例 12-26】 创建文件 demo-main.ts 的代码,设置启动项目的入口点文件。

```
import { platformBrowserDynamic } from '@angular/platform-browser-dynamic';
import { DemoModule } from './demo';
platformBrowserDynamic().bootstrapModule(DemoModule);
```

修改 src\examples 目录下的文件 examplesmodules1.module.ts,代码如例 12-27 所示。

【例 12-27】 修改文件 examplesmodules1.module.ts 的代码,设置启动组件。

```
import {NgModule} from '@angular/core';
import {TesthomeComponent} from "./testexample/testhome.component";
import {AppTestExampleModule} from "./testexample/app-test-example.module";
@NgModule({
  imports: [
    AppTestExampleModule
  ],
  bootstrap: [TesthomeComponent]
})
export class ExamplesmodulesModule1 {}
```

保持其他文件不变并成功运行程序后,在浏览器地址栏中输入 localhost:4200,自动跳转到 localhost:4200/home,结果如图 12-3 所示。

例 12-26 中的代码用文件 main.ts 的代码替换后,保持其他文件不变,可以成功运行。在浏览器地址栏中输入 localhost:4200,部分结果如图 12-4 所示。请读者自己参考源代码进行更多结果的验证。

图 12-3 成功运行程序后在浏览器地址栏中输入 localhost:4200 的结果

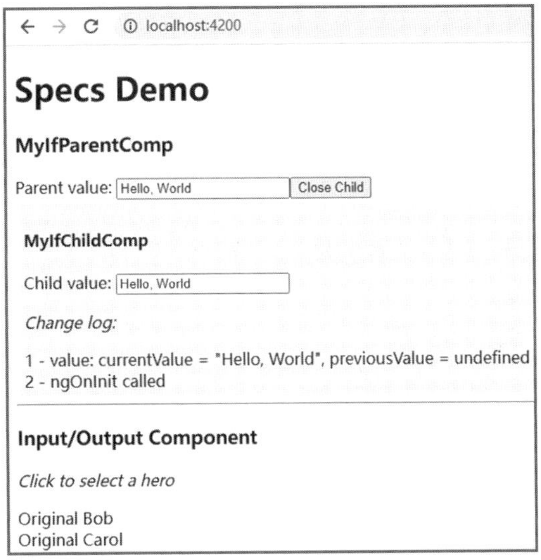

图 12-4 成功运行程序后在浏览器地址栏中输入 localhost:4200 的部分结果(从上往下)

习题 12

一、简答题

简述对测试的理解。

二、实验题

完成测试的应用开发。

第13章

高阶技术

13.1 Angular 统一平台

1. 含义

Angular 统一平台(Universal)是一项在服务端运行 Angular 下应用的技术。标准的 Angular 应用程序会运行在浏览器中,它会通过对 DOM 的处理和渲染页面,以响应用户的操作。统一平台会在服务端运行,生成一些静态的应用页面,然后再通过客户端启动静态页面。这意味着,应用程序的渲染会更快,让用户可以在应用程序变得完全可交互之前先查看应用程序的布局。使用 Angular CLI 为应用程序做好服务端渲染的准备。Angular Universal 需要用到活跃 LTS(长期支持版本)或维护 LTS 版本的 Node.js。从服务端渲染的应用过渡到客户端应用的过程会很快,但还是应该在实际场景中测试一下应用程序。

2. 应用

有三个主要的理由来为应用创建一个 Universal 版本,可以通过搜索引擎优化(SEO)来帮助网络爬虫;可以提升在手机和低功耗设备上的性能;还可以迅速显示出第一个支持首次内容绘制(FCP)的页面。

Google、Bing、Facebook、Twitter 和其他社交媒体网站都依赖网络爬虫去索引应用内容,并且让它的内容可以被用户通过网络搜索到。这些网络爬虫可能不会像人类那样导航到具有高度交互性的 Angular 应用程序,并为其建立索引。Angular Universal 可以生成应用的静态版本,它易搜索、可链接,浏览时也不必借助 JavaScript;它使站点可以被预览,因为每个 URL 返回的都是一个完全渲染好的页面。

有些设备不支持 JavaScript 或 JavaScript 执行得很差,导致用户体验较差。对于这些情况,可能会需要该应用程序的服务端渲染的、无 JavaScript 的版本。虽然有一些限制,但是这个版本可能是完全没办法使用该应用程序的用户的唯一选择。

快速显示用户界面第一页对于吸引用户是至关重要的。加载速度越快的页面,其效果会越好。应用程序要启动得更快一点,以便用户在决定做别的事情之前吸引他们的注意力。

使用 Angular Universal,可以为应用程序生成着陆页,它们看起来就和完整的应用一样。这些着陆页是纯 HTML,并且即使 JavaScript 被禁用了也能显示。这些页面不会处理浏览器事件,不过它们可以在这个网站中进行导航。实践中,可能要使用一个着陆页的静态版本来保持用户的注意力,同时,也会在幕后加载完整的 Angular 应用。用户会觉得着陆页几乎是立即出现的，而当完整的应用程序加载完之后，又可以获得完整的交互体验。

3. 服务器

Universal Web 服务器使用 Universal 模板引擎渲染出的静态 HTML 来响应对用户应用页面的请求。服务器接收并响应来自客户端(通常是浏览器)的 HTTP 请求，并回复静态文件，如脚本、CSS 和图片。它可以直接响应数据请求，也可以作为独立数据服务器的代理进行响应。任何一种 Web 服务器技术都可以作为 Universal 应用的服务器，只要它能调用 Universal 的 renderModule( )方法。Universal 应用程序使用 platform-server 包(而不是 platform-browser)，它提供了 DOM 的服务端实现、XMLHttpRequest 以及其他不依赖浏览器的底层特性。

服务器会把客户端对应用页面的请求传给 NgUniversal 的 ngExpressEngine。在内部实现上，它会调用 Universal 的 renderModule( )方法，它还提供了具有缓存等功能的 renderModule()方法接收一个模板 HTML 页面(通常是 index.html)、一个包含组件的 Angular 模块和一个用于决定该显示哪些组件的路由。路由从客户端的请求中传给服务器，每次请求都会给出所请求路由的一个适当的视图。最后，服务器把渲染好的页面返回给客户端。

由于 Universal 应用程序并没有运行在浏览器中，因此服务器上可能会缺少浏览器的某些 API 和其他能力，如服务端应用不能引用浏览器独有的全局对象(如 Window、Document、Navigator 或 Location)。Angular 提供了这些对象的可注入的抽象层，如 location 或 document，它可以作为所调用的 API 的等效替身。如果 Angular 没有提供它，也可以写一个自己的抽象层，当在浏览器中运行时，就把它委托给浏览器 API，当它在服务器中运行时，就提供一个符合要求的代为实现(或称垫片，shimming)。由于没有鼠标或键盘事件，所以 Universal 应用程序也不能依赖于用户单击某个按钮来显示某个组件。Universal 应用程序必须根据客户端发过来的请求决定要渲染的内容。把该应用程序做成可路由的，是一种好的设计方案。

Web 服务器必须把对应用程序页面的请求和其他类型的请求区分开。有三种常见类型的请求：数据请求、应用导航(请求的 URL 不带扩展名)和静态资源(所有其他请求)。如果应用程序只会通过服务器渲染，那么单击应用程序中任何一个链接都会发到服务器，就像导航时的地址会发到路由器一样。

在服务器上运行时请求的 URL 必须以某种方式转换为绝对 URL，在浏览器中运行时它们是相对 URL。如果使用@nguniversal/*-engine 包之一(如@nguniversal/express-engine)，就会自动做这件事。无须再做任何事情来让相对 URL 在服务器上运行。如果没有使用@nguniversal/*-engine 包，就需要手动进行处理：将完整的请求 URL 传给 renderModule()方法或 renderModuleFactory()方法的 options 参数。此选项的侵入性最小，因为它不需要对应用程序进行任何更改。

4. 预渲染

Angular Universal 允许预先渲染应用程序的页面。预先渲染是在构建时处理动态页面生成静态 HTML 的过程。要预先渲染静态页面，要先向应用程序添加 SSR 功能。向应用程序添加预先渲染时，可以使用不同构建选项：browserTarget 指定要构建的目标；serverTarget 指定用于预先渲染的应用的服务器目标；routes 定义要预先渲染的额外路由数组；guessRoutes 指定构建器是否应该提取路由并猜测要渲染的路径，默认为 true；routesFile 指定一个文件，其中包含要预先渲染的所有路由的列表，以换行符分隔，如果有大量路由，则此选项很有用；numProcesses 指定在运行预先渲染命令时要使用的 CPU 数量。还可以预先渲染动态路由。可以在命令行中提供额外的路由，使用文件来提供路由，预先渲染指定路由。

13.2 Angular CLI

1. 含义

很多 Angular CLI 命令都要在代码上执行一些复杂的处理，如风格检查(lint)构建或测试。这些命令会通过一个叫作架构师(architect)的内部工具来运行 Angular CLI 构建器，而这些构建器会运用一些第三方工具来完成目标任务。在 Angular 8 中，Angular CLI 构建器的 API 是稳定的，适合通过添加或修改命令来自定义 Angular CLI 的开发人员使用。

内部架构师工具会把工作委托给名叫构建器的处理器函数。处理器函数接收一组 options 对象(JSON 对象)和一个 context 对象(BuilderContext 对象)。options 对象是由 Angular CLI 的用户提供的，而 context 对象则由 Angular CLI 构建器的 API 提供。除了上下文信息之外，context 对象还允许访问调度方法 context.scheduleTarget()方法。调度器会用指定的目标配置来执行构建器处理函数。这个构建器处理函数可以是同步的(返回一个值)或异步的(返回一个 promise)，也可以监视并返回多个值(返回一个 Observable)。最终返回的值全都是 BuilderOutput 类型的。该对象包含一个逻辑字段 success 和一个可以包含错误信息的可选字段 error。

2. 内置构建器

Angular 提供了一些内置构建器，供 Angular CLI 命令使用(如 ng build 和 ng test 等)。这些内置 Angular CLI 构建器的默认目标配置可以在工作区配置文件 angular.json 的 architect 部分找到(并进行自定义)；也可以通过创建自定义构建器来扩展和自定义 Angular；还可以使用 Angular CLI 命令 ng run 来运行自定义构建器。

构建器必须有一个已定义的目标，此目标会把构建器与特定的输入配置和项目关联起来。目标是在 Angular CLI 配置文件 angular.json 中定义。目标用于指定要使用的构建器、默认的选项配置，以及指定的备用配置。架构师工具使用目标定义来为一次特定的执行解析输入选项。每个项目的 architect 部分都会为 Angular CLI 命令(如 build、test 和 lint)配置构建器目标。例如，默认情况下，build 命令会运行@angular-devkit/build-angular:browser 构建器来执行 build 任务，并传入配置文件 angular.json 中为 build 目标指定的默认选项值。

通用的 Angular CLI 命令 ng run 的第一个参数是形如 project:target[:configuration]的目标字符串。其中，project 是与目标关联的项目的名称；target 是文件 angular.json 中 architect 下的指定构建器配置；可选的 configuration 用于覆盖指定目标的具体配置名称。如果构建器调用另一个构建器，它可能需要读取一个传入的目标字符串。使用@angular-devkit/architect 中的工具函数 targetFromTargetString()可以把这个字符串解析成一个对象。

架构师工具会异步运行构建器。要调用某个构建器，就要在所有配置解析完成之后安排一个要运行的任务。在调度器返回 BuilderRun 控件对象之前，不会执行该构建器函数。Angular CLI 通常会通过调用 context.scheduleTarget()方法来调度任务，然后使用文件 angular.json 中的目标定义来解析输入选项。如果输入有效，架构师工具会创建上下文并执行该构建器；还可以通过调用 context.scheduleBuilder()方法从另一个构建器或测试中调用某个构建器；直接把 options 对象传给该方法，并且这些选项值会根据这个构建器的模式进行验证，而无须进一步调整。

架构师工具希望构建器运行一次(默认情况下)后返回。这种行为与需要监视文件更改的构建器(如 Webpack)并不完全兼容。如果需要在监视模式下使用，那么构建器处理函数应返回一个 Observable。架构师工具会订阅 Observable，直到 Observable 完成(complete)为止。如果使用相同的参数再次调度这个构建器，架构师工具还能复用这个 Observable。这个构建器应该总是在每次执行后发出一个 BuilderOutput 对象。一旦它被执行，就会进入一个由外部事件触发的监视模式。如果一个事件导致它被重启，那么此构建器应该执行 context.reportRunning()方法来告诉架构师工具再次运行它。如果调度器还计划了另一次运行，就会阻止架构师工具停掉这个构建器。

当构建器通过调用 BuilderRun.stop()方法来退出监视模式时，架构师工具可以从构建器的 Observable 中取消订阅，并调用构建器的退出逻辑进行清理(这种行为也允许停止和清理运行时间过长的构建)。

3. 自定义构建器

Angular CLI 构建器 API 提供了一种通过构建器执行自定义逻辑，以改变 Angular CLI 行为的方法。构建器既可以是同步的，也可以是异步的；它可以只执行一次，也可以监视外部事件，还可以调度其他构建器或目标。构建器在配置文件 angular.json 中指定了选项的默认值，它可以被目标的备用配置覆盖，还可以进一步被命令行标志所覆盖。建议用户使用集成测试来测试架构师工具的构建器；还可以用单元测试来验证这个构建器的执行逻辑。如果构建器返回一个 Observable，应该在 Observable 的退出逻辑中进行清理。

13.3 Angular 语言服务

1. 含义

Angular 语言服务为代码编辑器提供了一种在 Angular 模板中获取自动补全、错误、提示和导航的方法。它支持位于独立 HTML 文件中的外部模板以及内部模板。要启用最新的语言服务功能，在 tsconfig.json 文件中将 strictTemplates 选项设置为 true。

自动补全可以在输入时提供当前情境下的候选内容和提示，从而提高开发速度。

Angular 语言服务能对代码中存在的错误进行预警。快捷信息功能使用户可以悬停以查看组件、指令、模块等的来源。单击“转到定义”按钮或按 F12 键直接转到定义。

2. 支持工具

Angular 语言服务目前在 Visual Studio Code 和 WebStorm、Sublime Text 和 Eclipse IDE 中都有可用的扩展。在 WebStorm 中，启用 Angular 与 AngularJS 插件，从 WebStorm 2019.1 开始，@angular/language-service 已经不再需要了，应该从 package.json 文件中移除。

将编辑器与语言服务一起使用时，编辑器将启动一个单独的语言服务进程，并使用语言服务协议通过 RPC 与之通信。输入编辑器时，编辑器会将信息发送到语言服务流程，以跟踪项目状态。触发模板中的完成列表时，编辑器首先将模板解析为 HTML 抽象语法树(AST)。Angular 编译器解释这棵树以确定上下文：模板属于哪个模块，如当前作用域、组件选择器以及光标在模板 AST 中的位置。然后，编译器就可以确定可能位于该位置的符号。如果要进行插值，则需要更多的操作步骤。如果在 div 有{{data.---}}的插值，并且在 data.---之后需要自动补全列表，则编译器可能无法使用 HTML AST 查找答案。因为 HTML AST 只能告诉编译器某些文本带有字符{{data.---}}。模板解析器会生成一个表达式 AST，该表达式位于模板 AST 中。然后，Angular 语言服务会在其上下文中查找 data.---，询问 TypeScript 语言服务 data 的成员是什么并返回可能性列表。

13.4 AOT 编译器

1. 作用

Angular 应用程序需要先进行编译才能在浏览器中运行。在浏览器下载和运行代码之前的编译阶段，Angular 预先(AOT)编译器会先把 HTML 和 TypeScript 代码转换成高效的 JavaScript 代码。在构建期间，编译应用可以使浏览器中的渲染更快速。

使用 AOT 编译的部分原因包括以下 5 点。

(1) 更快的渲染，借助 AOT 浏览器可以下载应用程序的预编译版本。浏览器加载的是可执行代码，因此它可以立即渲染应用程序，而无须等待先编译好应用程序。

(2) 更少的异步请求。编译器会在应用 JavaScript 中内联外部 HTML 模板和 CSS 样式表，从而消除对源文件的单独 AJAX 请求。

(3) 较小的下载。如果已编译应用程序，则无须下载 Angular 编译器。编译器荷载大约占 Angular 的 1/2，因此省略编译器会大大减少应用的有效荷载。

(4) 尽早检测模板错误。AOT 编译器会在构建步骤中检测并报告模板绑定错误，然后用户才能看到它们。

(5) 更高的安全性。AOT 在将模板和组件提供给客户端之前就将其编译为 JavaScript 文件。没有要读取的模板，没有潜藏风险的客户端 HTML 或 JavaScript eval(求值)，受到注入攻击的机会就更少了。

2. JIT 和 AOT

Angular 提供了即时(JIT)编译、AOT 编译两种方式来编译应用。在运行期间，即时编译器在浏览器中编译应用程序，是 Angular 8 及更早版本的默认工作方式。AOT 编译器在构建时编译应用程序和库，是 Angular 9 及后续版本的默认工作方式。当运行 Angular CLI 命令的 ng build 或 ng serve 时，编译类型取决于配置文件 angular.json 中 aot 属性。默认情况下，新 Angular CLI 应用 aot 为 true。

AOT 编译器会提取元数据来解释应由 Angular 管理的应用部分。可以在装饰器(如@Component()装饰器和@Input()装饰器)中显式指定元数据，也可以在被装饰类的构造函数声明中隐式指定元数据。元数据告诉 Angular 要如何构造应用类的实例并在运行时与它们进行交互。

3. AOT 编译阶段

AOT 编译分为代码分析、代码生成、模板类型检查(可选)三个阶段。

(1) 在代码分析阶段,TypeScript 编译器和 AOT 收集器会创建源码的表现层。收集器不会尝试解释其收集到的元数据,它只是尽可能地表达元数据,并在检测到元数据语法冲突时记录错误。AOT 收集器会记录 Angular 装饰器中的元数据,并把它们输出到文件. metadata. json 中和每个类型定义文件. d. ts 相对应。可以把文件. metadata. json 看做一个包括全部装饰器的元数据的全景图,就像抽象语法树(AST)一样。Angular 的 schema. ts 把这个 JSON 格式表示成了一组 TypeScript 接口。AOT 收集器只能理解 JavaScript 的一个子集。定义元数据对象时要遵循一些语法限制。如果表达式使用了不支持的语法,收集器就会往文件. metadata. json 中写入一个错误节点。如果编译器用到元数据中的这部分内容来生成应用代码,它就会报告这个错误。AOT 编译中的错误通常是由元数据不符合编译器的要求而引起的。AOT 编译器不支持函数表达式和箭头函数(也称为 Lambda 函数)。编译器只会解析到已导出符号的引用。收集器可以在收集期间执行表达式,并将其结果记录到文件. metadata. json 中。编译器不能引用模板常量,因为它是未导出的。但是收集器可以通过内部模板常量的方式把它写进元数据定义中。

(2) 在代码生成阶段,编译器的 StaticReflector 会解释代码分析阶段中收集的元数据,对元数据执行附加验证,如果检测到元数据违反了限制,就抛出错误。解释文件. metadata. json 是编译器在代码生成阶段要承担的工作。只要语法有效,收集器就可以用 new 来表示函数调用或对象创建。但是,编译器在后面的步骤中可以拒绝生成对特定函数的调用或对特定对象的创建。编译器只能创建某些类的实例,仅支持核心装饰器,并且仅支持对返回表达式的宏(函数或静态方法)的调用。编译器只允许创建来自@angular/core 的 InjectionToken 类创建实例,并只支持来自@angular/core 模块的 Angular 装饰器的元数据。

(3) 在模板类型检查阶段,Angular 模板编译器使用 TypeScript 编译器来验证模板中的绑定表达式。通过往该项目的 TypeScript 配置文件 angularCompilerOptions 中添加编译器选项 fullTemplateTypeCheck,可以显式启用本阶段。当模板绑定表达式中检测到类型错误时,进行模板验证时就会生成错误。这与 TypeScript 编译器在处理文件. ts 中的代码时的报告错误很相似。

只能使用 TypeScript 的一个子集书写元数据,它必须满足下列限制:表达式语法只支持 JavaScript 的一个有限的子集;只能引用代码收缩后导出的符号;只能调用编译器支持的那些函数;被装饰和用于数据绑定的类成员必须是公共的。

13.5 Angular 应用的运行

13.5.1 不同配置方式

可以用不同的默认值来为项目定义出不同的命名配置项,如 stage,每个命名配置项都可以具有某些选项的默认值,并应用于各种构建目标,如 build。

项目基础环境文件 environment. ts 包含了默认的环境设置,如没有指定环境时,build 命令就会用它作为构建目标。可以添加其他变量,可以用该环境对象附加属性的形式,也可以用独立对象的形式。

Angular CLI 的主配置文件 angular. json 中的每个构建目标下都包含了一个 fileReplacements 区段,可以把 TypeScript 程序中的任何文件替换为针对特定目标的版本。当构建目标需要包含针对特定环境(如生产或预生产)的代码或变量时,这非常有用。默认情况下不会替换任何文件。

当应用程序的功能不断增长时,其文件大小也会同步增长。Angular CLI 允许通过配置项来限制文件大小,以确保应用的各个部分都处于定义的范围内。可以在 Angular CLI 配置文件 angular. json 的 budgets 区段为每个所配置的环境定义这些范围。如果配置了边界范围,构建系统就会在发现应用的某个部分达到或超过设置的边界范围时发出警告或报错。

Angular 应用中避免依赖 CommonJS 模块。对 CommonJS 模块的依赖会阻止打包器和压缩器优化应用,这会导致更大的打包尺寸。建议在应用程序中都使用 ECMAScript 模块来实现较小的打包尺寸。

Angular CLI 使用 Autoprefixer 来确保对不同浏览器及其版本的兼容性。当要从构建中针对特定的目标浏览器或排除指定的浏览器版本时,这是很有必要的。在内部,Autoprefixer 依赖一个名叫 Browserslist 的库来指出需要为哪些浏览器加前缀。Browserlist 会在文件 package. json 的 browserlist 属性或文件. browserslistrc 中配置这些选项。

用 Webpack 开发服务器中的代理支持可以把特定的 URL 转发给后端服务器,只要传入--proxy-config 选项就可以了。如果定义了环境变量 http_proxy 或 HTTP_PROXY,当运行 npm start 时,就会自动添加一个 agent 来通过企业代理转发网络调用。

13.5.2 开发者工具 DevTools

Angular DevTools(简称 DevTools)是一个 Chrome 扩展程序,可为 Angular 应用程序提供调试和剖析功能。DevTools 支持 Angular 9 及更高版本,并支持 Ivy。安装 DevTools 后,可以在 DevTools 的 Angular 标签下找到本扩展程序。

打开扩展程序时,还会看到 Components、Profiler 两个选项卡。Components 可以浏览应用中的组件和指令并预览或编辑它们的状态。通过 Profiler 可以剖析应用程序并了解变更检测执行期间的性能瓶颈。在 DevTools 应用界面的右上角,将找到页面上正在运行哪个版本的 Angular 以及该扩展的最后一次提交的哈希串。

13.5.3 开发、构建和布置

在开发过程中,可以使用 ng serve 命令来借助 webpack-dev-server 在本地内存中构建、监控和提供服务。打算部署应用程序(服务)时,就必须使用 ng build 命令来构建应用程序并在其他地方部署这些构建结果。ng build 命令会把生成的构建结果写入输出文件夹中。默认情况下,输出目录是 dist\project-name。要输出到其他文件夹,就要修改文件 angular. json 中的 outputPath。

同时使用两个终端才能体验到实时刷新的特性。在第一个终端上,在监控(watch)模式下执行 ng build 命令把该应用程序编译进 dist 文件夹。当源文件发生变化时,ng build --watch 命令就会重新生成输出文件。在第二个终端上,安装一个 Web 服务器,使用输出文件夹中的内容运行它。每当输出了新文件,服务器就会自动刷新浏览器。该方法只能用于开发和测试，在部署应用程序时，不支持该特性，也不是安全的方式。

Angular CLI 的 ng deploy 命令(在 Angular 8.3.0 版中引入)执行与项目关联的配置工作。有许多第三方构建器实现了到不同平台的部署功能。可以通过运行 ng add[package name]命令把它们中的任何一个添加到项目中。添加具有部署功能的程序包时，它将为所选项目自动更新配置文件 angular.json 中 deploy 部分。然后，就可以使用 ng deploy 命令来部署项目了。ng add @angular/fire 后执行 ng deploy 命令是交互式的。

如果要将应用程序部署到自己管理的服务器上，或者缺少针对云平台的构建器，则可以创建支持使用 ng deploy 命令的构建器。最简化的部署方式就是在开发环境构建下，把其输出复制到 Web 服务器上。

对服务器或准备部署到服务器的文件要做的一些修改。带路由的应用必须以 index.html 作为后备页面。Angular 应用很适合用简单的静态 HTML 服务器提供服务。不需要服务端引擎来动态合成应用页面，因为 Angular 会在客户端完成这件事。

如果应用程序使用 Angular 路由器，就必须配置服务器，让它对不存在的文件返回应用程序的宿主页。带路由的应用程序应该支持深链接。所谓深链接就是指一个 URL 用于指定到应用程序内某个组件的路径。当用户从运行中的客户端应用导航到这个 URL 时，Angular 路由器会拦截这个 URL，并且把它路由到正确的页面。但是，当从邮件中单击链接或在浏览器地址栏中输入它或仅仅在详情页刷新浏览器时，所有这些操作都是由浏览器本身处理的。在应用程序的控制范围之外，浏览器会直接向服务器请求 URL，而路由器没机会插手。静态服务器在收到请求时会拒绝请求，并返回一个 404-Not Found 错误，除非，它被配置成了返回 index.html。

没有一种配置可以适用于所有服务器。Angular 应用程序在向与该应用程序的宿主服务器不同域的服务器发起请求时，可能会遇到一种 CORS(跨域资源共享)错误。浏览器会阻止该请求，除非得到服务器的明确许可。客户端应用对这种错误无能为力。服务器必须配置成可以接受来自该应用程序的请求，才能使客户端得到响应。

13.5.4 生产环境

除了构建期优化之外，Angular 支持运行期的生产模式。生产模式通过禁用仅供开发用的安全检查和调试工具(如 expression-changed-after-checked 检测)来提高应用性能。使用生产配置构建应用时会自动启用 Angular 的运行时生产模式。

通过只加载应用启动时绝对必须的那些模块，可以极大缩短应用启动的时间。配置 Angular 的路由器可以推迟所有模块(及其相关代码)的加载时机，如一直等到应用启动完毕后加载，或者当用到时才按需惰性加载。

不要急性导入来自惰性加载模块中的任何东西。如果要惰性加载某个模块，就不要在应用启动时急性加载的模块(如根模块 AppModule)中导入它；否则，该模块就会立刻加载。配置打包方式时必须考虑惰性加载。因为在默认情况下，惰性加载的模块没有在 JavaScript 中导入过，因此打包器默认会排除它们。打包器不认识路由器配置，也就不能为惰性加载的模块创建独立的包，必须手动创建这些包。Angular CLI 会运行 Angular AOT Webpack 插件，它会自动识别出惰性加载的模块并为它们创建独立的包。

ng build 命令选项 deploy url 用于指定在编译时解析图片、脚本和样式表等资产

(assets)的相对 URL 的基础路径。deploy url 和 base href 这两个定义的作用有所重叠。两者都可用于初始脚本、样式表、惰性脚本和 CSS 资源。定义 base href 可用于定位相对路径模板资产和针对相对路径的 fetch/XMLHttpRequests,也可用于定义 Angular 路由器的默认基础 URL。需要进行更复杂设置的用户可能需要在应用程序中手动配置 APP_BASE_HREF 令牌。与可以只在一个地方定义的 base href 不同,deploy url 需要在构建时硬编码到应用中。这意味着指定 deploy url 会降低构建速度,但这是使用在整个应用程序中嵌入自己的工作方式的代价。这也是为什么说 base href 通常是更好的选择。

13.6 Angular 库的开发

13.6.1 含义

许多应用程序都需要解决一些同样的常见问题,例如提供统一的用户界面、渲染数据,以及允许数据输入。开发人员可以为特定的领域创建一些通用解决方案,以便在不同的应用中重复使用。像这样的解决方案就可以构建成可复用的 Angular 库,这些库可以作为 npm 包进行发布和共享。Angular 库是一个 Angular 项目,它与一般应用程序的不同之处在于它本身是不能运行的,必须在某个应用中导入库才能运行。这些库可以扩展 Angular 的基本功能。例如,Angular Material 是一个大型通用库的典范,它提供了一些复杂、可复用,兼具高度适应性的 UI 组件。同样,把 Service Worker 库添加到 Angular 应用中是将应用转换为渐进式 Web 应用的步骤之一。要向应用添加响应式表单,使用 ng add @angular/forms 命令添加该库的 npm 包,可在应用代码中从@angular/forms 库中导入 ReactiveFormsModule。

任何 Angular 应用开发者都可以使用 Angular 库,它们都已经由 Angular 团队或第三方发布为 npm 包。如果已经开发出了适合复用的功能,就可以创建自己的库。开发人员自己创建的库可以在本地使用,也可以把它们发布成 npm 包,共享给其他项目或其他开发者。这些包可以发布到 npm 服务器、私有的 npm 企业版服务器,或支持 npm 包的私有包管理系统。是否把一些功能打包成库是一种架构决策,类似于决定一个功能应该做成组件还是服务,或决定一个组件的范围该有多大。

把功能打包成库会强迫库中的工件与应用程序的业务逻辑分离。这有助于避免各种不良实践或架构失误,这些失误会导致将来很难解耦和复用代码。把代码放到一个单独的库中比简单地把所有内容都放在一个应用程序中要复杂得多。它需要更多的时间投入,并且需要管理、维护和更新这个库。不过,当把该库用在多个应用程序中时,这种复杂性就会得到回报。注意,这里所说的库是为了供 Angular 应用程序使用的库。如果想把 Angular 的功能添加到非 Angular 应用程序中,可以使用 Angular 自定义元素。

13.6.2 使用库

开发 Angular 应用程序时,可以选用 Angular 库,也可以使用丰富的第三方库。这些库都是作为 npm 包发布的,它们通常都带有一些与 Angular CLI 集成好的 schematic。要把可复用的库代码集成到应用中,需要安装该软件包并在使用时导入它提供的功能。对于大多数已发布的 Angular 库,可以使用 Angular CLI 的 ng add 命令来安装库包,并调用该包中的 schematic 在项目代码中添加脚手架。

通常，Angular 库包会在文件.d.ts 中包含类型信息。如果库包中没有包含类型信息并且 IDE 报错，则可能需要安装与该库关联的@types/<lib_name>包(<lib_name>为包名)。已安装到工作区的@types/包中所定义的类型，会自动添加到使用该库的项目的 TypeScript 配置文件中。TypeScript 默认就会在 node_modules\@types 文件夹中查找类型，不必单独添加每一个类型包。如果某个库没有@types/类型信息，仍然可以手动为它添加一些类型信息。

库的发布者可以对这些库进行更新，而这些库也有自己的依赖，所有依赖都需要保持最新。可以使用 ng update 命令检查、更新某个库的版本。Angular CLI 会检查库中最新发布的版本，如果最新版本比已安装的版本新，就会下载它并更新配置文件 package.json 以匹配最新版本。如果要把 Angular 更新到新版本，需要确保所用的库都是最新的。如果库之间相互依赖，可能还要按特定的顺序更新它们。

如果未将老式 JavaScript 库导入应用程序，可以将其添加到运行时全局范围并加载它，就像将其添加到 script 标记中一样。使用配置文件 angular.json 中构建目标的 scripts 和 styles 选项，配置 Angular CLI 以便在构建时执行此操作。

定义运行时全局库的类型信息。如果要用的全局库没有全局类型信息，就可以在 src 目录下的 typings.d.ts 中手动声明它们。如果不为由脚本定义的扩展添加接口，IDE 就会显示错误。

13.6.3　创建库

如果要在多个应用中解决同样的问题(或者要把解决方案分享给其他开发者)，就可以创建(修改成)库。例如，按钮会包含在构建的所有应用中。如果想在公共包注册表(如 npm)中发布它，要小心选择库名称。避免使用以 ng-为前缀的名称，ng-前缀是 Angular 框架及其库中使用的保留关键字。ng generate 命令会创建一个组件和一个服务的 NgModule。生成一个新库时，配置文件 angular.json 中增加了 library 类型的项目。

要让库代码可以复用，必须为它定义一个公共的 API。这个用户层定义了库中用户可用内容。该库的用户应该可以通过单个的导入路径来访问公共功能(如 NgModules、服务提供者和工具函数)。库的公共 API 在文件 public-api.ts 中维护。当库被导入应用程序时，从该文件导出的所有内容都会被公开。

为了让解决方案可供复用，需要先对它进行调整，再将其创建成一个库，以免它依赖应用特有的代码。将应用程序的功能迁移到库中时需要注意以下五点。

(1) 组件和管道等的可声明对象应该被设计成无状态的，这意味着它们不能被修改(如通过外部变量修改)。如果它们需要依赖于状态，就要进行分析、评估后，再决定是把它设计成 Angular 应用程序中的状态(即在 Angular 应用程序中修改)还是 Angular 库要管理的状态(即在 Angular 库中修改)。

(2) 组件内部订阅的所有可观察对象都应该在这些组件的生命周期内进行清理和释放。

(3) 组件对外暴露交互方式时，应该通过输入参数来提供上下文，通过输出参数来将事件传给其他组件。

(4) 检查所有内部依赖。对于在组件或服务中使用的自定义类或接口，检查它们是否依赖于其他类或接口，且需要将它们一起迁移。如果库代码依赖于某个服务，则需要迁移该服务；如果库代码或其模板依赖于其他库(如 Angular Material)，必须把它们配置为新创建库的依赖。

(5) 考虑如何为客户端应用程序提供服务。服务应该自己声明提供者(而不是在 NgModule 或组件中声明提供者)以便它们进行树抖优化。如果服务器从未被注入并导入该库的应用程序中，编译器就会把该服务从该 bundle 中删除；如果由多个 NgModules 注册全局服务提供者或提供者共享，使用 forRoot()方法和 forChild()方法设计模式由 RouterModule 提供服务。如果库中提供的可选服务没有被所有的客户端应用程序所使用，那么就可以通过轻量级令牌设计模式为这种情况支持正确的树状结构了。

一个库通常都包含可复用的代码，用于定义组件、服务以及导入项目中的其他 Angular 部件(如管道、指令等)。库被打包成一个 npm 包(或其他包)用于发布和共享。对于 Angular 库，可分发文件中可包含一些额外的资产，如主题文件、Sass mixins 或文档(变更日志)。各种 Angular 库应该把自己依赖的所有@angular/* 都列为同级依赖。这确保了当各个模块请求 Angular 时，都会得到完全相同的模块。如果某个库在 dependencies 列出 @angular/core 而不是用 peerDependencies，它可能会得到一个不同的 Angular 模块，这会破坏应用程序。如果要在同一个工作空间中使用某个库，不必把它发布到 npm 包管理器，但还是得先构建它。

13.6.4 构建、发布和编译库

要想在应用程序中使用库，先需要构建该库。用 Angular CLI 生成库时，会自动把它的路径添加到文件 tsconfig.json 中，告诉构建系统在哪里寻找这个库。依赖于库的应用程序应该只使用指向内置库的 TypeScript 路径映射。

发布库时可以使用 partial-Ivy(部分 Ivy)(推荐)和完全 Ivy 两种分发格式。部分 Ivy 包含可移植代码，使用任何版本(从 Angular 12 开始)的 Angular 构建的 Ivy 应用都可以使用这些可移植代码。完全 Ivy 包含专用的 Angular Ivy 指令，不能保证它们可在 Angular 的不同版本中使用。对于发布到 npm 的库，使用部分 Ivy 格式，因为它在 Angular 的各个补丁版本之间是稳定的。要发布到 npm，避免使用完全 Ivy 的方式编译库，因为生成的 Ivy 指令不属于 Angular 公共 API 的一部分，因此在补丁版本之间可能会有所不同。用于构建应用的 Angular 版本应始终与用于构建其任何依赖库的 Angular 版本相同或采用更大的版本。

在 Angular CLI 之外使用部分 Ivy 代码，可以实现分发格式。应用程序项目将 npm 中的许多 Angular 库安装到其 node_modules 目录中。但是，这些库中的代码不能与已编译的应用程序直接捆绑在一起，因为它尚未完全编译。要完成编译，可以使用 Angular 链接器。对于不使用 Angular CLI 的应用程序，此链接器可用作 Babel 插件。该插件要从@angular/compiler-cli/linker/babel 导入。Angular 链接器的 Babel 插件支持构建缓存，这意味着链接器只需一次处理库，而与其他 npm 操作无关。Angular CLI 自动集成了链接器插件，因此如果库的用户也在使用 Angular CLI，则他们可以从 npm 安装 Ivy 原生库，而无须其他任何配置。

13.6.5 Angular 包格式规范

Angular 包格式(APF)是针对 npm 包结构和格式的 Angular 专用规范，所有 Angular

官方生态系统(第一方)包(如@angular/core、@angular/material 等)和大多数第三方 Angular 库都使用了该规范。

APF 能让包在使用 Angular 的大多数常见场景下无缝工作。使用 APF 的包与 Angular 团队提供的工具以及更广泛的 JavaScript 生态系统兼容。建议开发人员也都遵循这种格式。APF 与 Angular 的其余部分一起进行版本控制,每个主要版本都改进了包格式。

在目前的 JavaScript 环境中,开发人员将使用多种不同的工具链(Webpack、Rollup、等)以多种不同的方式使用包。这些工具可能理解并需要不同的输入(一些工具能处理最新的 ES 语言版本,而其他工具也许要直接使用较旧的 ES 版本)。这种 Angular 分发格式支持所有常用的开发工具和工作流,并着重于优化,从而缩小应用有效负载大小或缩短开发迭代周期(构建时间)。开发人员可以依靠 Angular CLI 和 ng-packagr 命令来生成 APF 格式的包。

@angular/core 包中顶层配置文件 package. json 包含重要的包元数据。例如,把此包声明为 EcmaScript 模块(ESM)格式;包含一个 exports 字段用于定义所有入口点的可用源码格式;包含定义主入口点@angular/core 的可用源代码格式的键(这些键将被删除)供不理解 exports 的工具使用;声明此包是否包含副作用。除了顶层配置 package. json 之外,二级入口点有自己的文件 package. json。例如,在@angular/core/testing 入口点有 testing/package. json。不支持 exports 的旧解析器要用到二级配置文件 package. json。

APF 中的包含有一个主要入口点和零到多个次要入口点(如@angular/common/http)。入口点有多种功能,它们定义了用户要从中导入代码的模块说明符。用户通常将这些入口点视为具有不同用途或功能的不同符号组。特定入口点可能仅用于特殊目的,例如测试。此类 API 可以与主入口点分离,以减少它们被意外或错误使用的机会。它们定义了可以惰性加载代码的粒度。许多现代构建工具只能在 ES 模块级别进行代码拆分(又名惰性加载)。由于 APF 主要为每个入口点使用一个扁平 ES 模块,这意味着大多数构建工具无法将单个入口点中的代码拆分为多个输出块。APF 包的一般规则是为尽可能小的逻辑相关代码集使用入口点。大多数具有单一逻辑目的的库应该作为单一入口点发布。例如,@angular/core 为运行时使用单个入口点,Angular 运行时通常用作单个实体。次要入口点可以通过包的配置文件 package. json 的 exports 字段,通过节点模块解析规则和对应于入口点模块 ID 的子目录中的文件 package. json 来解析。

APF 指定代码要以扁平化的 ES 模块格式发布。这显著减少了 Angular 应用程序的构建时间以及最终应用包的下载和解析时间。Angular 编译器支持生成索引 ES 模块文件,然后可以让这些文件借助 Rollup 等工具生成扁平化模块,从而生成扁平化 ES 模块或 FESM 的文件格式模块。FESM 是一种文件格式,它会将所有可从入口点访问的 ES 模块扁平化为单个 ES 模块。它是通过跟踪包中的所有导入并将该代码复制到单个文件中而生成的,同时保留所有公共 ES 导出并删除所有私有导入。

默认情况下,ES 模块是有副作用的,即从模块导入可确保该模块顶层的任何代码都将执行。这通常是不可取的,因为典型模块中的大多数副作用代码并不是真正的副作用,而是仅影响特定符号。如果没有导入和使用这些符号,通常需要在称为树抖(tree-shaking)的优化过程中将它们删除,而副作用代码可以防止这种情况发生。

13.7 原理图

13.7.1 含义

原理图是一个基于模板的支持复杂逻辑的代码生成器。它是一组通过生成代码或修改代码来转换应用程序项目的指令；会将代码打包成集合并用 npm 安装；其集合可以作为一个强大的工具，以创建、修改和维护任何应用程序项目，特别是当要自定义 Angular 应用程序以满足特定需求时。例如，可以借助原理图来用预定义的模板或布局生成常用的 UI 模式或特定的组件；也可以用原理图来强制执行架构规则和约定，让整个项目保持一致性和互操作性。

Angular CLI 中的原理图是 Angular 生态系统的一部分。Angular CLI 使用原理图对 Web 应用项目进行转换。可以修改这些原理图，并定义新的原理图；如更新代码以修复依赖中的重大变更或者把新的配置项或框架添加到现有的项目中。

@schematics/angular 集合中的原理图是 ng generate 命令和 ng add 命令的默认原理图。此包里包含一些有名字的原理图，可用于配置 ng generate 命令的子命令选项，如 ng generate component 和 ng generate service。

与原理图相关联的 JSON 模式会告诉 Angular CLI 命令和子命令都有哪些选项以及默认值。这些默认值可以通过在命令行中为该选项提供不同的值来进行覆盖。在 Angular CLI 中，用来生成项目及其部件的默认原理图的 JSON 模式收集在@schematics/angular 包中。该模式描述了 Angular CLI 中每个可用的 ng generate 子命令选项。

13.7.2 自定义原理图

作为 Angular 库开发人员，可以创建自己的自定义原理图集合，以便把自己的库与 Angular CLI 集成在一起。添加原理图允许开发人员使用 ng add 命令安装库。生成原理图可以告诉 ng generate 命令如何修改项目、添加配置和脚本以及为库中定义的工件提供脚手架。更新原理图可以告诉 ng update 命令如何更新库的依赖，并在发布新版本时调整其中的重大变更。如果创建的新版本的库引入了潜在的重大更改，可以提供一个更新原理图，让 ng update 命令能够自动解决所更新项目中的任何重大修改；如果包中包含了涵盖从现有版本到新版本的迁移规则的更新原理图，那么该命令就会运行这个原理图。

开发人员可以创建自己的原理图来对 Angular 项目进行操作。库开发人员通常会把这些原理图与他们的库打包在一起；以便把它们与 Angular CLI 集成在一起；也可以创建独立的原理图来操作 Angular 应用程序中的文件和目录结构，以便为开发环境定制它们。多个原理图还可以串联起来，通过运行其他原理图来完成复杂的操作。

13.7.3 原理图的工作原理

在应用程序中，操作代码可能既强大又危险。Angular 原理图工具通过创建虚拟文件系统来防止副作用和错误。原理图描述了一个可应用于虚拟文件系统的转换管道。当原理图运行时，转换就会被记录在内存中，只有当这些更改被确认有效时，才会应用到实际的文件系统中。

原理图的公共 API 定义了表达其基本概念的类。虚拟文件系统用 Tree 表示。Tree 数据结构包含一个基础状态 base(一组已经存在的文件)和一个暂存区 staging(需要应用到 base 的更改列表)。在进行修改的过程中,并没有真正改变 base,而是把那些修改添加到了暂存区。Rule 对象定义了一个函数,它接受 Tree 进行转换,并返回一个新的 Tree。原理图的主文件 index.ts 定义了一组实现原理图逻辑的规则。转换由动作表示,有 Create、Rename、Overwrite 和 Delete 四种动作类型。每个原理图都在一个上下文中运行,上下文由一个 SchematicContext 对象表示。传给规则的上下文对象可以访问该原理图可能会用到的工具函数和元数据,包括一个帮助调试的日志 API。上下文还定义了一个合并策略,用于确定如何将这些更改从暂存树合并到基础树中,可以接受或忽略某个更改,也可以抛出异常。

使用 Schematics CLI 创建一个新的空白原理图时,它所生成的入口函数就是一个规则工厂。RuleFactory 对象定义了一个用于创建 Rule 的高阶函数。这些规则可以通过调用外部工具和实现逻辑来修改项目;可以从调用者那里收集选项值,并把它们注入模板中。规则可用的选项及其允许的值和默认值是在原理图的 JSON 模式文件 schema.json 中定义的。用 TypeScript 接口可以为这个模式定义变量或枚举数据类型。该模式定义了原理图中使用的变量的类型和默认值。定义原理图选项的 JSON 模式支持功能扩展,以允许对提示及其相应行为进行声明式定义。无须其他逻辑或更改原理图代码即可支持提示。

要把一个原理图添加到现有的集合中,可使用和新建原理图项目相同的命令。使用 Schematics CLI 创建空白原理图项目时,该集合的第一个成员是一张与该集合同名的空白原理图。把这个新的命名原理图添加到本集合中时,它就会自动添加到文件 collection.json 模式中。除了名称和描述外,每个原理图还有一个 factory 属性,用于标识此原理图的入口点。原理图可以在文件 index.ts 中提供它全部的逻辑,不需要额外的模板;也可以提供组件和模板来为 Angular 创建动态原理图,如独立的 Angular 项目;可以通过定义一些用来注入数据和修改变量的规则来配置这些模板。

13.7.4 库的原理图

创建 Angular 库时,可以将其打包进一组原理图里,并把它与 Angular CLI 集成在一起。借助原理图,用户可以用 ng add 命令来安装这个库的初始版本,可以用 ng generate 命令来创建在库中定义的一些部件,可以用 ng update 命令来调整项目,以支持在库的新版本中引入的重大变更。

要开始一个原理图集合,先要创建一些原理图文件,再用 ng add 命令帮助原理图增强用户的初始安装过程。提供初始命令 ng add 支持所需的唯一步骤是使用 SchematicContext 来触发安装任务。该任务会借助用户首选的包管理器将该库添加到宿主项目的配置文件 package.json 中,并将其安装到该项目的 node_modules 目录下。要把用户的原理图和库打包到一起,就必须把这个库配置成可单独构建的原理图,然后再把它们添加到发布包中。

可以把一个命名原理图添加到集合中,让 Angular 库的用户可以使用 ng generate 命令来创建在库中定义的部件。要把部件添加到项目中,原理图就需要有自己的模板文件。原理图模板支持特殊的语法来执行代码和替换变量。这里的模板会生成一个已把 Angular 的

HttpClient 注入其构造函数中的服务。有了基础设施，就可以开始定义一个 main()函数来执行要对用户项目做的各种修改了。Schematics 框架提供了一个文件模板系统，它支持路径和内容模板。系统会操作在这个输入文件 Tree 中加载的文件内或路径中定义的占位符，用传给规则的值来填充它们。Schematics 框架提供了许多实用函数来创建规则或在执行原理图时使用规则。导入已定义的模式接口，它会为原理图选项提供类型信息。

要想构建生成器原理图，就从一个空白的规则工厂开始。这个规则工厂返回 Tree 而不做任何修改。这些选项都是从 ng generate 命令传过来的选项值。有了一个框架，就可用来创建一些真正修改用户程序的代码，以便对库中定义的服务进行设置。用户安装的 Angular 库中会包含多个项目(应用程序和库)。用户可以在命令行中指定一个项目，也可以使用它的默认值。在任何情况下，代码都需要知道应该在哪个项目上应用此原理图，这样才能从该项目的配置中检索信息。可以使用传给工厂函数的 Tree 对象来做到这一点。通过 Tree 的一些方法，可以访问完整文件 Tree，以便在运行原理图时读/写文件。

要确定目标项目，可以使用 workspaces. readWorkspace()方法读取配置文件 angular. json 的内容。有了项目名称，可用它来检索指定项目的配置信息。workspace. projects 对象包含指定项目的全部配置信息。规则可以使用外部模板文件对它们进行转换，并使用转换后的模板返回另一个规则对象；可以用模板来生成原理图所需的任意自定义文件。规则工厂必须返回一个规则，并允许把多个规则组合到一个规则中，这样就可以在一个原理图中执行多个操作。在构建库和原理图之后，就可以安装一个原理图集合来运行项目了。

13.8 Angular 发布信息

13.8.1 版本发布

稳定性可以确保组件与库、教程、工具和现有实践不会突然被弃用。稳定性是让基于 Angular 的生态系统变得繁荣的基石。除了稳定性，还希望 Angular 能持续演进。随着 Angular 的不断演进，渴望新功能的开发者就可以使用新功能，同时为那些喜欢在新功能经过 Google 公司或开发人员的验证后才采纳新版本的用户，保持平台的稳定性和可靠性。

Angular 希望每个用户都明白在何时添加以及如何添加新特性，并且为那些将要移除的、准备弃用的特性提前做好准备。Angular 的版本号表明版本中所引入的变更级别。其中，重大变更(如移除特定的 API 和特性)有时候是必需的，如创新、让最佳实践与时俱进、变更依赖关系甚至来自 Web 平台自身的变化。Angular 是很多包、子项目和工具的集合。为了防止意外使用私有 API，Angular 官方文档给出公共 API 的说明。

13.8.2 路线图

Angular 以具有最小向后不兼容影响的方式对响应式表单实施更严格的类型检查。通过这种方式，可以让开发人员在开发期间发现更多问题，启用更好的文本编辑器和 IDE 支持，并改进响应式表单的类型检查。将 Angular 编译器的诊断扩展到类型检查之外，引入其他正确性和一致性检查，以进一步保证正确性和最佳实践。NgModules 作为可选的解决方案，使开发人员可以开发独立组件并实现用于声明组件编译范围的替代 API。

将 Angular 编译器作为 TypeScript 编译器的插件进行发布，可以大大提高开发者的开发效率，降低维护成本。在所有内部工具向 Ivy 的转换完成后，移除旧的 View Engine，以减少 Angular 的概念开销、获得更小的包大小、更低的维护成本和更低的代码库复杂度。

在组件级别对应用程序进行更细粒度的代码拆分。Angular 的开发工具提供用于调试和性能分析的实用程序，帮助开发人员了解 Angular 应用中的组件结构和变更检测。

为提高宿主元素添加指令的能力，允许开发人员使用额外的行为来扩展自己的组件，而不必使用继承机制。加载外部样式表是一个阻塞型操作，这意味着浏览器在加载所有引用的 CSS 之前无法开始渲染应用。为了使应用更快运行，通过在 Universal 应用中内联关键样式来提速。

MDC Web 是一个由 Google 公司的 Material Design 团队创建的库，它为构建 Material Design 组件提供了可复用的源语。使用 MDC Web 可以使 Angular Material 与 Material Design 规范更紧密地对齐，扩展无障碍性，提高组件质量。

使用 Angular 中的本机可信类型提高安全性，增加对新 Trusted Types API 的支持。此 Web 平台 API 可帮助开发人员构建更安全的 Web 应用；使用更好的 Angular 错误消息改进调试。错误消息通常会带来有限的行动指南来帮助开发人员解决它们。通过添加相关代码、开发指南和其他资料来使错误消息更易于发现，以确保更顺畅的调试体验。测试工具能让组件开发人员创建支持的 API 来测试组件交互。

13.8.3 浏览器支持

Angular Angular 构建于 Web 平台的最新标准之上。要支持这么多浏览器是一个不小的挑战，因为它们不支持现代浏览器的所有特性。可以通过加载腻子脚本(polyfills)来为支持的浏览器弥补这些特性。这些腻子脚本并没有神奇的魔力将老旧、慢速的浏览器变成现代、快速的浏览器，它只是填充了 API。

Angular CLI 提供了对腻子脚本的支持。使用 ng new 命令创建项目时，会在项目文件夹中创建一个配置文件 polyfills. ts。该文件包含许多强制性和可选腻子脚本的 JavaScript import 语句。使用 ng new 命令创建项目时，会自动安装一些强制性腻子脚本(如 zone. js)，并且它对应的 import 语句已在配置文件 polyfills. ts 中启用。如果需要一个可选的填充库，就必须安装它们的 npm 包，然后在配置文件 polyfills. ts 中反注释或创建一个对应的导入语句。如果不使用 Angular CLI，就要直接把腻子脚本添加到宿主页(index. html)中。

习题 13

简答题

1. 简述对 Angular 统一平台的理解。
2. 简述对开发工作流和工具的理解。
3. 简述对 Angular 的库开发的理解。
4. 简述对原理图的理解。
5. 简述对 Angular 发布信息的理解。

第14章

最佳实践

14.1 安全的最佳实践

安全的最佳实践包括及时把 Angular 包更新到最新版本以修复之前版本中发现的安全漏洞；不要修改 Angular 副本，私有的、定制版的 Angular 往往跟不上最新版本，这可能导致重要的安全修复与增强被忽略，应该在社区共享对 Angular 所做的改进并创建 Pull Request；避免使用官方文档中带"安全风险"标记的 Angular API。

14.1.1 XXS

跨站脚本(cross-site scripting,XSS)允许攻击者将恶意代码注入页面中。这些代码可以偷取用户数据(特别是用户的登录数据)，还可以冒充用户执行操作。它是 Web 上最常见的攻击方式之一。为了防范 XSS 攻击，必须阻止恶意代码进入 DOM。如，如果某个攻击者尝试把< script >标签插入 DOM，就可以在网站上运行任何代码；除了< script >，攻击者还可以使用很多 DOM 元素和属性来执行代码；如果攻击者所控制的数据混进了 DOM，就会导致安全漏洞。

为了系统性地防范 XSS 问题，Angular 默认把所有值都当作不可信任的。当值从模板中以属性(property)、DOM 元素特性(attribute)、CSS 类绑定或插值等途径插入 DOM 中的时候，Angular 将对这些值进行无害化处理(sanitize)，对不可信的值进行编码。如果某个值已经在 Angular 之外进行过无害化处理，可以确信是安全的，可以把这个值标记为安全的并通知 Angular。默认情况下，Angular 模板被认为是受信任的，应被视为可执行代码。切勿通过串联用户输入和模板语法来生成模板。这样做会使攻击者能够将任意代码注入你的应用程序中。为避免这些漏洞，始终在生产部署中使用默认的 AOT 模板编译器。借助内容安全策略和可信类型，可以提供额外的保护层。这些 Web 平台特性会在 DOM 级别运行，这是用来防范 XSS 问题的最有效位置，因为即使使用其他低级的 API，也无法绕过它们。因此，开发人员通过为其应用配置内容安全策略并启用强制可信类型来利用这些特性。

无害化处理会审查不可信的值，并将它们转换成可以安全插入 DOM 的形式。多数情况下，这些值并不会在处理过程中发生任何变化。无害化处理的方式取决于所在的环境。例如，一个在 CSS 里面无害的值，可能在 URL 里很危险。Angular 定义了 HTML、样式、

URL 和资源 URL 四个安全环境。其中，在 HTML 环境中，值需要在被解释为 HTML 时使用；在样式环境中，值需要作为 CSS 绑定到 style 属性时使用；在 URL 环境中，值需要被用作 URL 属性时使用；在资源 URL 中，值需要作为代码进行加载并执行。Angular 会对前三项中不可信的值进行无害化处理，但不能对第四种资源 URL 进行无害化处理，因为它们可能包含任何代码。在开发模式下，如果在进行无害化处理时需要被迫改变一个值，Angular 就会在控制台上输出一个警告。

除非强制使用可信类型(Trusted Types)，否则浏览器内置的 DOM API 不会自动保护程序免受安全漏洞的侵害。例如，document 通过 ElementRef 拿到的节点和很多第三方 API，都可能包含不安全的方法。如果使用能操纵 DOM 的其他库，也同样无法借助像 Angular 插值那样的自动清理功能。所以，要避免直接和 DOM 打交道，并尽可能地使用 Angular 模板。浏览器内置的 DOM API 不会自动针对安全漏洞进行防护。

有时候，应用程序确实需要包含可执行的代码，比如使用 URL 显示<iframe>，或者构造出有潜在危险的 URL。为了防止在这种情况下被自动无害化，可以告诉 Angular 已经审查了这个值，检查了它是怎么生成的，并确信它总是安全的。如果信任了一个可能是恶意的值，就会在应用中引入一个安全漏洞。注入 DomSanitizer 服务，然后调用 bypassSecurityTrustHtml()方法、bypassSecurityTrustScript()方法、bypassSecurityTrustStyle()方法、bypassSecurityTrustUrl()方法、bypassSecurityTrustResourceUrl()方法等之一，就可以把一个值标记为可信任的。一个值是否安全取决于它所在的环境，所以要为这个值按预定的用法选择正确的环境。通常，Angular 会自动无害化 URL 并禁止危险的代码。为了防止这种行为，可以调用 bypassSecurityTrustUrl()方法把这个 URL 值标记为一个可信任的 URL。如果需要把用户输入转换为一个可信任的值，可以在组件方法中处理。要调用一个组件方法来构造一个新的、可信任的视频 URL，Angular 就会允许把它绑定到<iframe src>标签上。

内容安全策略(CSP)是防止 XSS 的深度防御技术。启用 CSP，须将 Web 服务器配置为返回适当的 Content-Security-Policy HTTP 请求头。

使用可信类型可以帮助保护应用程序免受跨站脚本攻击。它是一项 Web 平台功能，可通过实施更安全的编码实践来防范跨站脚本攻击，还可以帮助简化应用代码的审计。可信类型尚未在应用目标的所有浏览器中可用。如果在不支持可信类型的浏览器中运行了可信类型的应用，那么应用的功能将被保留，并且应用将通过 Angular 的 DomSanitizer 防范 XSS。

要为应用强制实施可信类型，必须将应用的 Web 服务器配置为使用 angular、angular#unsafe-bypass、angular#unsafe-jit 等策略之一后，再发出 HTTP 请求头。其中，angular 策略用于 Angular 内部经过安全审查的代码，并且当强制执行可信类型时，Angular 需要此策略才能正常运行，且任何由 Angular 清理的内部模板值或内容都被此政策视为安全的；angular#unsafe-bypass 策略用于要使用 Angular 的 DomSanitizer 的各个方法来绕过安全性的应用，任何使用了这些方法的应用都必须启用此策略；JIT 编译器使用 angular#unsafe-jit 策略，如果应用直接与 JIT 编译器交互或使用 platformBrowserDynamic 在 JIT 模式下运行，则必须启用此策略。

应该在生产环境基础设施服务器、Angular CLI、Karma 等为可信类型配置 HTTP 请求头。Angular CLI(ng serve)使用文件 angular.json 中的 headers 属性，用于本地开发和端到端测试。Karma(ng test)使用文件 karma.config.js 中的 customHeaders 属性,进行单元测试。

AOT 模板编译器可防止称为模板注入的一类漏洞,并大大提高了应用性能。AOT 模板编译器是 Angular CLI 应用使用的默认编译器,应该在所有生产部署中使用它。AOT 编译器的替代方法是 JIT 编译器,它可以在运行时将模板编译为浏览器中的可执行模板代码。Angular 信任这些模板代码,因此动态生成模板并进行编译(尤其是包含用户数据的模板)可以规避 Angular 的内置保护,并且是一种安全性方面的反模式。

在服务器上构造的 HTML 容易受到注入攻击。将模板代码注入 Angular 应用中与注入可执行代码是一样的:它使攻击者可以完全控制该应用。为避免这种情况,请使用一种模板语言来自动转义值以防止服务器上的 XSS 漏洞。不要在服务器端使用模板语言生成 Angular 模板,否则,会带来引入模板注入漏洞的高风险。

14.1.2 XSRF 和 XSSI

Angular 内置了一些支持来防范跨站请求伪造(XSRF)和跨站脚本包含(XSSI)两个常见的 HTTP 漏洞。这两个漏洞主要在服务器端防范,但是 Angular 也自带了一些辅助特性,可以让防范代码在客户端的集成变得更容易。

在跨站请求伪造(XSRF 或 CSFR)中,攻击者欺骗用户,让他们访问一个假冒页面(该页面带有恶意代码并秘密向应用服务器发送恶意请求)。假设用户登录后打开一个邮件,单击里面的链接,在新页面中打开假冒页面,假冒页面立刻发送恶意请求到应用服务器,这个请求可能是从用户账户转账到攻击者的账户。与该请求一起,浏览器自动发出应用程序的 cookie。如果应用服务器缺乏 XSRF 保护,就无法辨识请求是从应用程序发来的合法请求还是从假冒页面来的假请求。为了防止这种情况,必须确保每个用户的请求都是从应用程序中发出的,而不是从另一个网站发出的。客户端和服务器必须合作来抵挡这种攻击。常见的反 XSRF 技术是服务器随机生成一个用户认证令牌到 cookie 中。客户端代码获取这个 cookie,并在接下来所有的请求中将其添加到自定义请求页头。服务器将收到的 cookie 值与请求页头的值进行比较,如果它们不匹配,便拒绝请求。这个技术之所以有效,是因为所有浏览器都实现了同源策略。只有设置 cookie 的网站的代码可以访问该站的 cookie,并为该站的请求设置自定义请求页头。也就是说,只有应用程序可以获取这个 cookie 令牌和设置自定义请求页头,而假冒页面的恶意代码不能。

跨站脚本包含(XSSI,也被称为 JSON 漏洞)可以允许一个攻击者的网站从 JSON API 读取数据。这种攻击发生在旧的浏览器上,它重写原生 JavaScript 对象的构造函数,然后使用< script >标签包含一个 API 的 URL。只有在返回的 JSON 能像 JavaScript 一样可以被执行时,这种攻击才会生效。所以服务端会约定给所有 JSON 响应体加上前缀")]}',\n",来把它们标记为不可执行的,以防范这种攻击。Angular 的 HttpClient 服务会识别这种约定,并在进一步解析之前,自动把字符串")]}',\n"从所有响应中去掉。

Angular 应用程序应该遵循和常规 Web 应用一样的安全原则并按照这些原则进行审计。Angular 中某些应该在安全评审中被审计的 API(如 bypassSecurityTrust()方法)都在文档中被明确标记为安全性敏感。

14.2　无障碍性

1. 含义

Web 会被各种各样的人使用，包括有视觉或运动障碍的人。有多种辅助技术能使这些人更轻松地和基于 Web 的应用程序进行交互。另外，将应用程序设计得更易于访问通常也能改善所有用户的体验。

建立无障碍的 Web 体验通常会涉及设置 ARIA 特性(attribute)以提供可能会丢失的语义。使用 attribute 绑定模板语法来控制与无障碍性相关的特性(attribute)值。在 Angular 中绑定 ARIA 特性(attribute)时，必须使用 attr. 前缀，因为 ARIA 规范针对的是 HTML 特性(attribute)，而不是 DOM 元素的属性(property)。此语法仅对于特性(attribute)绑定是必需的。静态 ARIA 特性(attribute)不需要额外的语法。按照约定，HTML 特性(attribute)使用小写名称，而 property 使用小驼峰方法命名。

由 Angular 团队维护的 Angular Material 库旨在提供完全无障碍的一组可复用 UI 组件。

2. 设计与实现

原生 HTML 元素捕获了许多对无障碍性很重要的标准交互模式。在制作 Angular 组件时，应尽可能直接复用这些原生元素，而不是重新实现已获良好支持的行为。有时要使用的原生元素需要一个容器元素。例如，原生<input>元素不能有子元素，因此任何自定义的文本输入组件都需要用其他元素来包装<input>。尽管可能只在自定义组件的模板中包含<input>，但这将使该组件的用户无法为 input 元素设置任意 property 和 attribute。相反，可以创建一个使用内容投影的容器组件，以将原生控件包含在组件的 API 中。

在设计无障碍性时，在 UI 中跟踪和控制焦点是很重要的考虑因素。使用 Angular 路由时，需要确定页面焦点在导航上的位置。为了避免仅依靠视觉提示，需要确保路由代码在页面导航之后更新焦点。使用 Router 服务中的 NavigationEnd 事件可以知道何时该更新焦点。

在实际的应用中，哪些元素获得焦点将取决于该应用特有的结构和布局。获得焦点的元素应使用户能够立即移动到刚刚进入视野的主要内容；应该避免在路由发生变化后的焦点重新回到 body 元素的情况。

Angular ESLint 提供了整理(linting)规则，可以帮助确保代码符合无障碍性标准。

14.3　保持最新和属性绑定

1. 保持最新

就像 Web 及其整个生态系统一样，Angular 也在持续改进中。Angular 平衡了持续改进与强调稳定性之间的冲突，并努力让升级变得更简单。让 Angular 始终保持最新，可以使开发者享受前沿的新特性所带来的好处，从而有利于进行各种优化和 Bug 修复等工作。

2. 属性绑定的最佳实践

通过遵循一些指导原则，可以使用属性绑定最大限度地减少错误并让代码保持可读性。

避免副作用模板表达式的计算应该没有明显的副作用。使用模板表达式的语法可避免产生副作用。通常,正确的语法会阻止开发者为属性绑定表达式中的任何东西赋值。该语法还会阻止开发者使用递增和递减运算符。如果开发者的表达式改变了所绑定的其他东西的值,那么这种更改就会产生副作用。Angular 可能显示也可能不显示更改后的值。如果 Angular 确实检测到了这个变化,就会抛出一个错误。作为一项最佳实践,只使用属性和返回值的方法。

模板表达式应该为目标属性所期望的值类型求值。例如,如果目标属性需要一个字符串,就返回一个字符串;如果需要一个数值,就返回一个数值;如果需要一个对象,就返回一个对象。

14.4 惰性加载

1. 惰性加载特性模块

默认情况下,NgModule 都是急性加载的。也就是说,它会在应用程序加载时尽快加载,所有模块都是如此,无论是否立即要用。对于带有很多路由的大型应用,考虑使用惰性加载(一种按需加载 NgModule 的模式)。惰性加载可以减小初始包的尺寸,从而减少加载时间。

建立惰性加载的特性模块有两个主要步骤:使用 route 标志用 Angular CLI 创建特性模块;配置相关路由。

Angular CLI 会将每个特性模块自动添加到应用级的路由映射表中。通过添加默认路由来最终完成这些步骤。

Angular CLI 会把 RouterModule.forRoot(routes)方法添加到 AppRoutingModule 的 imports 数组中。这会让 Angular 知道 AppRoutingModule 是一个路由模块,而 forRoot()方法表示这是一个根路由模块。它会配置传入的所有路由、让开发者能访问路由器指令并注册 Router。forRoot()方法在应用中只能使用一次,也就是这个 AppRoutingModule。Angular CLI 还会把 RouterModule.forChild(routes)方法添加到各个特性模块中。这种方式下的 Angular 就会知道这个路由列表只负责提供额外的路由并且其设计意图是作为特性模块使用。可以在多个模块中使用 forChild()方法。forRoot()方法为路由器管理全局性的注入器配置。forChild()方法中没有注入器配置,只有像 RouterOutlet 和 RouterLink 这样的指令。

2. 预加载

预加载通过在后台加载部分应用程序来改进用户体验,即预加载模块或组件数据。其中,预加载模块通过在后台加载部分应用程序来改善用户体验,这样用户在激活路由时就无须等待下载这些元素。如果想要启用所有惰性加载模块的预加载,就要确定从 Angular 的 router 导入的是 PreloadAllModules 还是 AppRoutingModule,可通过 forRoot()方法指定预加载策略;要预加载组件数据,可以用 resolver 守卫。解析器通过阻止页面加载来改进用户体验,直到显示页面时的全部必要数据都可用。

惰性加载模块时常见的错误之一,就是在应用程序中的多个位置导入通用模块。可以先用 Angular CLI 生成模块并包括 route route-name 参数来测试这种情况。其中,route-name 是模块的名称。接下来,生成不带 route 参数的模块。如果用了 route 参数,Angular

CLI 就会生成错误，但如果不使用 Angular CLI 便可以正确运行，则可能是在多个位置导入了相同的模块。许多常见的 Angular 模块都应该导入应用程序的基础模块中。

14.5 令牌

14.5.1 轻量级注入令牌

使用轻量级注入令牌设计 Angular 库，有助于优化用到库的客户应用程序发布包的体积。可以使用可摇树优化的提供者来管理组件和可注入服务之间的依赖结构，以优化发布包体积。这通常会确保如果提供的组件或服务从未被应用实际使用过，那么编译器就可以从发布包中删除它的代码。但是，由于 Angular 存储注入令牌的方式可能导致未用到的组件或服务最终进入发布包中。它通过使用轻量级注入令牌来支持正确的摇树优化。这种轻量级注入令牌设计模式对于 Angular 库开发者来说尤其重要：它可以确保当应用程序只用到了库中的某些功能时，可以从客户应用的发布包中删除未使用过的代码；当某应用程序用到了库时，库中可能会提供一些客户应用未用到的服务。在这种情况下，开发者会期望该服务是可摇树优化的，不让这部分代码增加应用的编译后大小。由于开发者既无法了解也无法解决库的摇树优化问题，因此这是库开发者的责任。为了防止未使用的组件被保留下来，开发者的库应该使用轻量级注入令牌这种设计模式。

编译器在从 TypeScript 转换完后会删除这些类型位置上的引用，所以它们对于摇树优化没什么影响。编译器必须在运行时保留值位置上的引用，这就会阻止该组件被摇树优化掉。例如，编译器保留了 LibHeaderComponent 令牌，它出现在了值位置上，这就会防止所引用的组件被摇树优化掉，即使开发者实际上没有在任何地方用过< lib-header >。如果 LibHeaderComponent 很大（代码、模板和样式），那么把它包含进来就会大大增加客户应用的大小。

当一个组件被用作注入令牌时，就会出现摇树优化的问题。有两种情况可能会发生：令牌用在内容查询中值的位置上；该令牌用作构造函数注入的类型说明符。

轻量级注入令牌设计模式包括：使用一个小的抽象类作为注入令牌，并在稍后为它提供实际实现。该抽象类固然会被留下（不会被摇树优化掉），但它很小，对应用的大小没有任何重大影响。

综上所述，轻量级注入令牌模式由以下几部分组成：一个轻量级的注入令牌，它表现为一个抽象类；一个实现该抽象类的组件定义；注入这种轻量级模式时使用@ContentChild()或者@ContentChildren()；实现轻量级注入令牌的提供者，它将轻量级注入令牌及其实现关联起来；使用轻量级注入令牌进行 API 定义。

14.5.2 注入令牌的应用

那些注入了轻量级注入令牌的组件可能要调用注入的类中的方法。因为令牌现在是一个抽象类，并且可注入组件实现了抽象类，所以开发者还必须在作为轻量级注入令牌的抽象类中声明一个抽象方法。该方法的实现代码（及其所有相关代码）都会留在可注入组件中，但这个组件本身仍可被摇树优化。这样就能让父组件以类型安全的方式与子组件（如果存在）进行通信。

轻量级注入令牌只对组件有用。Angular 风格指南中建议使用 Component 后缀命名组件。例如,LibHeaderComponent 就遵循这个约定。

为了维护组件及其令牌之间的对应关系,同时又要区分它们,推荐的写法是使用组件基本名加上后缀 Token 来命名轻量级注入令牌：LibHeaderToken。

微课视频

14.6 安全的应用开发

14.6.1 创建组件

在项目 src\examples 根目录下创建 securityexamples 子目录,在 src\examples\securityexamples 目录下创建文件 inner-html-binding. component. ts,代码如例 14-1 所示。

【例 14-1】 创建文件 inner-html-binding. component. ts 的代码,定义组件。

```
import {Component} from '@angular/core';
@Component({
  selector: 'app - inner - html - binding',
  template: `
    < h3 > Binding innerHTML </h3 >
    < p > Bound value:</p >
    < p class = "e2e - inner - html - interpolated">{{htmlSnippet}}</p >
    < p > Result of binding to innerHTML:</p >
    < p class = "e2e - inner - html - bound" [innerHTML] = "htmlSnippet"></p >
  `,
})
export class InnerHtmlBindingComponent {
  htmlSnippet  =  'Template < script > alert("Owned")</script > < b > Syntax </b >';
}
```

在 src\examples\securityexamples 目录下创建文件 bypass-security. component. ts,代码如例 14-2 所示。

【例 14-2】 创建文件 bypass-security. component. ts 的代码,定义组件。

```
import {Component} from '@angular/core';
import {DomSanitizer, SafeResourceUrl, SafeUrl} from '@angular/platform - browser';
@Component({
  selector: 'app - bypass - security',
  template: `
    < h4 > Bypass Security Component </h4 >
    < h5 > An untrusted URL:</h5 >
    < p >< a class = "e2e - dangerous - url" [href] = "dangerousUrl"> Click me </a ></p >
    < h5 > A trusted URL:</h5 >
    < p >< a class = "e2e - trusted - url" [href] = "trustedUrl"> Click me </a ></p >
    < h5 > Resource URL:</h5 >
    < p > Showing: {{dangerousVideoUrl}}</p >
    < p > Trusted:</p >
    < iframe class = " e2e - iframe - trusted - src" width = " 640" height = " 390" [src] =
"videoUrl"></iframe >
    < p > Untrusted:</p >
    < iframe class = " e2e - iframe - untrusted - src" width = " 640" height = " 390" [src] =
```

```
"dangerousVideoUrl"></iframe>
  `,
})
export class BypassSecurityComponent {
  dangerousUrl: string;
  trustedUrl: SafeUrl;
  dangerousVideoUrl!: string;
  videoUrl!: SafeResourceUrl;
  constructor(private sanitizer: DomSanitizer) {
    this.dangerousUrl = 'javascript:alert("Hi there")';
    this.trustedUrl = sanitizer.bypassSecurityTrustUrl(this.dangerousUrl);
this.updateVideoUrl('courseId = 1209722814&share = 2&shareId = 480000001936408 #/learn/
video?lessonId = 1280314488&courseId = 1209722814');
  }
  updateVideoUrl(id: string) {
    this.dangerousVideoUrl = 'https://study.163.com/course/courseLearn.htm?' + id;
    this.videoUrl =
        this.sanitizer.bypassSecurityTrustResourceUrl(this.dangerousVideoUrl);
  }
}
```

在 src\examples\securityexamples 目录下创建文件 securityhome.component.ts，代码如例 14-3 所示。

【例 14-3】 创建文件 securityhome.component.ts 的代码，定义组件。

```
import {Component} from '@angular/core';
@Component({
  selector: 'root',
  template: `
  <h3>Security</h3>
  <app-inner-html-binding></app-inner-html-binding>
  <app-bypass-security></app-bypass-security>
  `
})
export class SecurityhomeComponent {
}
```

14.6.2 模块和运行结果

在 src\examples\securityexamples 目录下创建文件 app-security.module.ts，代码如例 14-4 所示。

【例 14-4】 创建文件 app-security.module.ts 的代码，定义路由并声明组件。

```
import {NgModule} from '@angular/core';
import {BrowserModule} from '@angular/platform-browser';
import {SecurityhomeComponent} from './securityhome.component';
import {BypassSecurityComponent} from './bypass-security.component';
import {InnerHtmlBindingComponent} from './inner-html-binding.component';
import {RouterModule} from "@angular/router";
@NgModule({
  imports: [ BrowserModule,
    RouterModule.forRoot([
```

```
      {path: 'security', component: SecurityhomeComponent},
    ]),
  ],
  declarations: [
    SecurityhomeComponent,
    BypassSecurityComponent,
    InnerHtmlBindingComponent
  ],
})
export class AppSecurityModule { }
```

修改 src\examples 目录下的文件 examplesmodules1. module. ts，代码如例 14-5 所示。

【例 14-5】 修改文件 examplesmodules1. module. ts 的代码，设置启动模板。

```
import {NgModule} from '@angular/core';
import {AppSecurityModule} from "./securityexamples/app-security.module";
import {SecurityhomeComponent} from "./securityexamples/securityhome.component";
@NgModule({
  imports: [
    AppSecurityModule,
  ],
  bootstrap: [SecurityhomeComponent]
})
export class ExamplesmodulesModule1 {}
```

保持其他文件不变并成功运行程序后，在浏览器地址栏中输入 localhost:4200，自动跳转到 localhost:4200/home；也可以在浏览器地址栏中输入 localhost:4200/security，部分结果如图 14-1 所示。

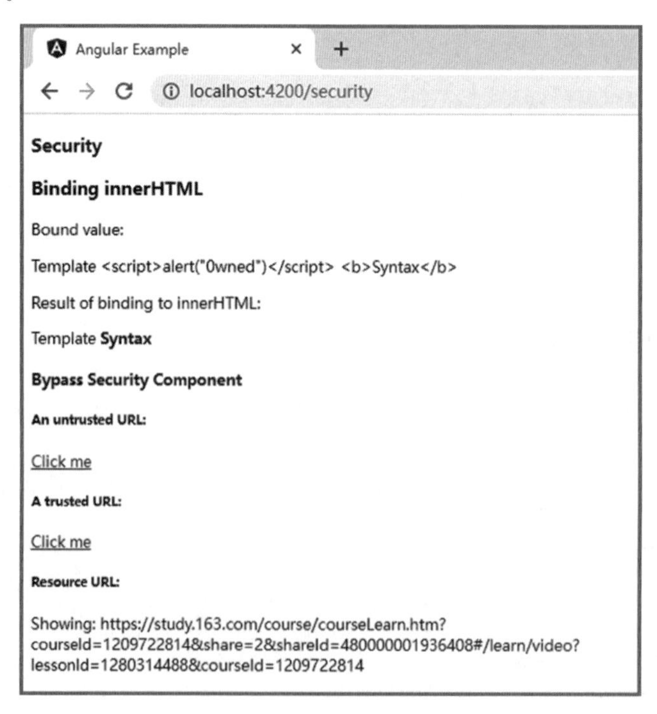

图 14-1　成功运行程序后在浏览器地址栏中输入 localhost:4200/security 的部分结果(从上往下)

14.7　无障碍性的应用开发

14.7.1　创建组件

在项目 src\examples 根目录下创建 accessibilityexamples 子目录，在 src\examples\accessibilityexamples 目录下创建文件 progress-bar. component. ts，代码如例 14-6 所示。

【例 14-6】 创建文件 progress-bar. component. ts 的代码，定义组件。

```
import {Component, Input} from '@angular/core';
@Component({
  selector: 'app - example - progressbar',
  template: '< div class = "bar" [style.width. % ] = "value"></div >',
  styles: ['      :host { \n' +
  '        display: block; \n' +
  '        width: 300px; \n' +
  '        height: 25px; \n' +
  '        border: 1px solid black; \n' +
  '        margin - top: 16px;\n' +
  '      }\n' +
  '      \n' +
  '      .bar { \n' +
  '        background: blue; \n' +
  '        height: 100 % ; \n' +
  '      }'],
  host: {
    role: 'progressbar',
    'aria - valuemin': '0',
    'aria - valuemax': '100',
    '[attr.aria - valuenow]': 'value',
  }
})
export class ExampleProgressbarComponent  {
  /** Current value of the progressbar. */
  @Input() value = 0;
}
```

在 src\examples\accessibilityexamples 目录下创建文件 accessibilityhome. component. ts，代码如例 14-7 所示。

【例 14-7】 创建文件 accessibilityhome. component. ts 的代码，定义组件。

```
import {Component} from '@angular/core';
@Component({
  selector: 'root',
  template: `
    < h1 > Accessibility Example 无障碍访问示例</h1 >
    < label >
      Enter an example progress value 输入(或选择)进度条的数值
      < input type = "number" min = "0" max = "100"
              [value] = "progress" (input) = "setProgress( $ event)">
    </label >
    <! -- The user of the progressbar sets an aria - label to communicate what the progress
```

```
means. -->
    <app-example-progressbar [value]="progress" aria-label="Example of a progress bar">
    </app-example-progressbar>
  `,
})
export class AccessibilityhomeComponent {
  progress = 0;
  setProgress($event: Event) {
    this.progress = +($event.target as HTMLInputElement).value;
  }
}
```

14.7.2 模块和运行结果

在 src\examples\accessibilityexamples 目录下创建文件 app-accessibility. module. ts，代码如例 14-8 所示。

【例 14-8】 创建文件 app-accessibility. module. ts 的代码，定义路由并声明组件。

```
import {NgModule} from '@angular/core';
import {BrowserModule} from '@angular/platform-browser';
import {AccessibilityhomeComponent} from './accessibilityhome.component';
import {ExampleProgressbarComponent} from './progress-bar.component';
import {RouterModule} from "@angular/router";
@NgModule({
  imports:      [
    BrowserModule,
    RouterModule.forRoot([
      {path: 'accessibility', component: AccessibilityhomeComponent},
    ]),
  ],
  declarations: [ AccessibilityhomeComponent, ExampleProgressbarComponent ],
})
export class AppAccessibilityModule { }
```

修改 src\examples 目录下的文件 examplesmodules1. module. ts，代码如例 14-9 所示。

【例 14-9】 修改文件 examplesmodules1. module. ts 的代码，设置启动组件。

```
import {NgModule} from '@angular/core';
import {AppAccessibilityModule} from "./propertybindingexamples/app-property-binding.
module";
import {AccessibilityhomeComponent} from './accessibilityexamples/accessibilityhome.component';
@NgModule({
  imports: [
    AppAccessibilityModule,
  ],
  bootstrap:[AccessibilityhomeComponent]
})
export class ExamplesmodulesModule1 {}
```

保持其他文件不变并成功运行程序后，在浏览器地址栏中输入 localhost:4200，自动跳转到 localhost:4200/home；也可以在浏览器地址栏中输入 localhost:4200/accessibility，结果如图 14-2 所示。在图 14-2 中的文本框输入(选择)一个数值，将在下方的进度条显示对应的进度，如图 14-3 所示。

图 14-2　成功运行程序后在浏览器地址栏中输入 localhost:4200 的结果

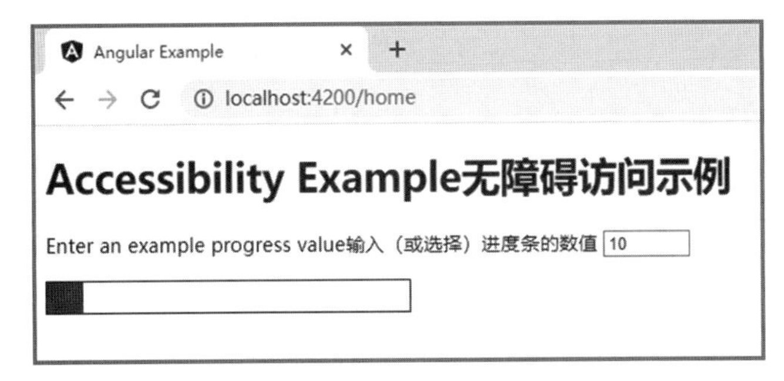

图 14-3　在图 14-2 中的文本框输入(选择)一个数值，将在进度条显示对应进度的结果

14.8　属性绑定的应用

14.8.1　创建组件

在项目 src\examples 根目录下创建 propertybindingexamples 子目录，在 src\examples\propertybindingexamples 目录下创建文件 item-detail. component. ts，代码如例 14-10 所示。

【例 14-10】　创建文件 item-detail. component. ts 的代码，定义组件。

```
import {Component, OnInit, Input} from '@angular/core';
@Component({
  selector: 'app-item-detail',
  template: `
    <p>Your item is: {{childItem}} </p>
  `,
})
export class ItemDetailComponent implements OnInit {
  @Input() childItem = '';
  constructor() { }
  ngOnInit() {
  }
}
```

在 src\examples\propertybindingexamples 目录下创建文件 item-list. component. ts，代码如例 14-11 所示。

【例 14-11】　创建文件 item-list. component. ts 的代码，定义组件。

```
import {Component, Input} from '@angular/core';
```

```
import {Item, ITEMS} from './mock-items';
@Component({
  selector: 'app-item-list',
  template: `
    <h4>Nested component's list of items:</h4>
    <ul>
      <li *ngFor="let item of listItems">{{item.id}} {{item.name}}</li>
    </ul>
    <h4>Pass an object from parent to nested component:</h4>
    <ul>
      <li *ngFor="let item of items">{{item.id}} {{item.name}}</li>
    </ul>
  `,
})
export class ItemListComponent {
  listItems = ITEMS;
  @Input() items: Item[] = [];
  constructor() {
  }
}
```

在src\examples\propertybindingexamples目录下创建文件mock-items.ts,代码如例14-12所示。

【例14-12】 创建文件mock-items.ts的代码,定义接口和数组。

```
export interface Item {
  id: number;
  name: string;
}
export const ITEMS: Item[] = [
  {id: 11, name: 'bottle'},
  {id: 12, name: 'boombox'},
  {id: 13, name: 'chair'},
  {id: 14, name: 'fishbowl'},
];
```

在src\examples\propertybindingexamples目录下创建文件property-binding.component.ts,代码如例14-13所示。

【例14-13】 创建文件property-binding.component.ts的代码,定义组件。

```
import {Component} from '@angular/core';
@Component({
  selector: 'root',
  template: `
    <div>
      <h4>绑定到 img 的 src 属性</h4>
      <img [src]="itemImageUrl">
      <hr/>
      <h4>绑定到 colSpan 属性</h4>
      <table border=1>
        <tr>
          <td>Column 1</td>
          <td>Column 2</td>
        </tr>
```

```
        <!-- Notice the colSpan property is camel case -->
        <tr>
          <td [colSpan]="2">Span 2 columns</td>
        </tr>
      </table>
      <hr/>
      <h4>Button disabled state bound to isUnchanged property:</h4>
      <!-- Bind button disabled state to \`isUnchanged\` property -->
      <button [disabled]="isUnchanged">Disabled Button</button>
      <hr/>
      <h4>绑定到属性 directive 的属性</h4>
      <p [ngClass]="classes">[ngClass] binding to the classes property making this blue</p>
      <hr/>
      <h2>Model property of a custom component:</h2>
      <app-item-detail [childItem]="parentItem"></app-item-detail>
      <app-item-detail childItem="parentItem"></app-item-detail>
      <h3>Pass objects:</h3>
      <app-item-list [items]="currentItems"></app-item-list>
      <hr/>
      <h2>Property binding and interpolation</h2>
      <p><img src="{{itemImageUrl}}"> is the <i>interpolated</i> image.</p>
      <p><img [src]="itemImageUrl"> is the <i>property bound</i> image.</p>
      <p><span>"{{interpolationTitle}}" is the <i>interpolated</i> title.</span></p>
      <p>"<span [innerHTML]="propertyTitle"></span>" is the <i>property bound</i>
title.</p>
      <hr/>
      <h2>Malicious content</h2>
      <p><span>"{{evilTitle}}" is the <i>interpolated</i> evil title.</span></p>
      <!--
      Angular generates a warning for the following line as it sanitizes them
      WARNING: sanitizing HTML stripped some content (see https://g.co/ng/security#xss).
      -->
      <p>"<span [innerHTML]="evilTitle"></span>" is the <i>property bound</i> evil
title.</p>
    </div>
  `,
  styles: ['div {\n' +
  '  margin: 1rem auto;\n' +
  '  width: 90%\n' +
  '}\n' +
  '.special {\n' +
  '  background-color: #1976d2;\n' +
  '  color: #fff;\n' +
  '  padding: 1rem;\n' +
  '}\n']
})
export class PropertyBindingComponent {
  itemImageUrl = '../../assets/phone.png';    //准备一张图片
  isUnchanged = true;
  classes = 'special';
  parentItem = 'lamp';
  currentItems = [{
    id: 21,
    name: 'phone'
```

```
  }];
  interpolationTitle = 'Interpolation';
  propertyTitle = 'Property binding';
  evilTitle = 'Template <script>alert("evil never sleeps")</script> Syntax';
}
```

14.8.2 模块和运行结果

在 src\examples\propertybindingexamples 目录下创建文件 app-property-binding.module.ts，代码如例 14-14 所示。

【例 14-14】 创建文件 app-property-binding.module.ts 的代码，定义路由并声明组件。

```
import {BrowserModule} from '@angular/platform-browser';
import {NgModule} from '@angular/core';
import {PropertyBindingComponent} from './property-binding.component';
import {ItemDetailComponent} from './item-detail.component';
import {ItemListComponent} from './item-list.component';
import {RouterModule} from "@angular/router";
@NgModule({
  declarations: [
    PropertyBindingComponent,
    ItemDetailComponent,
    ItemListComponent
  ],
  imports: [
    BrowserModule,
    RouterModule.forRoot([
      {path: 'property-binding', component: PropertyBindingComponent},
    ]),
  ],
})
export class AppPropertyBindingModule { }
```

修改 src\examples 目录下的文件 examplesmodules1.module.ts，代码如例 14-15 所示。

【例 14-15】 修改文件 examplesmodules1.module.ts 的代码，设置启动组件。

```
import {NgModule} from '@angular/core';
import {AppPropertyBindingModule} from "./propertybindingexamples/app-property-binding.
module";
import {PropertyBindingComponent} from './propertybindingexamples/property-binding.component';
@NgModule({
  imports: [
    AppProperty BindingModule,
  ],
  bootstrap: [PropertyBindingComponent]
})
export class ExamplesmodulesModule1 {}
```

准备图片 phone.png 后，将其存放在 src\examples 目录下。保持其他文件不变并成功运行程序后，在浏览器地址栏中输入 localhost:4200，自动跳转到 localhost:4200/home，也可在浏览器地址栏中输入 localhost:4200/property-binding，结果如图 14-4（上半部分）和图 14-5 所示（下半部分）。注意，图 14-4（最后两行文字）和图 14-5（最上两行文字）有重复（体现了衔接关系）。

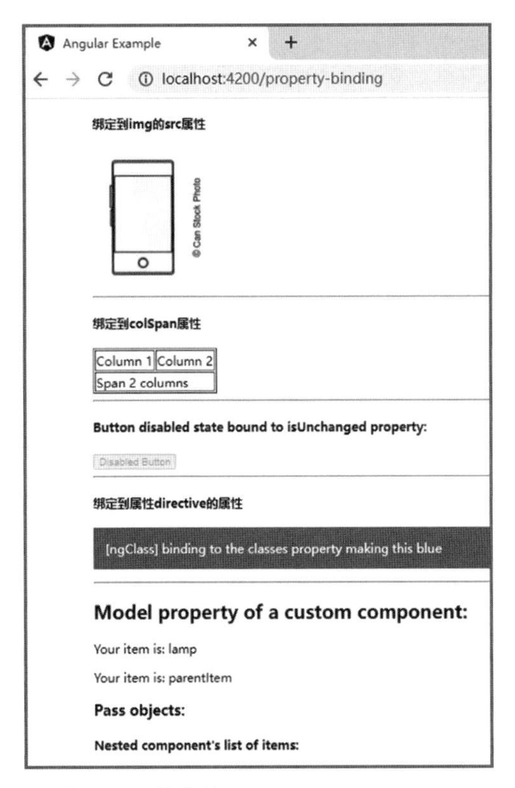

图 14-4 成功运行程序后在浏览器地址栏中输入 localhost:4200/property-binding 的结果(上半部分)

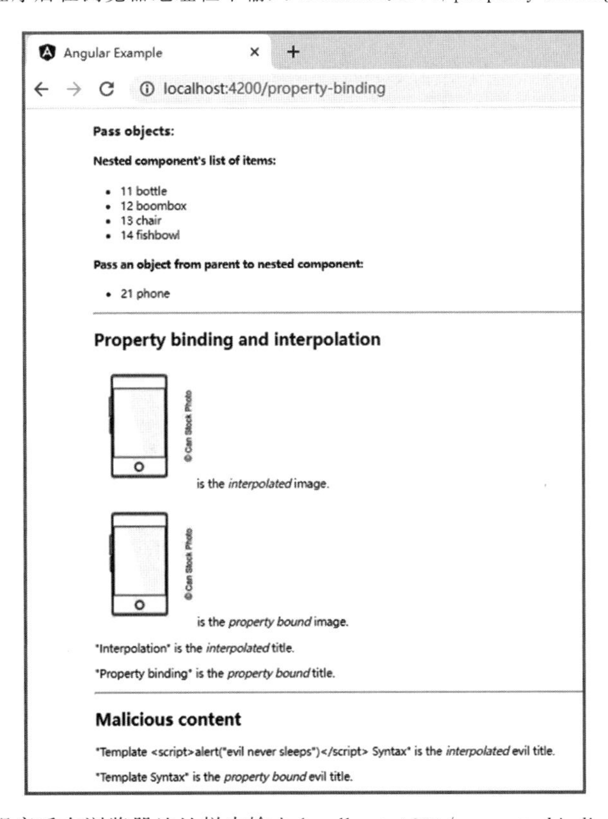

图 14-5 成功运行程序后在浏览器地址栏中输入 localhost:4200/property-binding 的结果(下半部分)

微课视频

14.9 惰性加载特性模块的应用

14.9.1 创建组件

在项目 src\examples 根目录下创建 lazyloadingexamples 子目录，在 src\examples\lazyloadingexamples 目录下创建文件 customers. component. ts，代码如例 14-16 所示。

【例 14-16】 创建文件 customers. component. ts 的代码，定义组件。

```
import {Component, OnInit} from '@angular/core';
@Component({
  selector: 'app-customers',
  template: `
    <p>
      customers works!
    </p>
  `,
})
export class CustomersComponent implements OnInit {
   ngOnInit() {   }
}
```

在 src\examples\lazyloadingexamples 目录下创建文件 orders. component. ts，代码如例 14-17 所示。

【例 14-17】 创建文件 orders. component. ts 的代码，定义组件。

```
import {Component, OnInit} from '@angular/core';
@Component({
  selector: 'app-orders',
  template: `
    <p>
      orders works!
    </p>
  `,
})
export class OrdersComponent implements OnInit {
  ngOnInit() {   }
}
```

在 src\examples\lazyloadingexamples 目录下创建文件 lazyloading. component. ts，代码如例 14-18 所示。

【例 14-18】 创建文件 lazyloading. component. ts 的代码，定义组件。

```
import {Component} from '@angular/core';
@Component({
  selector: 'root',
  template: `
    <h1>
      {{title}}
    </h1>
    <button routerLink="/customers">customers</button>
    <button routerLink="/orders">orders</button>
```

```
    <button routerLink = "/lazyloading"> home </button>
    <router - outlet></router - outlet>
  `,
})
export class LazyloadingComponent {
  title = 'Lazy loading feature modules';
}
```

14.9.2 模块和运行结果

在 src\examples\lazyloadingexamples 目录下创建文件 app-lazyloading. module. ts,代码如例 14-19 所示。

【例 14-19】 创建文件 app-lazyloading. module. ts 的代码,定义路由并声明组件。

```
import {HttpClientModule} from '@angular/common/http';
import {NgModule} from '@angular/core';
import {FormsModule} from '@angular/forms';
import {BrowserModule} from '@angular/platform - browser';
import {LazyloadingComponent} from './lazyloading.component';
import {CommonModule} from "@angular/common";
import {CustomersComponent} from "./customers.component";
import {OrdersComponent} from "./orders.component";
import {RouterModule, Routes} from "@angular/router";
const routes: Routes = [
  {   path: 'customers', component:CustomersComponent,   },
  {     path: 'orders',  component:OrdersComponent,   },
  {     path: 'lazyloading', component:LazyloadingComponent}
  ]
@NgModule({
  declarations: [
    LazyloadingComponent,
    CustomersComponent,
    OrdersComponent
  ],
  imports: [
    BrowserModule,
    FormsModule,
    HttpClientModule,
    CommonModule,
    RouterModule.forRoot(routes)
  ],
})
export class AppLazyloadingModule { }
```

修改 src\examples 目录下的文件 examplesmodules1. module. ts,代码如例 14-20 所示。

【例 14-20】 修改文件 examplesmodules1. module. ts 的代码,设置启动组件。

```
import {NgModule} from '@angular/core';
import {LazyloadingComponent} from './lazyloadingexamples/lazyloading.component';
import {AppLazyloadingModule} from "./lazyloadingexamples/app - lazyloading.module";
@NgModule({
  imports: [
    AppLazyloadingModule,
  ],
```

```
  bootstrap: [LazyloadingComponent]
})
export class ExamplesmodulesModule1 {}
```

保持其他文件不变并成功运行程序后，在浏览器地址栏中输入 localhost:4200，自动跳转到 localhost:4200/home，结果如图 14-6 所示。在浏览器地址栏中输入 localhost:4200/lazyloading(或单击图 14-6 中 home 按钮)，结果如图 14-7 所示。单击图 14-6(或图 14-7)中 customers 按钮，或者在浏览器地址栏中输入 localhost:4200/customers，结果如图 14-7 所示。单击图 14-6(或图 14-7)中的 orders 按钮，或者在浏览器地址栏中输入 localhost:4200/orders，结果如图 14-8 所示。单击图 14-6 中的 home 按钮，结果如图 14-6 所示。

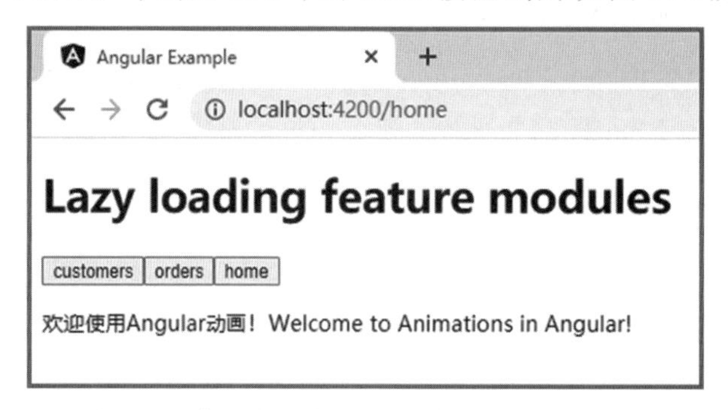

图 14-6　成功运行程序后在浏览器地址栏中输入 localhost:4200 的结果

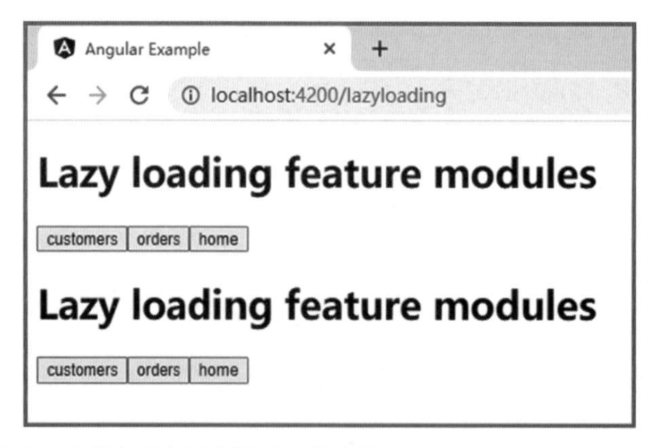

图 14-7　成功运行程序后在浏览器地址栏中输入 localhost:4200/lazyloading 的结果

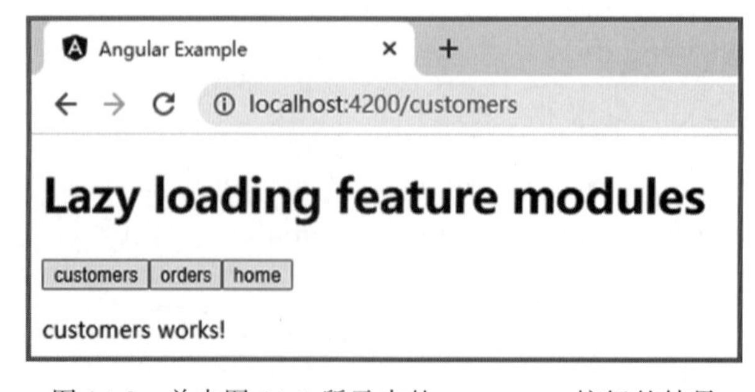

图 14-8　单击图 14-6 所示中的 customers 按钮的结果

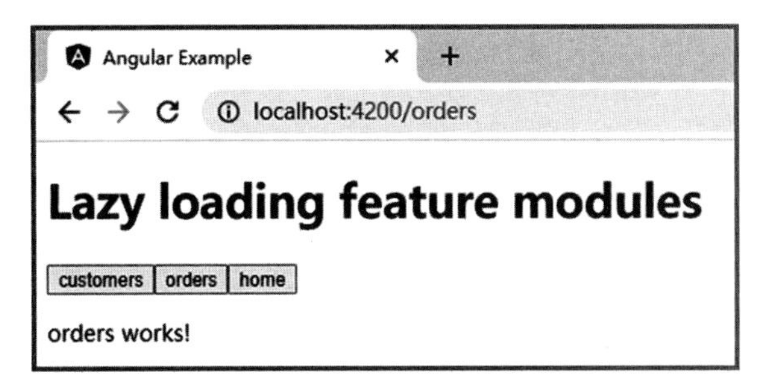

图 14-9 单击图 14-6 中的 orders 按钮的结果

习题 14

一、简答题

1. 简述对安全的理解。
2. 简述对无障碍性的理解。
3. 简述对属性绑定的理解。
4. 简述对惰性加载特性模块的理解。

二、实验题

1. 实现安全的应用开发。
2. 实现无障碍性的应用开发。
3. 实现属性绑定的应用开发。
4. 实现惰性加载特性模块的应用开发。

第 15 章

综合案例

微课视频

15.1 英雄信息

15.1.1 创建文件

在项目 src 根目录下创建 caseofhero 子目录，在 src\caseofhero 目录下创建文件 hero.ts，代码如例 15-1 所示。

【例 15-1】 创建文件 hero.ts 的代码，定义接口和数组。

```
export interface Hero {
  id: number;
  name: string;
}
export const HEROES: Hero[] = [
  { id: 1, name: '张三丰' },
  { id: 2, name: '李斯' },
  { id: 3, name: '王重阳' },
];
```

在 src\caseofhero 目录下创建文件 message.service.ts，代码如例 15-2 所示。

【例 15-2】 创建文件 message.service.ts 的代码，定义类。

```
import {Injectable} from '@angular/core';
@Injectable({ providedIn: 'root' })
export class MessageService {
  messages: string[] = [];
  add(message: string) {
    this.messages.push(message);
  }
  clear() {
    this.messages = [];
  }
}
```

在 src\caseofhero 目录下创建文件 hero.service.ts，代码如例 15-3 所示。

【例 15-3】 创建文件 hero.service.ts 的代码，定义类。

```
import {Injectable} from '@angular/core';
import {HttpClient, HttpHeaders} from '@angular/common/http';
```

```
import {Observable, of} from 'rxjs';
import {catchError, map, tap} from 'rxjs/operators';
import {Hero} from './hero';
import {MessageService} from "./message.service";
@Injectable({providedIn: 'root'})
export class HeroService {
  private heroesUrl = 'api/heroes';
  httpOptions = {
    headers: new HttpHeaders({'Content-Type': 'application/json'})
  };
  constructor(
    private http: HttpClient,
    private messageService: MessageService) { }
  /** GET 方法 */
  getHeroes(): Observable<Hero[]> {
    return this.http.get<Hero[]>(this.heroesUrl)
      .pipe(
        tap(_ => this.log('fetched heroes')),
        catchError(this.handleError<Hero[]>('getHeroes', []))
      );
  }
  /** GET 方法,根据 id 查找,找不到时返回 undefined */
  getHeroNo404<Data>(id: number): Observable<Hero> {
    const url = `${this.heroesUrl}/?id=${id}`;
    return this.http.get<Hero[]>(url)
      .pipe(
        map(heroes => heroes[0])
        tap(h => {
          const outcome = h ? 'fetched' : 'did not find';
          this.log(`${outcome} hero id=${id}`);
        }),
        catchError(this.handleError<Hero>(`getHero id=${id}`))
      );
  }
  /** GET 方法,根据 id 查找,找不到时返回 404 */
  getHero(id: number): Observable<Hero> {
    const url = `${this.heroesUrl}/${id}`;
    return this.http.get<Hero>(url).pipe(
      tap(_ => this.log(`fetched hero id=${id}`)),
      catchError(this.handleError<Hero>(`getHero id=${id}`))
    );
  }
  /* GET 方法,根据名字查找 */
  searchHeroes(term: string): Observable<Hero[]> {
    if (!term.trim()) {
      // 没有找到时返回空数组
      return of([]);
    }
    return this.http.get<Hero[]>(`${this.heroesUrl}/?name=${term}`).pipe(
      tap(x => x.length ?
         this.log(`found heroes matching "${term}"`) :
         this.log(`no heroes matching "${term}"`)),
      catchError(this.handleError<Hero[]>('searchHeroes', []))
    );
```

```
  }
  /** POST 方法 */
  addHero(hero: Hero): Observable<Hero> {
    return this.http.post<Hero>(this.heroesUrl, hero, this.httpOptions).pipe(
      tap((newHero: Hero) => this.log(`added hero w/ id=${newHero.id}`)),
      catchError(this.handleError<Hero>('addHero'))
    );
  }
  /** DELETE 方法 */
  deleteHero(id: number): Observable<Hero> {
    const url = `${this.heroesUrl}/${id}`;
    return this.http.delete<Hero>(url, this.httpOptions).pipe(
      tap(_ => this.log(`deleted hero id=${id}`)),
      catchError(this.handleError<Hero>('deleteHero'))
    );
  }
  /** PUT 方法 */
  updateHero(hero: Hero): Observable<any> {
    return this.http.put(this.heroesUrl, hero, this.httpOptions).pipe(
      tap(_ => this.log(`updated hero id=${hero.id}`)),
      catchError(this.handleError<any>('updateHero'))
    );
  }
//处理 HTTP 操作失败
  private handleError<T>(operation = 'operation', result?: T) {
    return (error: any): Observable<T> => {
      console.error(error); // log to console instead
      this.log(`${operation} failed: ${error.message}`);
      return of(result as T);
    };
  }
  private log(message: string) {
    this.messageService.add(`HeroService: ${message}`);
  }
}
```

在 src\caseofhero 目录下创建文件 messages.component.ts，代码如例 15-4 所示。

【例 15-4】 创建文件 messages.component.ts 的代码，定义组件。

```
import {Component, OnInit} from '@angular/core';
import {MessageService} from './message.service';
@Component({
  selector: 'app-messages',
  template: `
    <div *ngIf="messageService.messages.length">
      <h2>Messages</h2>
      <button class="clear"
              (click)="messageService.clear()">Clear messages</button>
      <div *ngFor='let message of messageService.messages'> {{message}} </div>
    </div>
  `,
  styles: ['h2 {\n' +
  '  color: #A80000;\n' +
  '  font-family: Arial, Helvetica, sans-serif;\n' +
```

```
  '  font-weight: lighter;\n' +
  '}\n' +
  '\n' +
  '.clear {\n' +
  '  color: #333;\n' +
  '  background-color: #eee;\n' +
  '  margin-bottom: 12px;\n' +
  '  padding: 1rem;\n' +
  '  border-radius: 4px;\n' +
  '  font-size: 1rem;\n' +
  '}\n' +
  '.clear:hover {\n' +
  '  color: #fff;\n' +
  '  background-color: #42545C;\n' +
  '}\n']
})
export class MessagesComponent implements OnInit {
  constructor(public messageService: MessageService) {}
  ngOnInit() {  }
}
```

在 src\caseofhero 目录下创建文件 heroes. component. ts，代码如例 15-5 所示。

【例 15-5】 创建文件 heroes. component. ts 的代码，定义组件。

```
import {Component, OnInit} from '@angular/core';
import {Hero} from './hero';
import {HeroService} from './hero.service';
@Component({
  selector: 'app-heroes',
  template: `
    <h2>My Heroes</h2>
    <div>
      <label for="new-hero">Hero name: </label>
      <input id="new-hero" #heroName />
      <!-- (click) passes input value to add() and then clears the input -->
      <button class="add-button" (click)="add(heroName.value); heroName.value=''">
        Add hero
      </button>
    </div>
    <ul class="heroes">
      <li *ngFor="let hero of heroes">
        <a routerLink="/detail/{{hero.id}}">
          <span class="badge">{{hero.id}}</span> {{hero.name}}
        </a>
        <button class="delete" title="delete hero"
                (click)="delete(hero)">x</button>
      </li>
    </ul>
  `,
  styles: ['.heroes {\n' +
  '  margin: 0 0 2em 0;\n' +
  '  list-style-type: none;\n' +
  '  padding: 0;\n' +
  '  width: 15em;\n' +
```

```
'}\n' +
'\n' +
'input {\n' +
'  display: block;\n' +
'  width: 100 % ;\n' +
'  padding: .5rem;\n' +
'  margin: 1rem 0;\n' +
'  box - sizing: border - box;\n' +
'}\n' +
'\n' +
'.heroes li {\n' +
'  position: relative;\n' +
'  cursor: pointer;\n' +
'}\n' +
'\n' +
'.heroes li:hover {\n' +
'  left: .1em;\n' +
'}\n' +
'\n' +
'.heroes a {\n' +
'  color: #333;\n' +
'  text - decoration: none;\n' +
'  background - color: #EEE;\n' +
'  margin: .5em;\n' +
'  padding: .3em 0;\n' +
'  height: 1.6em;\n' +
'  border - radius: 4px;\n' +
'  display: block;\n' +
'  width: 100 % ;\n' +
'}\n' +
'\n' +
'.heroes a:hover {\n' +
'  color: #2c3a41;\n' +
'  background - color: #e6e6e6;\n' +
'}\n' +
'\n' +
'.heroes a:active {\n' +
'  background - color: #525252;\n' +
'  color: #fafafa;\n' +
'}\n' +
'\n' +
'.heroes .badge {\n' +
'  display: inline - block;\n' +
'  font - size: small;\n' +
'  color: white;\n' +
'  padding: 0.8em 0.7em 0 0.7em;\n' +
'  background - color: #405061;\n' +
'  line - height: 1em;\n' +
'  position: relative;\n' +
'  left: -1px;\n' +
'  top: -4px;\n' +
'  height: 1.8em;\n' +
'  min - width: 16px;\n' +
'  text - align: right;\n' +
```

```
  '  margin-right: .8em;\n' +
  '  border-radius: 4px 0 0 4px;\n' +
  '}\n' +
  '\n' +
  '.add-button {\n' +
  'padding: .5rem 1.5rem;\n' +
  'font-size: 1rem;\n' +
  'margin-bottom: 2rem;\n' +
  '}\n' +
  '\n' +
  '.add-button:hover {\n' +
  '  color: white;\n' +
  '  background-color: #42545C;\n' +
  '}\n' +
  '\n' +
  'button.delete {\n' +
  '  position: absolute;\n' +
  '  left: 210px;\n' +
  '  top: 5px;\n' +
  '  background-color: white;\n' +
  '  color:   #525252;\n' +
  '  font-size: 1.1rem;\n' +
  '  padding: 1px 10px 3px 10px;\n' +
  '}\n' +
  '\n' +
  'button.delete:hover {\n' +
  '  background-color: #525252;\n' +
  '  color: white;\n' +
  '}\n']
})
export class HeroesComponent implements OnInit {
  heroes: Hero[] = [];
  constructor(private heroService: HeroService) { }
  ngOnInit(): void {
    this.getHeroes();
  }
  getHeroes(): void {
    this.heroService.getHeroes()
    .subscribe(heroes => this.heroes = heroes);
  }
  add(name: string): void {
    name = name.trim();
    if (!name) { return; }
    this.heroService.addHero({name} as Hero)
      .subscribe(hero => {
        this.heroes.push(hero);
      });
  }
  delete(hero: Hero): void {
    this.heroes = this.heroes.filter(h => h !== hero);
    this.heroService.deleteHero(hero.id).subscribe();
  }
}
```

在 src\caseofhero 目录下创建文件 hero-detail.component.ts,代码如例 15-6 所示。

【例 15-6】 创建文件 hero-detail. component. ts 的代码，定义组件。

```
import {Component, OnInit} from '@angular/core';
import {ActivatedRoute} from '@angular/router';
import {Location} from '@angular/common';
import {Hero} from './hero';
import {HeroService} from './hero.service';
@Component({
  selector: 'app-hero-detail',
  template: `
    <div *ngIf="hero">
      <h2>{{hero.name | uppercase}} Details</h2>
      <div><span>id: </span>{{hero.id}}</div>
      <div>
        <label for="hero-name">Hero name: </label>
        <input id="hero-name" [(ngModel)]="hero.name" placeholder="Hero name"/>
      </div>
      <button (click)="goBack()">go back</button>
      <button (click)="save()">save</button>
    </div>
  `,
  styles: ['label {\n' +
  '  color: #435960;\n' +
  '  font-weight: bold;\n' +
  '}\n' +
  'input {\n' +
  '  font-size: 1em;\n' +
  '  padding: .5rem;\n' +
  '}\n' +
  'button {\n' +
  '  margin-top: 20px;\n' +
  '  margin-right: .5rem;\n' +
  '  background-color: #eee;\n' +
  '  padding: 1rem;\n' +
  '  border-radius: 4px;\n' +
  '  font-size: 1rem;\n' +
  '}\n' +
  'button:hover {\n' +
  '  background-color: #cfd8dc;\n' +
  '}\n' +
  'button:disabled {\n' +
  '  background-color: #eee;\n' +
  '  color: #ccc;\n' +
  '  cursor: auto;\n' +
  '}\n']
})
export class HeroDetailComponent implements OnInit {
  hero: Hero | undefined;
  constructor(
    private route: ActivatedRoute,
    private heroService: HeroService,
    private location: Location
  ) {}
  ngOnInit(): void {
```

```
    this.getHero();
  }
  getHero(): void {
    const id = parseInt(this.route.snapshot.paramMap.get('id')!, 10);
    this.heroService.getHero(id)
      .subscribe(hero => this.hero = hero);
  }
  goBack(): void {
    this.location.back();
  }
  save(): void {
    if (this.hero) {
      this.heroService.updateHero(this.hero)
        .subscribe(() => this.goBack());
    }
  }
}
```

在 src\caseofhero 目录下创建文件 hero-search.component.ts，代码如例 15-7 所示。

【例 15-7】 创建文件 hero-search.component.ts 的代码，定义组件。

```
import {Component, OnInit} from '@angular/core';
import {Observable, Subject} from 'rxjs';
import {
   debounceTime, distinctUntilChanged, switchMap
 } from 'rxjs/operators';
import {Hero} from './hero';
import {HeroService} from './hero.service';
@Component({
  selector: 'app-hero-search',
  template: `
    <div id="search-component">
      <label for="search-box">Hero Search</label>
      <input #searchBox id="search-box" (input)="search(searchBox.value)" />
      <ul class="search-result">
        <li *ngFor="let hero of heroes$ | async" >
          <a routerLink="/detail/{{hero.id}}">
            {{hero.name}}
          </a>
        </li>
      </ul>
    </div>
  `,
  styles: [ 'label {\n' +
  '  display: block;\n' +
  '  font-weight: bold;\n' +
  '  font-size: 1.2rem;\n' +
  '  margin-top: 1rem;\n' +
  '  margin-bottom: .5rem;\n' +
  '\n' +
  '}\n' +
  'input {\n' +
  '  padding: .5rem;\n' +
  '  width: 100%;\n' +
  '  max-width: 600px;\n' +
  '  box-sizing: border-box;\n' +
```

```
  '  display: block;\n' +
  '}\n' +
  '\n' +
  'input:focus {\n' +
  '  outline: #336699 auto 1px;\n' +
  '}\n' +
  '\n' +
  'li {\n' +
  '  list-style-type: none;\n' +
  '}\n' +
  '.search-result li a {\n' +
  '  border-bottom: 1px solid gray;\n' +
  '  border-left: 1px solid gray;\n' +
  '  border-right: 1px solid gray;\n' +
  '  display: inline-block;\n' +
  '  width: 100%;\n' +
  '  max-width: 600px;\n' +
  '  padding: .5rem;\n' +
  '  box-sizing: border-box;\n' +
  '  text-decoration: none;\n' +
  '  color: black;\n' +
  '}\n' +
  '\n' +
  '.search-result li a:hover {\n' +
  '  background-color: #435A60;\n' +
  '  color: white;\n' +
  '}\n' +
  '\n' +
  'ul.search-result {\n' +
  '  margin-top: 0;\n' +
  '  padding-left: 0;\n' +
  '}\n']
})
export class HeroSearchComponent implements OnInit {
  heroes$!: Observable<Hero[]>;
  private searchTerms = new Subject<string>();
  constructor(private heroService: HeroService) {}
  search(term: string): void {
    this.searchTerms.next(term);
  }
  ngOnInit(): void {
    this.heroes$ = this.searchTerms.pipe(
      debounceTime(300),
      distinctUntilChanged(),
      switchMap((term: string) => this.heroService.searchHeroes(term)),
    );
  }
}
```

在 src\caseofhero 目录下创建文件 dashboard.component.ts，代码如例 15-8 所示。

【例 15-8】 创建文件 dashboard.component.ts 的代码，定义组件。

```
import {Component, OnInit} from '@angular/core';
import {Hero} from './hero';
import {HeroService} from "./hero.service";
@Component({
  selector: 'app-dashboard',
```

```
template: `
  <h2>Top Heroes</h2>
  <div class="heroes-menu">
    <a *ngFor="let hero of heroes"
       routerLink="/detail/{{hero.id}}">
      {{hero.name}}
    </a>
  </div>
  <app-hero-search></app-hero-search>
`,
styles: [ 'h2 {\n' +
'  text-align: center;\n' +
'}\n' +
'\n' +
'.heroes-menu {\n' +
'  padding: 0;\n' +
'  margin: auto;\n' +
'  max-width: 1000px;\n' +
'\n' +
'  /* flexbox */\n' +
'  display: -webkit-box;\n' +
'  display: -moz-box;\n' +
'  display: -ms-flexbox;\n' +
'  display: -webkit-flex;\n' +
'  display: flex;\n' +
'  flex-direction: row;\n' +
'  flex-wrap: wrap;\n' +
'  justify-content: space-around;\n' +
'  align-content: flex-start;\n' +
'  align-items: flex-start;\n' +
'}\n' +
'\n' +
'a {\n' +
'  background-color: #3f525c;\n' +
'  border-radius: 2px;\n' +
'  padding: 1rem;\n' +
'  font-size: 1.2rem;\n' +
'  text-decoration: none;\n' +
'  display: inline-block;\n' +
'  color: #fff;\n' +
'  text-align: center;\n' +
'  width: 100%;\n' +
'  min-width: 70px;\n' +
'  margin: .5rem auto;\n' +
'  box-sizing: border-box;\n' +
'\n' +
'  /* flexbox */\n' +
'  order: 0;\n' +
'  flex: 0 1 auto;\n' +
'  align-self: auto;\n' +
'}\n' +
'\n' +
'@media (min-width: 600px) {\n' +
'  a {\n' +
'    width: 18%;\n' +
'    box-sizing: content-box;\n' +
'  }\n' +
```

```
  '}\n' +
  '\n' +
  'a:hover {\n' +
  '  background-color: black;\n' +
  '}\n' ]
})
export class DashboardComponent implements OnInit {
  heroes: Hero[] = [];
  constructor(private heroService: HeroService) { }
  ngOnInit(): void {
    this.getHeroes();
  }
  getHeroes(): void {
    this.heroService.getHeroes()
      .subscribe(heroes => this.heroes = heroes.slice(1, 5));
  }
}
```

在 src\caseofhero 目录下创建文件 app-case.component.ts，代码如例 15-9 所示。

【例 15-9】 创建文件 app-case.component.ts 的代码，定义组件。

```
import {Component} from '@angular/core';
@Component({
  selector: 'root',
  template: `
    <h1>{{title}}</h1>
    <nav>
      <a routerLink="/dashboard">Dashboard</a>
      <a routerLink="/heroesofcase">Heroes</a>
    </nav>
    <router-outlet></router-outlet>
    <app-messages></app-messages>
  `,
  styles: [ 'h1 {\n' +
  '  margin-bottom: 0;\n' +
  '}\n' +
  'nav a {\n' +
  '  padding: 1rem;\n' +
  '  text-decoration: none;\n' +
  '  margin-top: 10px;\n' +
  '  display: inline-block;\n' +
  '  background-color: #e8e8e8;\n' +
  '  color: #3d3d3d;\n' +
  '  border-radius: 4px;\n' +
  '}\n' +
  'nav a:hover {\n' +
  '  color: white;\n' +
  '  background-color: #42545C;\n' +
  '}\n' +
  'nav a.active {\n' +
  '  background-color: black;\n' +
  '}']
})
export class AppCaseComponent {
  title = 'Tour of Heroes';
}
```

在 src\caseofhero 目录下创建文件 app-case. module. ts，代码如例 15-10 所示。

【例 15-10】 创建文件 app-case. module. ts 的代码，定义路由并声明组件等。

```
import {BrowserModule} from '@angular/platform-browser';
import {NgModule} from '@angular/core';
import {AppCaseComponent} from './app-case.component';
import {FormsModule} from "@angular/forms";
import {RouterModule, Routes} from "@angular/router";
import {DashboardComponent} from "./dashboard.component";
import {HeroDetailComponent} from "./hero-detail.component";
import {HeroesComponent} from "./heroes.component";
import {MessagesComponent} from "./messages.component";
import {HeroSearchComponent} from "./hero-search.component";
const routes: Routes = [
  { path: '', redirectTo: '/dashboard', pathMatch: 'full' },
  { path: 'dashboard', component: DashboardComponent },
  { path: 'detail/:id', component: HeroDetailComponent },
  { path: 'heroesofcase', component: HeroesComponent }
];
@NgModule({
  declarations: [
    AppCaseComponent,
    HeroesComponent,
    HeroDetailComponent,
    MessagesComponent,
    HeroSearchComponent,
    DashboardComponent
  ],
  imports: [
    BrowserModule,
    FormsModule,
    RouterModule.forRoot(routes)
  ],
})
export class AppCaseModule { }
```

15.1.2 修改文件

修改 src\examples 目录下的文件 examplesmodules1. module. ts，代码如例 15-11 所示。可以在 src\caseofhero 目录下创建与例 15-11 类似功能的模块文件（而不是修改文件 examplesmodules1. module. ts）并修改 main. ts，这样做更符合一个项目开发的实际。而此处只修改文件 examplesmodules1. module. ts 是为了节省篇幅。

【例 15-11】 修改文件 examplesmodules1. module. ts 的代码，设置启动组件。

```
import {NgModule} from '@angular/core';
import {AppCaseComponent} from "../caseofhero/app-case.component";
import {AppCaseModule} from "../caseofhero/app-case.module";
@NgModule({
  imports: [
    AppCaseModule
  ],
  bootstrap: [AppCaseComponent]
})
export class ExamplesmodulesModule1 {}
```

15.1.3 运行结果

保持其他文件不变并成功运行程序后，在浏览器地址栏中输入 localhost:4200，自动跳转到 localhost:4200/dashboard，结果如图 15-1 所示。单击图 15-1 中的 Heroes 链接，结果如图 15-2 所示。

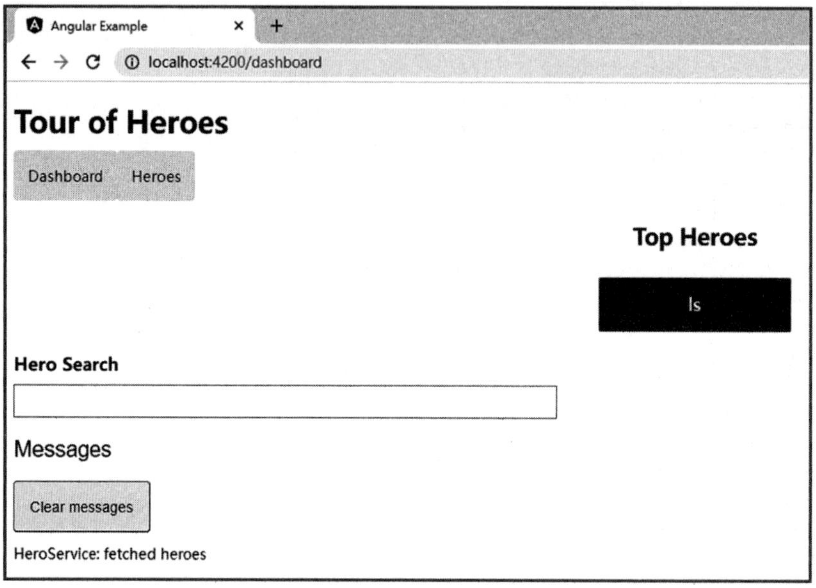

图 15-1 成功运行程序后在浏览器地址栏中输入 localhost:4200 的结果

图 15-2 单击图 15-1 中的 Heroes 链接的结果

15.2 简易通讯录

微课视频

15.2.1 创建文件

在项目 src 根目录下创建 caseofcontact 子目录，在 src\caseofcontact 目录下创建文件 contact. ts，代码如例 15-12 所示。

【例 15-12】 创建文件 contact. ts 的代码，定义接口。

```
export interface Contact {
  name: string;
  phone: string;
}
```

在 src\caseofcontact 目录下创建文件 comp. component. ts，代码如例 15-13 所示。

【例 15-13】 创建文件 comp. component. ts 的代码，定义组件。

```
import {Component, OnInit} from '@angular/core';
import {Contact} from "./contact";
@Component({
  selector: 'app-comp',
  template: `
    <div class = "card border-primary mb-3 text-center">
      <div class = "card-header">通讯录信息输入</div>
      <div  class  = "card-body">
        <form novalidate #Form = "ngForm" class = "form">
          <div class = "form-group" class  >
            <label>姓名</label>      
<input type = "text" placeholder = "请输入人名" min = "20" [(ngModel)] = "name" name =
"name" #userName = "ngModel" minlength = "2" required>
<div *ngIf = "userName.errors?.[minlength] && userName.touched" class = "alert alert-
danger">Name should be at least 2 characters</div>
          </div>
          <div class = "form-group" >
            <label>电话</label>      
<input type = "text" placeholder = "请输入电话" size = "20" [(ngModel)] = "phone" name =
"phone" #userPhone = "ngModel" minlength = "10">
<div *ngIf = "userPhone.errors?.[minlength] && userPhone.touched" class = "alert alert-
danger">Enter valid phone number</div>
          </div>
          <button type = "button" class = "btn btn-success" (click) = "get(); Form.reset()">
保存</button>
        </form>
      </div>
    </div>
    <div class = "row card-block">
      <div class = "col-4 card" *ngFor = "let contact of Contacts" >
        <div class = "card" style = "width:18rem;">
          <div class = "card border-primary mb-3 text-center">
            <div class = "card-header">联系人通讯录</div>
            <div class = "card-body">
              <p>人名: {{contact.name}}</p>
              <p>电话: {{contact.phone}}</p>
```

```
            </div>
          </div>
        </div>
      </div>
    </div>
  `,
  styles: ['label {\n' +
  '    color: blue;\n' +
  '  }\n' +
  '  table, tr, td {\n' +
  '    border: 1px solid green ;\n' +
  '    text-align: left;\n' +
  '\n' +
  '  }\n']
})
export class CompComponent implements OnInit {
  constructor() {
  }
  Contacts: Contact[] = [];
  name: string | undefined;
  phone: string | undefined;
  minlength: any;
  ngOnInit() {
  }
  get() {
    const contact = {name: this.name, phone: this.phone};
    this.Contacts.push(<Contact> contact);
  }
}
```

在 src\caseofcontact 目录下创建文件 app-contact. component. ts,代码如例 15-14 所示。

【例 15-14】 创建文件 app-contact. component. ts 的代码,定义组件。

```
import {NgModule} from '@angular/core';
import {RouterModule, Routes} from '@angular/router';
import {FormsModule} from '@angular/forms';
import {BrowserModule} from '@angular/platform-browser';
import {AppContactComponent} from './app-contact.component';
import {CompComponent} from "./comp.component";
export const routes: Routes = [
  {path: 'contacts', component: AppContactComponent},
];
@NgModule({
  declarations: [
    AppContactComponent,
    CompComponent
  ],
  imports: [
    BrowserModule,
    FormsModule,
    RouterModule.forRoot(routes)
  ],
})
export class AppContactModule {}
```

在 src\caseofcontact 目录下创建文件 app-contact. module. ts，代码如例 15-15 所示。

【例 15-15】 创建文件 app-contact. module. ts 的代码，声明组件。

```
import {NgModule} from '@angular/core';
import {RouterModule, Routes} from '@angular/router';
import {FormsModule} from '@angular/forms';
import {BrowserModule} from '@angular/platform-browser';
import {AppContactComponent} from './app-contact.component';
import {CompComponent} from "./comp.component";
export const routes: Routes = [
  {path: 'contacts', component: AppContactComponent},
];
@NgModule({
  declarations: [
    AppContactComponent,
    CompComponent
  ],
  imports: [
    BrowserModule,
    FormsModule,
    RouterModule.forRoot(routes)
  ],
})
export class AppContactModule {}
```

15.2.2 修改文件

修改 src\examples 目录下的文件 examplesmodules1. module. ts，代码如例 15-16 所示。可以在 src\caseofhero 目录下创建与例 15-16 类似功能的模块文件（而不是修改文件 examplesmodules1. module. ts）并修改 main. ts，这样做更符合一个项目开发的实际。而此处只修改文件 examplesmodules1. module. ts 是为了节省篇幅。

【例 15-16】 修改文件 examplesmodules1. module. ts 的代码，设置启动组件。

```
import {NgModule} from '@angular/core';
import {AppTestExampleModule} from "./testexample/app-test-example.module";
import {AppLazyloadingModule} from "./lazyloadingexamples/app-lazyloading.module";
import {AppCaseModule} from "../caseofhero/app-case.module";
import {AppContactModule} from "../caseofcontact/app-contact.module";
import {AppContactComponent} from "../caseofcontact/app-contact.component";
@NgModule({
  imports: [
    AppContactModule,
  ],
  bootstrap: [AppContactComponent]
})
export class ExamplesmodulesModule1 {}
```

15.2.3 运行结果

保持其他文件不变并成功运行程序后，在浏览器地址栏中输入 localhost:4200，也可以在浏览器地址栏中输入 localhost:4200/contacts，结果如图 15-3 所示。在图 15-3 所示的对应文本框中输入姓名(张三丰)、电话(12345678901)后单击“保存”按钮,结果如图 15-4 所示。

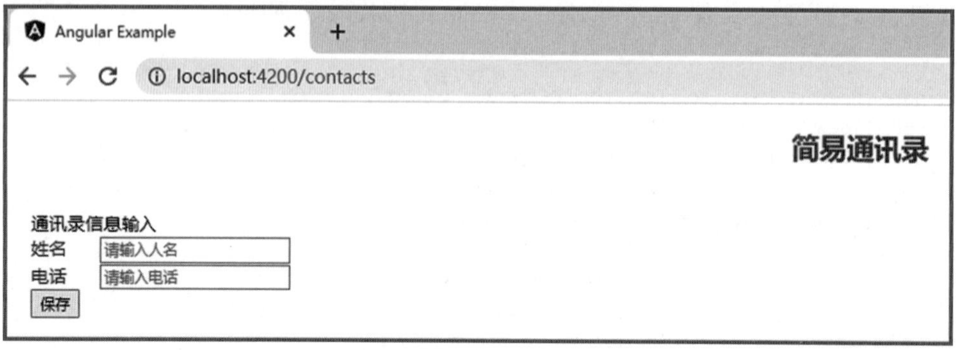

图 15-3 成功运行程序后在浏览器地址栏中输入 localhost:4200/contacts 的结果

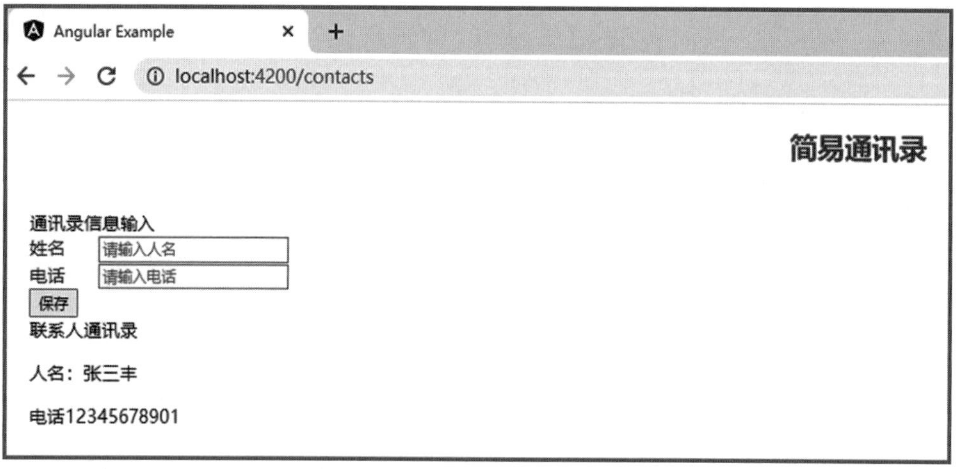

图 15-4 在图 15-3 所示的对应文本框中输入姓名、电话后单击“保存”按钮的结果

习题 15

实验题

1. 完成本章的英雄信息案例。
2. 完成本章的简易通讯录案例。

第16章

整合开发

16.1 与 Ant Design of Angular 的整合开发

16.1.1 创建文件

在项目 src 根目录下创建 ng-zorro-antd 子目录，在 src\ng-zorro-antd 目录下创建文件 local-storage. service. ts，代码如例 16-1 所示。

【例 16-1】 创建文件 local-storage. service. ts 的代码，定义类。

```
import {Injectable} from '@angular/core';
export const USER_NAME:string = 'user_name';
export const LOGIN_TIME:string = 'login_time'
@Injectable({
  providedIn: 'root'
})
export class LocalStorageService {
  constructor() { }
  public get<T>(key:string):any{
    // @ts-ignore
    return JSON.parse(localStorage.getItem(key))as T;
  }
  public set(key:string,value:any):void{
    if(!value && value === undefined ){ return ;}
    const arr = JSON.stringify(value);
    localStorage.setItem(key,arr);
  }
}
```

在 src\ng-zorro-antd 目录下创建文件 login. component. ts，代码如例 16-2 所示。

【例 16-2】 创建文件 login. component. ts 的代码，定义组件。

```
import {Router} from '@angular/router';
import {LocalStorageService, LOGIN_TIME, USER_NAME} from './local-storage.service';
import {Component, OnInit} from '@angular/core';
@Component({
  selector: 'app-login',
  template: `
    <div class="full-screen page-content">
      <div class="wrapper">
```

```
        <img src = "/assets/images/logo.png" alt = "">
        <div class = "text - wrapper">
          <h1 class = "text - text">输入用户名</h1>
        </div>
        <input nz - input placeholder = "用户名" #usernameInput [(ngModel)] = "username">
<button nz - button [nzType] = "'primary'" [disabled] = "!usernameInput.value" (click) = "login()">
开始</button>
        <div class = "copy - right">
          Copyright © 2022 Angular Book Code
        </div>
      </div>
    </div>
  `,
  styles: ['.full - screen{\n' +
  '    position: fixed;\n' +
  '    top: 0;\n' +
  '    bottom: 0;\n' +
  '    left: 0;\n' +
  '    right: 0;\n' +
  '    background - color: blue;\n' +
  '}\n' +
  'div.page - content{\n' +
  '    display: flex;\n' +
  '    justify - content: center;\n' +
  '    align - items: center;\n' +
  '    padding - top: 50px;\n' +
  '    .wrapper{\n' +
  '        display: flex;\n' +
  '        flex - direction: column;\n' +
  '        align - items: center;\n' +
  '        min - height: 400px;\n' +
  '        min - width: 300px;\n' +
  '        width: 30vw;\n' +
  '        max - width: 400px;\n' +
  '        padding: 40px 30px 10px;\n' +
  '        border - radius: 8px;\n' +
  '        background - color: white;\n' +
  '        img{\n' +
  '            width: 120px;\n' +
  '            flex: 0 0 120px;\n' +
  '            height: 120px;\n' +
  '        }\n' +
  '        .text - wrapper{\n' +
  '            margin - top: 20px;\n' +
  '        }\n' +
  '        button{\n' +
  '            width:100%;\n' +
  '            margin - top: 20px;\n' +
  '        }\n' +
  '        .copy - right{\n' +
  '            margin - top: 20px;\n' +
  '        }\n' +
  '    }\n' +
  '}']
```

```
})
export class LoginComponent implements OnInit {
  username: string | undefined;
  constructor(
    private store:LocalStorageService,
    private router:Router
  ) { }
  ngOnInit() {  }
  login():void{
    this.store.set(LOGIN_TIME,Date().toLocaleString());
    this.store.set(USER_NAME,this.username);
    this.router.navigateByUrl('main');
  }
}
```

在 src\ng-zorro-antd 目录下创建文件 header.component.ts，代码如例 16-3 所示。

【例 16-3】 创建文件 header.component.ts 的代码，定义组件。

```
import {Component} from '@angular/core';
@Component({
  selector: 'app-header',
  template: `
    <div class="header-wrapper">
      <img src="./assets/images/default-avatar.png" alt="">
    </div>
  `,
  styles: ['div{ color: blue; }\n' +
  '.header-wrapper{\n' +
  '   > img{\n' +
  '         width: 40px;\n' +
  '         height: 40px;\n' +
  '         margin:0 8px;\n' +
  '     }\n' +
  '}']
})
export class HeaderComponent {  }
```

在 src\ng-zorro-antd 目录下创建文件 content.component.ts，代码如例 16-4 所示。

【例 16-4】 创建文件 content.component.ts 的代码，定义组件。

```
import {Component} from '@angular/core';
@Component({
  selector: 'app-content',
  template: `
    <div>模拟主页内容</div>
  `,
})
export class ContentComponent {  }
```

在 src\ng-zorro-antd 目录下创建文件 main.component.ts，代码如例 16-5 所示。

【例 16-5】 创建文件 main.component.ts 的代码，定义组件。

```
import {Component} from '@angular/core';
@Component({
  selector: 'app-main',
```

```
    template: `
      <nz - layout class = "full - screen">
        <nz - sider nzCollapsible [(nzCollapsed)] = "isCollapsed" [nzWidth] = "260">
          <app - header></app - header>
        </nz - sider>
        <nz - content class = "container">
          <app - content></app - content>
        </nz - content>
      </nz - layout>
    `,
    styles: ['.full - screen{\n' +
    '    position: fixed;\n' +
    '    top: 0;\n' +
    '    bottom: 0;\n' +
    '    right: 0;\n' +
    '    left: 0;\n' +
    '}']
})
export class MainComponent {
  isCollapsed: any;
}
```

在 src\ng-zorro-antd 目录下创建文件 app-ng-zorro-antd.component.ts,代码如例 16-6 所示。

【例 16-6】 创建文件 app-ng-zorro-antd.component.ts 的代码,定义组件。

```
import {Component} from '@angular/core';
@Component({
  selector: 'root',
  template: `
    <router - outlet></router - outlet>
  `,
})
export class AppNgZorroAntdComponent {
}
```

在 src\ng-zorro-antd 目录下创建文件 ng-zorro-antd.module.ts,代码如例 16-7 所示。

【例 16-7】 创建文件 ng-zorro-antd.module.ts 的代码,定义路由并声明组件。

```
import {BrowserModule} from '@angular/platform - browser';
import {NgModule} from '@angular/core';
import {FormsModule} from '@angular/forms';
import {HttpClientModule} from '@angular/common/http';
import {BrowserAnimationsModule} from '@angular/platform - browser/animations';
import {registerLocaleData} from '@angular/common';
import en from '@angular/common/locales/en';
import {LoginComponent} from './login.component';
import {HeaderComponent} from './header.component';
import {ContentComponent} from './content.component';
import {RouterModule, Routes} from "@angular/router";
import {MainComponent} from "./main.component";
import {NzLayoutModule} from "ng - zorro - antd/layout";    //安装
import {NzButtonModule} from "ng - zorro - antd/button";
import {AppNgZorroAntdComponent} from "./app - ng - zorro - antd.component";
```

```
registerLocaleData(en);
const routes:Routes = [
  {path:'',redirectTo:'/login',pathMatch:'full'},
  {path:'login',component:LoginComponent},
  {path:'main',component:MainComponent,}
]
// @ts - ignore
@NgModule({
  declarations: [
    LoginComponent,
    MainComponent,
    HeaderComponent,
    ContentComponent,
    AppNgZorroAntdComponent
  ],
  imports: [
    BrowserModule,
    FormsModule,
    HttpClientModule,
    BrowserAnimationsModule,
    RouterModule.forRoot(routes),
    NzLayoutModule,
    NzButtonModule
  ],
  exports: [
    LoginComponent
  ],
})
export class NgZorroAntdModule {}
```

16.1.2 修改文件

修改 src\examples 目录下的文件 examplesmodules1. module. ts,代码如例 16-8 所示。可以在 src\caseofhero 目录下创建与例 16-8 类似功能的模块文件(而不是修改文件 examplesmodules1. module. ts)并修改 main. ts,这样做更符合一个项目开发的实际。而此处只修改文件 examplesmodules1. module. ts 是为了节省篇幅。

【例 16-8】 修改文件 examplesmodules1. module. ts 的代码,设置启动组件。

```
import {NgModule} from '@angular/core';
import {NgZorroAntdModule} from "../ng - zorro - antd/ng - zorro - antd.module";
import {AppNgZorroAntdComponent} from "../ng - zorro - antd/app - ng - zorro - antd.component";
@NgModule({
  imports: [
    NgZorroAntdModule,
  ],
  bootstrap: [AppNgZorroAntdComponent]
})
export class ExamplesmodulesModule1 {}
```

16.1.3 运行结果

保持其他文件不变并成功运行程序后,在浏览器地址栏中输入 localhost:4200,自动跳转到 localhost：4200/login，也可在浏览器地址栏中输入 localhost：4200/login，结果如图 16-1 所示。在图 16-1 所示的文本框中输入用户名(如张三丰)后单击“开始”按钮，或者在浏览器地址栏中输入 localhost:4200/login，结果如图 16-2 所示。

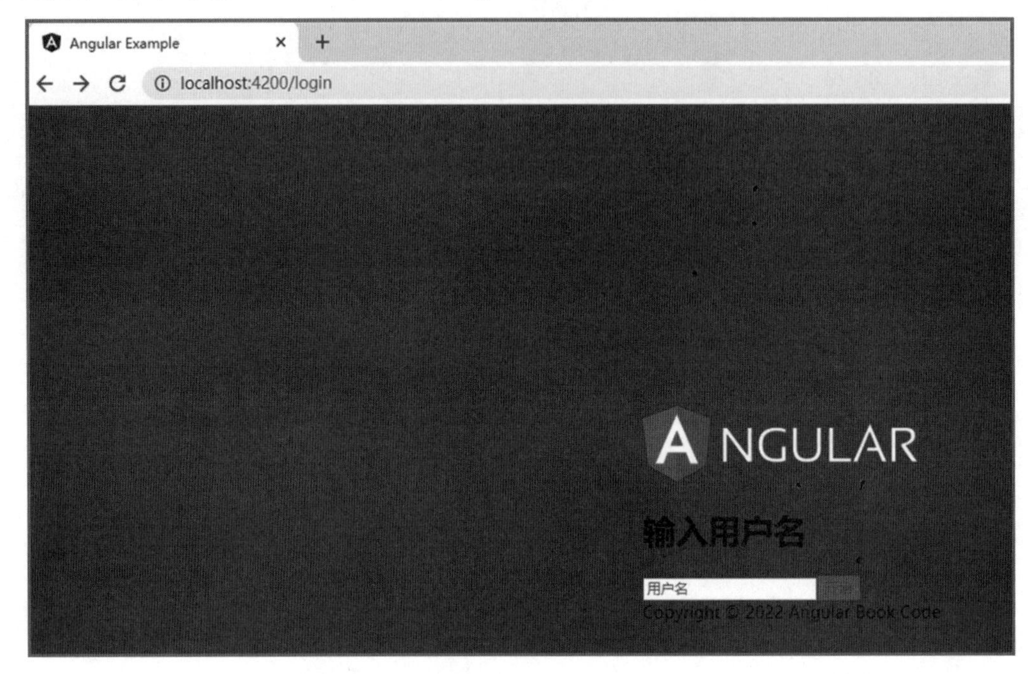

图 16-1　成功运行程序后在浏览器地址栏中输入 localhost:4200 的结果

图 16-2　在图 16-1 所示的文本框中输入用户名后单击“开始”按钮的结果

16.2　与 Spring Boot 的整合开发

16.2.1　创建 Spring Boot 项目 backendofangular

参考电子资源(或其他资料)，使用 IDEA 创建 Spring Boot 项目 backendofangular。

修改文件 pom. xml，代码如例 16-9 所示。

【例 16-9】 修改文件 pom. xml 的代码，实现项目元数据的设置。

```
<?xml version = "1.0" encoding = "UTF - 8"?>
< project xmlns = "http://maven.apache.org/POM/4.0.0" xmlns:xsi = "http://www.w3.org/2001/
```

```
XMLSchema - instance"
         xsi:schemaLocation = "http://maven.apache.org/POM/4.0.0
https://maven.apache.org/xsd/maven - 4.0.0.xsd">
    < modelVersion > 4.0.0 </modelVersion >
    < parent >
        < groupId > org.springframework.boot </groupId >
        < artifactId > spring - boot - starter - parent </artifactId >
        < version > 3.0.0 - M1 </version >
        < relativePath/> <!-- lookup parent from repository -- >
    </parent >
    < groupId > edu.bookcode </groupId >
    < artifactId > backendofangular </artifactId >
    < version > 0.0.1 - SNAPSHOT </version >
    < name > backendofangular </name >
    < description > backendofangular </description >
    < properties >
        < java.version > 17 </java.version >
    </properties >
    < dependencies >
        < dependency >
            < groupId > org.springframework.boot </groupId >
            < artifactId > spring - boot - starter - web </artifactId >
        </dependency >
        < dependency >
            < groupId > org.springframework.boot </groupId >
            < artifactId > spring - boot - starter - data - jpa </artifactId >
        </dependency >
        < dependency >
            < groupId > mysql </groupId >
            < artifactId > mysql - connector - java </artifactId >
            < scope > runtime </scope >
        </dependency >
        < dependency >
            < groupId > redis.clients </groupId >
            < artifactId > jedis </artifactId >
            < version > 2.4.2 </version >
        </dependency >
        < dependency >
            < groupId > org.springframework.boot </groupId >
            < artifactId > spring - boot - starter - data - rest </artifactId >
        </dependency >
        < dependency >
            < groupId > org.springframework.boot </groupId >
            < artifactId > spring - boot - starter - test </artifactId >
            < scope > test </scope >
        </dependency >
        < dependency >
            < groupId > jakarta.validation </groupId >
            < artifactId > jakarta.validation - api </artifactId >
        </dependency >
        < dependency >
            < groupId > org.projectlombok </groupId >
            < artifactId > lombok </artifactId >
            < optional > true </optional >
```

```
        </dependency>
    </dependencies>
    <build>
        <plugins>
            <plugin>
                <groupId>org.springframework.boot</groupId>
                <artifactId>spring-boot-maven-plugin</artifactId>
            </plugin>
        </plugins>
    </build>
    <repositories>
        <repository>
            <id>spring-milestones</id>
            <name>Spring Milestones</name>
            <url>https://repo.spring.io/milestone</url>
            <snapshots>
                <enabled>false</enabled>
            </snapshots>
        </repository>
        <repository>
            <id>spring-releases</id>
            <name>Spring Releases</name>
            <url>https://repo.spring.io/libs-release</url>
        </repository>
        <repository>
            <id>org.jboss.repository.releases</id>
            <name>JBoss Maven Release Repository</name>
            <url>https://repository.jboss.org/nexus/content/repositories/releases</url>
        </repository>
    </repositories>
    <pluginRepositories>
        <pluginRepository>
            <id>spring-milestones</id>
            <name>Spring Milestones</name>
            <url>https://repo.spring.io/milestone</url>
            <snapshots>
                <enabled>false</enabled>
            </snapshots>
        </pluginRepository>
        <pluginRepository>
            <id>spring-releases</id>
            <name>Spring Releases</name>
            <url>https://repo.spring.io/libs-release</url>
        </pluginRepository>
    </pluginRepositories>
</project>
```

16.2.2 创建类 Employee

在 Spring Boot 项目 backendofangular 的包 edu.bookcode 下创建类 Employee，代码如例 16-10 所示。

【例 16-10】 创建类 Employee 的代码，定义实体。

```
package edu.bookcode;
import jakarta.persistence.*;
@Entity
@Table(name = "employees")
public class Employee {
@Id
@GeneratedValue(strategy = GenerationType.IDENTITY)
    private long id;
    private String firstName;
    private String lastName;
    private String emailId;
    public Employee() {     }
    public long getId() {
        return id;
    }
    public void setId(long id) {
        this.id = id;
    }
    @Column(name = "first_name", nullable = false)
    public String getFirstName() {
        return firstName;
    }
    public void setFirstName(String firstName) {
        this.firstName = firstName;
    }
    @Column(name = "last_name", nullable = false)
    public String getLastName() {
        return lastName;
    }
    public void setLastName(String lastName) {
        this.lastName = lastName;
    }
    @Column(name = "email_address", nullable = false)
    public String getEmailId() {
        return emailId;
    }
    public void setEmailId(String emailId) {
        this.emailId = emailId;
    }
    @Override
    public String toString() {
        return "Employee [id = " + id + ", firstName = " + firstName + ", lastName = " +
lastName + ", emailId = " + emailId + "]";
    }
}
```

16.2.3 创建接口 EmployeeRepository

在 Spring Boot 项目 backendofangular 的包 edu.bookcode 下创建接口 EmployeeRepository,代码如例 16-11 所示。

【例 16-11】 创建接口 EmployeeRepository 的代码,定义访问数据库的接口。

```
package edu.bookcode;
```

```
import org.springframework.data.jpa.repository.JpaRepository;
import org.springframework.stereotype.Repository;
@Repository
public interface EmployeeRepository extends JpaRepository<Employee, Long>{
}
```

16.2.4 创建类 EmployeeController

在 Spring Boot 项目 backendofangular 的包 edu.bookcode 下创建类 EmployeeController，代码如例 16-12 所示。

【例 16-12】 创建类 EmployeeController 的代码，定义控制类。

```
package edu.bookcode;
import edu.bookcode.exception.ResourceNotFoundException;
import jakarta.validation.Valid;
import org.springframework.beans.factory.annotation.Autowired;
import org.springframework.http.ResponseEntity;
import org.springframework.web.bind.annotation.*;
import java.util.HashMap;
import java.util.List;
import java.util.Map;
//@CrossOrigin(origins = "http://localhost:4200")
@RestController
@RequestMapping("/api/v1")
public class EmployeeController {
    @Autowired
    private EmployeeRepository employeeRepository;
    @GetMapping("/employees")
    public List<Employee> getAllEmployees() {
        return employeeRepository.findAll();
    }
    @GetMapping("/employees/{id}")
    public ResponseEntity<Employee> getEmployeeById(@PathVariable(value = "id") Long
employeeId)
            throws ResourceNotFoundException {
        Employee employee = employeeRepository.findById(employeeId)
                .orElseThrow(() -> new ResourceNotFoundException("Employee not found for
this id :: " + employeeId));
        return ResponseEntity.ok().body(employee);
    }
    @PostMapping("/employees")
    public Employee createEmployee(@Valid @RequestBody Employee employee) {
        return employeeRepository.save(employee);
    }
    @PutMapping("/employees/{id}")
    public ResponseEntity<Employee> updateEmployee(@PathVariable(value = "id") Long
employeeId,
            @Valid @RequestBody Employee employeeDetails) throws ResourceNotFoundException {
        Employee employee = employeeRepository.findById(employeeId)
                .orElseThrow(() -> new ResourceNotFoundException("Employee not found for
this id :: " + employeeId));
        employee.setEmailId(employeeDetails.getEmailId());
        employee.setLastName(employeeDetails.getLastName());
```

```
            employee.setFirstName(employeeDetails.getFirstName());
            final Employee updatedEmployee = employeeRepository.save(employee);
            return ResponseEntity.ok(updatedEmployee);
        }
        @DeleteMapping("/employees/{id}")
        public Map<String, Boolean> deleteEmployee(@PathVariable(value = "id") Long employeeId)
                throws ResourceNotFoundException {
            Employee employee = employeeRepository.findById(employeeId)
                    .orElseThrow(() -> new ResourceNotFoundException("Employee not found for
    this id :: " + employeeId));
            employeeRepository.delete(employee);
            Map<String, Boolean> response = new HashMap<>();
            response.put("deleted", Boolean.TRUE);
            return response;
        }
    }
```

16.2.5 创建类 MvcConfig

在 Spring Boot 项目 backendofangular 的包 edu.bookcode 下创建类 MvcConfig，代码如例 16-13 所示。

【例 16-13】 创建类 MvcConfig 的代码，定义配置类。

```
package edu.bookcode;
import org.springframework.context.annotation.Configuration;
import org.springframework.web.servlet.config.annotation.CorsRegistry;
import org.springframework.web.servlet.config.annotation.WebMvcConfigurer;
@Configuration
public class MvcConfig implements WebMvcConfigurer {
    //解决跨域问题
    @Override
    public void addCorsMappings(CorsRegistry registry) {
        //所有请求都允许跨域 使用这种方法就不能在 interceptor 中再配置 header 了
        registry.addMapping("/**")
                .allowCredentials(true)
                .allowedOrigins("http://localhost:4200")
                .allowedMethods("POST","GET","PUT","OPTIONS","DELETE")
                .allowedHeaders("*")
                .maxAge(3600);
    }
}
```

16.2.6 修改后端配置文件

修改 src\main\resources 目录下的文件 application.properties，代码如例 16-14 所示。注意，本案例中使用的 MySQL 版本是 8.x，若使用其他版本 MySQL（如 5.x），则例 16-14 中的代码需要调整。

【例 16-14】 修改文件 application.properties 的代码，定义访问 MySQL 等配置信息。

```
spring.datasource.url = jdbc:mysql://localhost:3306/studywebsite?serverTimezone =
GMT&useUnicode = true&characterEncoding = UTF-8&useSSL = false
spring.datasource.username = root
```

```
spring.datasource.password = ws780125
spring.datasource.driver-class-name = com.mysql.cj.jdbc.Driver
spring.jpa.generate-ddl = true
spring.jpa.hibernate.ddl-auto = update
spring.jpa.show-sql = true
server.error.include-message = always
```

16.2.7 运行后端 Spring Boot 程序

在数据库 MySQL 中创建数据库 studywebsite，以便于存放自动生成的表 employees。

运行入口类 BackendofangularApplication，成功启动自带的内置 Tomcat。在浏览器地址栏中输入 localhost：8080/employees 后，部分结果如图 16-3 所示。

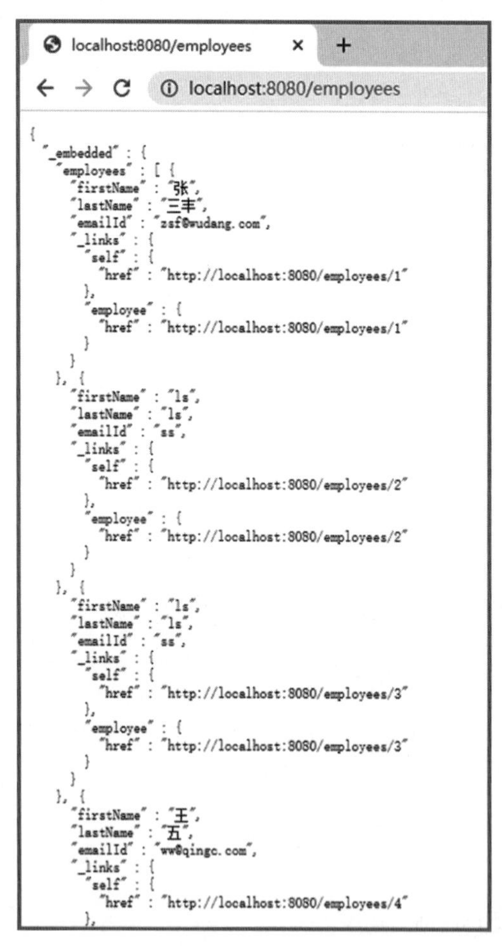

图 16-3 在浏览器地址栏中输入 localhost：8080/employees 后的部分结果(从上往下)

16.2.8 创建前端目录和文件

在 Angular 项目 src 根目录下创建 employee 子目录，在 src\employee 目录下创建 create-employee 子目录，在 src\employee\create-employee 目录下创建文件 create-employee.component.html，代码如例 16-15 所示。

【例 16-15】 创建文件 create-employee.component.html 的代码，实现增加员工信息

的界面。

```
<h3>增加员工 Create Employee</h3>
<div [hidden] = "submitted" style = "width: 400px;">
  <form (ngSubmit) = "onSubmit()">
    <div class = "form-group">
      <label for = "name">姓 First Name</label>
<input type = "text" class = "form-control" id = "firstName" required [(ngModel)] = "employee.
firstName" name = "firstName">
    </div>
    <div class = "form-group">
      <label for = "name">名 Last Name</label>
<input type = "text" class = "form-control" id = "lastName" required [(ngModel)] = "employee.
lastName" name = "lastName">
    </div>
    <div class = "form-group">
      <label for = "name">邮箱 Email Id</label>
<input type = "text" class = "form-control" id = "emailId" required [(ngModel)] = "employee.
emailId" name = "emailId">
    </div>
    <button type = "submit" class = "btn btn-success">保存 Submit</button>
  </form>
</div>
<div [hidden] = "!submitted">
  <h4>保存成功 You submitted successfully!</h4>
</div>
```

在 Angular 项目 src\employee\create-employee 目录下创建文件 create-employee.component.ts,代码如例 16-16 所示。本章为了更好地演示 Angular 项目分解方法,使用的是独立的外部模板文件 create-employee.component.html。

【例 16-16】 创建文件 create-employee.component.ts 的代码,定义组件。

```
import {EmployeeService} from '../employee.service';
import {Employee} from '../employee';
import {Component, OnInit} from '@angular/core';
import {Router} from '@angular/router';
@Component({
  selector: 'app-create-employee',
  templateUrl: './create-employee.component.html',
})
export class CreateEmployeeComponent implements OnInit {
  employee: Employee = new Employee();
  submitted = false;
  constructor(private employeeService: EmployeeService,
    private router: Router) { }
  ngOnInit() {
  }
  save() {
    this.employeeService
    .createEmployee(this.employee).subscribe(data => {
      console.log(data)
      this.employee = new Employee();
      this.gotoList();
    },
```

```
      error => console.log(error));
  }
  onSubmit() {
    this.submitted = true;
    this.save();
  }
  gotoList() {
    this.router.navigate(['/employees']);
  }
}
```

在 Angular 项目 src\employee 创建 employee-details 子目录，在 src\employee\employee-details 目录下创建文件 employee-details. component. html，代码如例 16-17 所示。

【例 16-17】 创建文件 employee-details. component. html 的代码，实现显示员工信息的界面。

```
<h2>Employee Details 员工详细信息</h2>
<hr/>
<div *ngIf="employee">
  <div>
    <label><b>姓 First Name: </b></label> {{employee.firstName}}
  </div>
  <div>
    <label><b>名 Last Name: </b></label> {{employee.lastName}}
  </div>
  <div>
    <label><b>邮箱 Email Id: </b></label> {{employee.emailId}}
  </div>
</div>
<br>
<br>
<button (click)="list()" class="btn btn-primary">返回到员工列表 Back to Employee List</
button><br>
```

在 Angular 项目 src\employee\employee-details 目录下创建文件 employee-details. component. ts，代码如例 16-18 所示。例 16-18 调用的是外部模板文件 employee-details. component. html。

【例 16-18】 创建文件 employee-details. component. ts 的代码，定义组件。

```
import {Employee} from '../employee';
import {Component, OnInit} from '@angular/core';
import {EmployeeService} from '../employee.service';
import {Router, ActivatedRoute} from '@angular/router';
@Component({
  selector: 'app-employee-details',
  templateUrl: './employee-details.component.html',
})
export class EmployeeDetailsComponent implements OnInit {
  // @ts-ignore
  id: number;
  // @ts-ignore
  employee: Employee;
  constructor(private route: ActivatedRoute,private router: Router,
```

```
      private employeeService: EmployeeService) { }
    ngOnInit() {
      this.employee = new Employee();
      this.id = this.route.snapshot.params['id'];
      this.employeeService.getEmployee(this.id)
        .subscribe(data => {
          console.log(data)
          this.employee = data;
        }, error => console.log(error));
    }
    list(){
      this.router.navigate(['employees']);
    }
  }
```

在 Angular 项目 src\employee 创建 employee-list 子目录，在 src\employee\employee-list 目录下创建文件 employee-list.component.html，代码如例 16-19 所示。

【例 16-19】 创建文件 employee-list.component.html 的代码，实现显示员工列表信息的界面。

```
<div class="panel panel-primary">
  <div class="panel-heading">
    <h2>Employee List 员工列表</h2>
  </div>
  <div class="panel-body">
    <table class="table table-striped">
      <thead>
        <tr>
          <th>Firstname 姓</th>
          <th>Lastname 名</th>
          <th>Email 邮箱</th>
          <th>Actions 操作</th>
        </tr>
      </thead>
      <tbody>
        <tr *ngFor="let employee of employees | async">
          <td>{{employee.firstName}}</td>
          <td>{{employee.lastName}}</td>
          <td>{{employee.emailId}}</td>
         <td><button (click)="deleteEmployee(employee.id)" class="btn btn-danger">Delete
删除</button>
<button (click)="updateEmployee(employee.id)" class="btn btn-info" style="margin-
left: 10px">Update 更新</button>
<button (click)="employeeDetails(employee.id)" class="btn btn-info" style="margin-
left: 10px">Details 详细</button>
          </td>
        </tr>
      </tbody>
    </table>
  </div>
</div>
```

在 Angular 项目 src\employee\employee-list 目录下创建文件 employee-list.component.ts，代码如例 16-20 所示。

【例 16-20】 创建文件 employee-list. component. ts 的代码,定义组件。

```
import {Observable} from "rxjs";
import {EmployeeService} from "./../employee.service";
import {Employee} from "./../employee";
import {Component, OnInit} from "@angular/core";
import {Router} from '@angular/router';
@Component({
  selector: "app-employee-list",
  templateUrl: "./employee-list.component.html",
})
export class EmployeeListComponent implements OnInit {
  // @ts-ignore
  employees: Observable<Employee[]>;
  constructor(private employeeService: EmployeeService,
    private router: Router) {}
  ngOnInit() {
    this.reloadData();
  }
  reloadData() {
    this.employees = this.employeeService.getEmployeesList();
  }
  deleteEmployee(id: number) {
    this.employeeService.deleteEmployee(id)
      .subscribe(
        data => {
          console.log(data);
          this.reloadData();
        },
        error => console.log(error));
  }
  employeeDetails(id: number){
    this.router.navigate(['details', id]);
  }
  updateEmployee(id: number){
    this.router.navigate(['update', id]);
  }
}
```

在 Angular 项目 src\employee 创建 employee-list 子目录,在 src\employee\update-employee 目录下创建文件 update-employee. component. html,代码如例 16-21 所示。

【例 16-21】 创建文件 update-employee. component. html 的代码,定义显示更新员工信息的界面。

```
<h3>更新员工信息 Update Employee</h3>
<div style="width: 400px;">
  <form (ngSubmit)="onSubmit()">
    <div class="form-group">
      <label for="name">姓 First Name</label>
<input type="text" class="form-control" id="firstName" required [(ngModel)]="employee.
firstName" name="firstName">
    </div>
    <div class="form-group">
      <label for="name">名 Last Name</label>
```

```
< input type = "text" class = "form - control" id = "lastName" required [(ngModel)] = "employee.
lastName" name = "lastName">
    </div >
    < div class = "form - group">
      < label for = "name">邮箱 Email Id </label >
< input type = "text" class = "form - control" id = "emailId" required [(ngModel)] = "employee.
emailId" name = "emailId">
    </div >
    < button type = "submit" class = "btn btn - success">保存 Submit </button >
  </form >
</div >
```

在 Angular 项目 src\employee\update-employee 目录下创建文件 update-employee.component.ts,代码如例 16-22 所示。

【例 16-22】 创建文件 update-employee.component.ts 的代码,定义组件。

```
import {Component, OnInit} from '@angular/core';
import {Employee} from '../employee';
import {ActivatedRoute, Router} from '@angular/router';
import {EmployeeService} from '../employee.service';
@Component({
  selector: 'app - update - employee',
  templateUrl: './update - employee.component.html',
})
export class UpdateEmployeeComponent implements OnInit {
  // @ts - ignore
  id: number;
  // @ts - ignore
  employee: Employee;
  constructor(private route: ActivatedRoute,private router: Router,
    private employeeService: EmployeeService) { }
  ngOnInit() {
    this.employee = new Employee();
    this.id = this.route.snapshot.params['id'];
    this.employeeService.getEmployee(this.id)
      .subscribe(data => {
        console.log(data)
        this.employee = data;
      }, error => console.log(error));
  }
  updateEmployee() {
    this.employeeService.updateEmployee(this.id, this.employee)
      .subscribe(data => {
        console.log(data);
        this.employee = new Employee();
        this.gotoList();
      }, error => console.log(error));
  }
  onSubmit() {
    this.updateEmployee();
  }
  gotoList() {
    this.router.navigate(['/employees']);
  }
}
```

在 Angular 项目 src\employee 目录下创建文件 employee. ts，代码如例 16-23 所示。

【例 16-23】 创建文件 employee. ts 的代码，定义类。

```
export class Employee {
  // @ts - ignore
  id: number;  // @ts - ignore
  firstName: string;  // @ts - ignore
  lastName: string;  // @ts - ignore
  emailId: string;  // @ts - ignore
  active: boolean;
}
```

在 Angular 项目 src\employee 目录下创建文件 employee. service. ts，代码如例 16-24 所示。

【例 16-24】 创建文件 employee. service. ts 的代码，定义类。

```
import {Injectable} from '@angular/core';
import {HttpClient} from '@angular/common/http';
import {Observable} from 'rxjs';
@Injectable({
  providedIn: 'root'
})
export class EmployeeService {
  private baseUrl = 'http://localhost:8080/api/v1/employees';
  constructor(private http: HttpClient) { }
  getEmployee(id: number): Observable<any> {
    return this.http.get(`${this.baseUrl}/${id}`);
  }
  createEmployee(employee: Object): Observable<Object> {
    return this.http.post(`${this.baseUrl}`, employee);
  }
  updateEmployee(id: number, value: any): Observable<Object> {
    return this.http.put(`${this.baseUrl}/${id}`, value);
  }
  deleteEmployee(id: number): Observable<any> {
    return this.http.delete(`${this.baseUrl}/${id}`, { responseType: 'text' });
  }
  getEmployeesList(): Observable<any> {
    return this.http.get(`${this.baseUrl}`);
  }
}
```

在 Angular 项目 src\employee 目录下创建文件 app-employee. component. html，代码如例 16-25 所示。

【例 16-25】 创建文件 app-employee. component. html 的代码，实现首页界面。

```
<nav class = "navbar navbar - expand - sm bg - primary navbar - dark">
  <!-- 链接部分 -->
  <ul class = "navbar - nav">
    <li class = "nav - item">
 <a routerLink = "employees" class = "nav - link" routerLinkActive = "active">员工列表 Employee
List </a>
    </li>
    <li class = "nav - item">
```

```
<a routerLink = "add" class = "nav - link" routerLinkActive = "active">增加员工 Add Employee </a>
      </li>
    </ul>
</nav>
<!-- 标题 -->
<div class = "container">
  <br>
  <h2 style = "text - align: center;">{{title}}</h2>
  <hr>
  <div class = "card">
    <div class = "card - body">
  <router - outlet></router - outlet>
    </div>
  </div>
</div>
<!-- 底部部分版权信息 -->
<footer class = "footer">
  <div class = "container">
      <span>版权 2022 @ Angular + JavaGuides【All Rights Reserved 2019 @JavaGuides】</span>
  </div>
</footer>
```

在 Angular 项目 src\employee 目录下创建文件 app-employee. component. ts，代码如例 16-26 所示。

【例 16-26】 创建文件 app-employee. component. ts 的代码，定义组件。

```
import {Component} from '@angular/core';
@Component({
  selector: 'root',
  templateUrl: './app - employee.component.html',
})
export class AppEmployeeComponent {
  title = 'Angular 与 Spring Boot 整合开发';
}
```

16.2.9 模块

在 Angular 项目 src\employee 目录下创建文件 app-routing. module. ts，代码如例 16-27 所示。

【例 16-27】 创建文件 app-routing. module. ts 的代码，定义路由等。

```
import {EmployeeDetailsComponent} from './employee - details/employee - details.component';
import {CreateEmployeeComponent} from './create - employee/create - employee.component';
import {NgModule} from '@angular/core';
import {Routes, RouterModule} from '@angular/router';
import {EmployeeListComponent} from './employee - list/employee - list.component';
import {UpdateEmployeeComponent} from './update - employee/update - employee.component';
const routes: Routes = [
  { path: '', redirectTo: 'employee', pathMatch: 'full' },
  { path: 'employees', component: EmployeeListComponent },
  { path: 'add', component: CreateEmployeeComponent },
  { path: 'update/:id', component: UpdateEmployeeComponent },
  { path: 'details/:id', component: EmployeeDetailsComponent },
```

```
];
@NgModule({
  imports: [RouterModule.forRoot(routes)],
  exports: [RouterModule]
})
export class AppRoutingModule { }
```

在 Angular 项目 src\employee 目录下创建文件 app-employee. module. ts，代码如例 16-28 所示。本章为了更好地演示 Angular 项目分解方法，使用的是独立的路由文件。例 16-28 调用的是外部路由文件 app-routing. module. ts。

【例 16-28】 创建文件 app-employee. module. ts 的代码，声明组件并设置启动组件。

```
import {BrowserModule} from '@angular/platform-browser';
import {NgModule} from '@angular/core';
import {FormsModule} from '@angular/forms';
import {AppRoutingModule} from './app-routing.module';
import {AppEmployeeComponent} from './app-employee.component';
import {CreateEmployeeComponent} from './create-employee/create-employee.component';
import {EmployeeDetailsComponent} from './employee-details/employee-details.component';
import {EmployeeListComponent} from './employee-list/employee-list.component';
import {HttpClientModule} from '@angular/common/http';
import {UpdateEmployeeComponent} from './update-employee/update-employee.component';
@NgModule({
  declarations: [
    AppEmployeeComponent,
    CreateEmployeeComponent,
    EmployeeDetailsComponent,
    EmployeeListComponent,
    UpdateEmployeeComponent
  ],
  imports: [
    BrowserModule,
    AppRoutingModule,
    FormsModule,
    HttpClientModule
  ],
  providers: [],
  bootstrap: [AppEmployeeComponent]
})
export class AppEmployeeModule { }
```

16.2.10 修改文件 main. ts

修改 src 目录下的文件 main. ts，代码如例 16-29 所示。

【例 16-29】 修改文件 main. ts 的代码，设置启动模块。

```
import {enableProdMode} from '@angular/core';
import {platformBrowserDynamic} from '@angular/platform-browser-dynamic';
import {AppEmployeeModule} from "./employee/app-employee.module";
import {environment} from './environments/environment';
if (environment.production) {
```

```
  enableProdMode();
}
platformBrowserDynamic().bootstrapModule(AppEmployeeModule)
   .catch(err => console.error(err));
```

16.2.11　运行结果

保持其他文件不变并成功运行程序后，在浏览器地址栏中输入 localhost:4200，结果如图 16-4 所示。单击图 16-4 所示中的超链接“员工列表 Employee List”，结果如图 16-5 所示。单击图 16-4 所示中的超链接“增加员工 Add Employee”，结果如图 16-6 所示。在图 16-6 所示中分别输入姓、名、邮箱，结果如图 16-7 所示。在图 16-7 所示的基础上单击“保存 Submit”按钮，结果如图 16-8 所示。单击图 16-8 所示中的超链接“员工列表 Employee List”（注意，在图 16-5 所示的基础上修改了第 2 个 Employee 信息后增加了第 3 个 Employee 信息），结果如图 16-9 所示。

图 16-4　成功运行程序后在浏览器地址栏中输入 localhost:4200 的结果

图 16-5　单击图 16-4 所示中的超链接“员工列表 Employee List”的结果

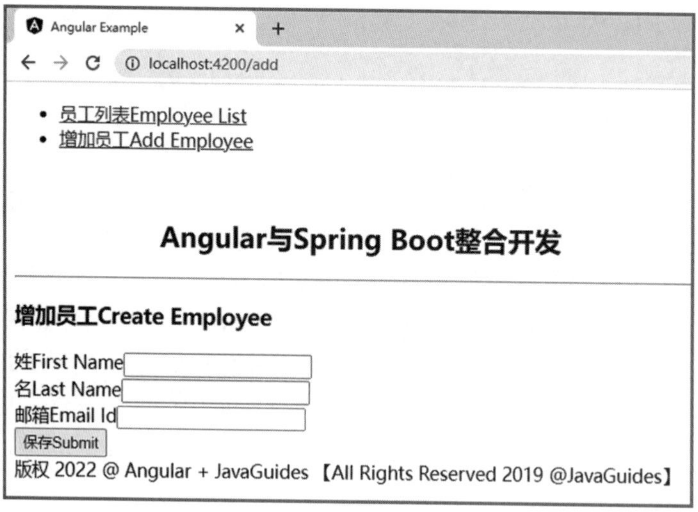

Angular Example
localhost:4200/add

- 员工列表Employee List
- 增加员工Add Employee

Angular与Spring Boot整合开发

增加员工Create Employee

姓First Name
名Last Name
邮箱Email Id
保存Submit
版权 2022 @ Angular + JavaGuides 【All Rights Reserved 2019 @JavaGuides】

图 16-6 单击图 16-4 所示中的超链接“增加员工 Add Employee”的结果

Angular Example
localhost:4200/add

- 员工列表Employee List
- 增加员工Add Employee

Angular与Spring Boot整合开发

增加员工Create Employee

姓First Name 王
名Last Name 五
邮箱Email Id ww@qingc.com
保存Submit
版权 2022 @ Angular + JavaGuides 【All Rights Reserved 2019 @JavaGuides】

图 16-7 在图 16-6 所示中分别输入姓、名、邮箱的结果

Angular Example
localhost:4200/add

- 员工列表Employee List
- 增加员工Add Employee

Angular与Spring Boot整合开发

增加员工Create Employee

保存成功You submitted successfully!

版权 2022 @ Angular + JavaGuides 【All Rights Reserved 2019 @JavaGuides】

图 16-8 在图 16-7 所示的基础上单击“保存 Submit”按钮的结果

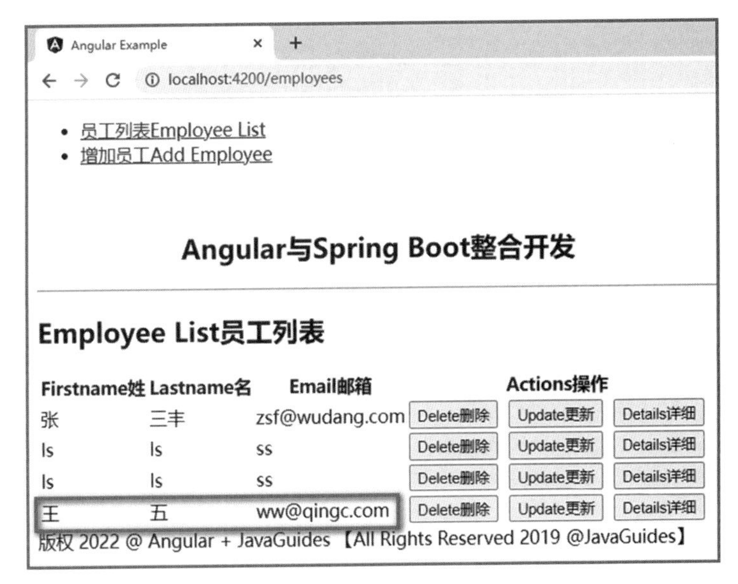

图 16-9　单击图 16-8 所示中的超链接"员工列表 Employee List"的结果

习题 16

实验题

1. 实现本章 Angular 与 Ant Design of Angular 整合开发的案例。
2. 实现本章 Angular 与 Spring Boot 整合开发的案例。

参考文献

[1] 兰泽军. Angular 开发入门与实战[M]. 北京：人民邮电出版社，2021.

[2] 广发证券互联网金融技术团队. 揭秘 Angular 2[M]. 北京：电子工业出版社，2017.

[3] Freeman A. Angular 高级编程[M]. 4 版. 陈磊，译. 北京：清华大学出版社，2021.

[4] 李一鸣. Angular 开发实战[M]. 北京：清华大学出版社，2019.

[5] 王芃. 全栈技能修炼：使用 Angular 和 Spring Boot 打造全栈应用[M]. 北京：电子工业出版社，2019.

[6] 王芃. Angular 从零到一[M]. 北京：机械工业出版社，2017.

[7] 柳伟卫. Angula 企业级应用开发实战[M]. 北京：电子工业出版社，2019.

[8] 成龙. Angular 应用开发指南[M]. 北京：人民邮电出版社，2020.

[9] 吴胜. Spring Boot 开发实战：微课视频版[M]. 北京：清华大学出版社，2019.

图书资源支持

感谢您一直以来对清华版图书的支持和爱护。为了配合本书的使用，本书提供配套的资源，有需求的读者请扫描下方的“书圈”微信公众号二维码，在图书专区下载，也可以拨打电话或发送电子邮件咨询。

如果您在使用本书的过程中遇到了什么问题，或者有相关图书出版计划，也请您发邮件告诉我们，以便我们更好地为您服务。

我们的联系方式：

地　　址：北京市海淀区双清路学研大厦 A 座 714

邮　　编：100084

电　　话：010-83470236　010-83470237

客服邮箱：2301891038@qq.com

QQ：2301891038（请写明您的单位和姓名）

资源下载：关注公众号“书圈”下载配套资源。

资源下载、样书申请

书圈

图书案例

清华计算机学堂

观看课程直播